Lean Sigma

A Practitioner's Guide

Lean Sigma

A Practitioner's Guide

Dr. Ian Wedgwood

Upper Saddle River, NJ • Boston • Indianapolis • San Francisco
New York • Toronto • Montreal • London • Munich • Paris • Madrid
Cape Town • Sydney • Tokyo • Singapore • Mexico City

ISBN 0-13-239078-7
Text printed in the United States on recycled paper at *R.R. Donnelley & Sons in Crawfordsville, Indiana.*
October 2006

Library of Congress Cataloging-in-Publication Data

Wedgwood, Ian D.
Lean sigma : a practitioner's guide / Ian D. Wedgwood.
p. cm.
ISBN 0-13-239078-7 (hardback : alk. paper) 1. Project management. 2. Six sigma (Quality control standard) 3. Total quality management. 4. Leadership. 5. Organizational effectiveness. I. Title.
HD69.P75W44 2006
658.4'013—dc22

2006021756

This book is dedicated to my wonderful wife, Veronica, for relentlessly urging me to write this book for the past eight years, but also for the tireless patience and support when I finally did.

Contents

Foreword xix

Preface xxiii

Acknowledgments xxvii

About the Author xxix

Chapter 1 Introduction 1

Overview 1

Intended Audience 2

Prerequisites 2

Basics 3

A Process 3

Entities 4

Deliverables 4

Methodologies 6

Lean Sigma Roadmap 7

How to Use This Book 9

Problem Categories 11

And Finally... 11

PART I PROJECT ROADMAPS TO SOLVE BUSINESS PROBLEMS

Chapter 2 Define—Tools Roadmap Applied to the Beginning of All Projects 13
Overview 13
Roadmap 14

Chapter 3 Global Process Problems 23
A: On-Time Delivery Issues 23
Overview 23
Examples 23
Measuring Performance 23
Tool Approach 24
B: Capacity of Process Is Too Low 27
Overview 27
Examples 27
Measuring Performance 27
Tool Approach 27
C: RTY, Defects, Accuracy, Quality, Scrap, and Rework Issues 29
Overview 29
Examples 29
Measuring Performance 29
Tool Approach 30
D: % Uptime Is Too Low 33
Overview 33
Examples 33
Measuring Performance 33
Tool Approach 34
E: Pace of Process Too Slow 36
Overview 36
Examples 36
Measuring Performance 36
Tool Approach 37
F: Process Has Enough Capacity, But Fails Intermittently 38
Overview 38
Examples 39
Measuring Performance 39
Tool Approach 39
G: Process Has Enough Capacity, But Process Lead Time Is Too Long 41
Overview 41
Examples 41

Measuring Performance 42
Tool Approach 42
H: Individual Steps Meet Takt, Global Process Does Not 44
Overview 44
Examples 44
Measuring Performance 44
Tool Approach 44
I: Demand from the Customer Is Too Variable 47
Overview 47
Examples 47
Measuring Performance 47
Tool Approach 48
J: Too Many Entity Types (Products) 51
Overview 51
Examples 51
Measuring Performance 52
Tool Approach 52
K: High Schedule Variation 54
Overview 54
Examples 54
Measuring Performance 55
Tool Approach 55
L: Measurement System Broken 60
Overview 60
Examples 61
Measuring Performance 61
Tool Approach 61
Other Considerations 62
M: Performance Characteristic Not Good Enough 63
Overview 63
Examples 63
Measuring Performance 63
Tool Approach 64
N: Planned Maintenance Takes Too Long 64
Overview 64
Examples 64
Measuring Performance 64
Tool Approach 65

O: Setup/Changeover Takes Too Long 65
- Overview 65
- Examples 65
- Measuring Performance 65
- Tool Approach 65

P: Too Much Unplanned Maintenance 68
- Overview 68
- Examples 68
- Measuring Performance 68
- Tool Approach 69

Q: Process Can't Make Product at All 72
- Overview 72
- Examples 73
- Measuring Performance 73
- Tool Approach 73

R: Resource Usage Is Too High (Headcount Reduction) 75
- Overview 75
- Examples 75
- Measuring Performance 76
- Tool Approach 76

S: Inventory Is Too High 78
- Overview 78
- Examples 79
- Measuring Performance 79
- Tool Approach 80

T: Waste/Process Loss Too High 82
- Overview 82
- Examples 83
- Measuring Performance 83
- Tool Approach 83

U: High Forecast Variation 85
- Overview 85
- Examples 86
- Measuring Performance 86
- Tool Approach 86

V: Not Enough Sales 89
- Overview 89
- Examples 89
- Measuring Performance 90
- Tool Approach 90

W: Backlog of Orders Is Too High 91
Overview 91
Examples 92
Measuring Performance 92
Tool Approach 92
X: Payments Made to Suppliers Not Optimized 93
Overview 93
Examples 93
Measuring Performance 93
Tool Approach 94
Y: Accounts Receivable Are Too High 95
Overview 95
Examples 95
Measuring Performance 95
Tool Approach 96

Chapter 4 Individual Step Process Problems 99
1: A Single Process Step Does Not Meet Takt 99
Overview 99
Measuring Performance 99
Tool Approach 100
2: The Pace for a Single Process Step Is Too Slow 102
Overview 102
Measuring Performance 102
Tool Approach 103
3: Too Much Variation in the Cycle Time of a Single Step 104
Overview 104
Measuring Performance 105
Tool Approach 105

Chapter 5 Control—Tools Used at the End of All Projects 107
Overview 107
Tool Approach 108

PART II ROADMAPS TO FIND PROJECTS WHERE NO OBVIOUS CANDIDATES EXIST

Chapter 6 Discovery—Tools Applied to Identify Projects 115
Overview 115
Tool Approach 115

Part III Roadmaps to Guide in the Practical Application of Each Lean Sigma Tool

Chapter 7 Tools 119

01: 5 Whys 119
- Overview 119
- Other Options 120

02: 5S 122
- Overview 122
- Logistics 122
- Roadmap 123

03: Affinity 130
- Overview 130
- Logistics 131
- Roadmap 131
- Interpreting the Output 133

04: ANOVA 133
- Overview 133
- Roadmap 135
- Interpreting the Output 138
- Other Options 140

05: Box Plot 141
- Overview 141
- Interpreting the Output 142

06: Capability—Attribute 143
- Overview 143
- Roadmap 145
- Interpreting the Output 146

07: Capability—Continuous 146
- Overview 146
- Roadmap 148
- Interpreting the Output 149
- Other Options 151

08: Cause & Effect (C&E) Matrix 153
- Overview 153
- Logistics 153
- Roadmap 153
- Interpreting the Output 155
- Other Options: 2-Phase C&E 155

09: Chi-Square 157
Overview 157
Roadmap 160
Interpreting the Output 161
10: Control Charts 163
Overview 163
Roadmap 167
Interpreting the Output 168
Other Options 169
11: Critical Path Analysis 171
Overview 171
Roadmap 172
Interpreting the Output 173
12: Customer Interviewing 174
Overview 174
Logistics 174
Roadmap 175
Interpreting the Output 183
13: Customer Requirements Tree 183
Overview 183
Logistics 184
Roadmap 184
Interpreting the Output 186
14: Customer Surveys 186
Overview 186
Logistics 186
Roadmap 186
Interpreting the Output 192
15: Demand Profiling 193
Overview 193
Logistics 193
Roadmap 194
Interpreting the Output 194
Other Options 196
16: Demand Segmentation 196
Overview 196
Logistics 197
Roadmap 198
Interpreting the Output 199

17: DOE—Introduction 202
Overview 202
Logistics 208
Roadmap 208
Other Considerations 212
18: DOE—Screening 213
Overview 213
Roadmap 216
Other Options 221
19: DOE—Characterizing 222
Overview 222
Roadmap 231
Other Options 236
20: DOE—Optimizing 237
Overview 237
Roadmap 246
Other Options 247
21: Fishbone Diagram 249
Overview 249
Logistics 251
Roadmap 251
Interpreting the Output 251
22: Handoff Map 253
Overview 253
Logistics 254
Roadmap 254
Interpreting the Output 256
23: KPOVs and Data 257
Overview 257
Roadmap 261
24: Load Chart 268
Overview 268
Logistics 269
Roadmap 269
Interpreting the Output 270
Other Options 271
25: MSA—Validity 272
Overview 272
Logistics 274

Roadmap 274
Interpreting the Output 275
26: MSA—Attribute 276
Overview 276
Logistics 278
Roadmap 278
Interpreting the Output 279
27: MSA—Continuous 284
Overview 284
Logistics 289
Roadmap 289
Interpreting the Output 290
Other Options 294
28: Multi-Cycle Analysis 294
Overview 294
Logistics 295
Roadmap 295
Interpreting the Output 298
29: Multi-Vari Studies 300
Overview 300
Logistics 301
Roadmap 301
30: Murphy's Analysis 306
Overview 306
Logistics 307
Roadmap 307
Interpreting the Output 308
31: Normality Test 308
Overview 308
Roadmap 309
Interpreting the Output 309
32: Overall Equipment Effectiveness (OEE) 311
Overview 311
Logistics 312
% Uptime 312
% Pace 314
% Quality 315
Special Cases 315
Interpreting the Output 316
Other Options 317

33: Pareto Chart 318
- Overview 318
- Roadmap 319
- Interpreting the Output 319

34: Poka Yoke (Mistake Proofing) 321
- Overview 321
- Logistics 323
- Roadmap 323

35: Process FMEA 325
- Overview 325
- Logistics 326
- Roadmap 326
- Interpreting the Output 329

36: Process Variables (Input/Output) Map 330
- Overview 330
- Logistics 330
- Roadmap 330
- Interpreting the Output 333

37: Project Charter 333
- Overview 333
- Logistics 335
- Roadmap 335

38: Pull Systems and Kanban 342
- Overview 342
- Logistics 348
- Roadmap 348
- Other Options 354

39: Rapid Changeover (SMED) 354
- Overview 354
- Logistics 356
- Roadmap 357

40: Regression 362
- Overview 362
- Roadmap 370
- Interpreting the Output 370
- Other Options 371

41: SIPOC 372
- Overview 372
- Logistics 373

Roadmap 374
Interpreting the Output 376
42: Spaghetti (Physical Process) Map 376
Overview 376
Roadmap 377
Interpreting the Output 379
43: Statistical Process Control (SPC) 380
Overview 380
Logistics 381
Roadmap 381
Interpreting the Output 382
44: Swimlane Map 384
Overview 384
Logistics 384
Roadmap 386
Interpreting the Output 387
Other Options 387
45: Test of Equal Variance 389
Overview 389
Roadmap 389
Interpreting the Output 391
46: Time—Global Process Cycle Time 392
Note 392
Overview 393
Logistics 393
Roadmap 394
Interpreting the Output 395
47: Time—Individual Step Cycle Time 395
Overview 395
Logistics 396
Roadmap 397
Interpreting the Output 398
48: Time—Process Lead Time 398
Note 398
Overview 399
Logistics 399
Roadmap 400
Interpreting the Output 400

49: Time—Replenishment Time 401
Overview 401
Logistics 402
Roadmap 403
Interpreting the Output 403
50: Time—Takt Time 404
Overview 404
Logistics 406
Roadmap 406
Interpreting the Output 407
51: Total Productive Maintenance 408
Overview 408
Logistics 409
Roadmap 409
52: t-Test—1-Sample 411
Overview 411
Roadmap 411
Interpreting The Output 414
Other Options 415
53: t-Test—2-Sample 416
Overview 416
Roadmap 417
Interpreting the Output 420
Other Options 421
54: t-Test—Paired 422
Overview 422
Roadmap 423
55: Value Stream Map 423
Overview 423
Logistics 426
Roadmap 426
Interpreting the Output 432

Index 435

Foreword

Six Sigma was established in 1987 by Motorola. It lay in limbo with sporadic deployment among relatively few companies. Then, in 1994, the tsunami began. AlliedSignal, a $14 Billion company lead by Larry Bossidy, applied Six Sigma to his mediocre company with a vengeance. With the help of a former Motorola quality expert, Rich Schroeder, Larry established Six Sigma to be clearly a business process and quite effective. He was able to track Six Sigma project activities directly to earnings per share for 1995.

When Larry convinced his friend, Jack Welsh, Six Sigma would work in GE, the rest is history. Since 1996, Six Sigma has spread like wildfire throughout every industry. Then companies realized that when Six Sigma was integrated with Lean Enterprise methods (Lean Sigma), the results were magnificent.

So, why have Six Sigma and Lean Sigma not gone the way of TQM, Business Process Re-engineering and other programs of the month? There are several reasons explaining the endurance of Six Sigma and Lean Sigma:

1. Lean Six Sigma moved from a quality function to a business function
2. Lean Six Sigma is roadmap based
3. Training innovation resulted from Lean Six Sigma
    a. Learning, accountability, execution, results
4. Lean Six Sigma dovetails effectively with other initiatives (Lean)
5. Precise Level of Accountability
6. Lean Six Sigma is expandable to any business process
7. Lean Six Sigma gets results

Of the above differentiators, the third addressing training innovations was probably a revolutionary feature. Companies had not previously considered sending students away for four weeks of training around a high impact project. Companies normally did not treat training as a business process where training was linked to strategic goals and performance criteria.

Nor was training directly accountable to attain significant business results and achieve a return-on-investment. In fact, the Chief Learning Officer of GM, Donnee Ramelli, has linked all the training from GM University to the business turnaround GM is currently attempting. In doing so with innovative training methods, he has saved GM millions of dollars in training funds and produced new capabilities in GM employees. It's no wonder that Donnee learned this approach while driving Six Sigma in AlliedSignal's Engineered Materials Sector.

The second differentiator, the process improvement roadmaps, is the second most profound difference between Lean Sigma and previous improvement methods. As Ian Wedgwood points out in this fine book that the number improvement tools required to integrate Lean and Six Sigma is vast but the tools are not new. Ian should know, as he is an expert in both Lean and Six Sigma tools as well as the statistics. And he learned the nuances of the Lean Six Sigma roadmaps by actually using them in a long series of process improvement projects. The complexity of many of the statistical tools is intimidating. The irony is that these tools have been around a long time, some since the 1920's. The integration and linking of these tools have been the keys to the success of Lean and Six Sigma.

To provide a student a set of proven process improvement tools and a clear order in which to provide them (the roadmap) had never been done effectively in other quality initiatives. Over the years, thousands of students have achieved success using the roadmap approach and aggressive results oriented training. A critical success factor for the best Black Belts is whether they receive mentoring from a Master Black Belt at critical times during the project. Because of the complexity of the roadmaps, having access to a roadmap navigator has paid dividends.

The goal is to make Black Belts and Green Belts self sustaining in their improvement efforts. Ian's book is the first step in that evolution from beginner to expert. While numerous books have been written on Six Sigma, Lean, and Lean Sigma, all pay lip service to the roadmaps and hand wave around the tools. A large set of technical books cover the statistical tools in egregious detail but never links those tools to other tools and methods or to a process improvement roadmap. I know as a Ph.D. statistician, I was an expert in statistical tools but not an expert problem solver. When I built the MAIC roadmap for Six Sigma, I finally became an expert problem solver.

So, it's not about the tool set (TQM has a similar toolset) but about project execution. By using the right tools at the right time, in the right order, and in the right place, dramatic improvement occurs. This book fills an important gap in the literature. This book facilitates a student moving smoothly and swiftly from project definition to project completion. And, uniquely, Ian has laid out roadmaps for different classes of problems such as accuracy, capacity, lead-time downtime and inventory. His roadmaps work in manufacturing and transactional processes. The Black Belts coach and mentor will always be by his or her side at the bookshelf. And the book itself will be seen as an integral part of any Black Belt's or Green Belt's training.

Welcome to the wonderful and rich world of roadmap-based process improvement. Your capable navigator will be Ian Wedgwood. Ian, at the Ph.D. level, has paid his dues in the corporate world by leading both Lean and Six Sigma deployments in a very large company. Through his consulting efforts with a large number of companies, he has also learned what always works and where the barriers are to project success. He has taken everything he has learned over the last 10 years and distilled the knowledge to a set of step-by-step improvement roadmaps that really work. With the requisite Black Belt or Green Belt training and this book, you will be ready to lead a team to solve some of the most complex problems in your business. My advice is to do great things, be safe, and have fun. For the next few years of your professional career will be some of the most rewarding you'll ever have. Welcome aboard!

—Dr. Stephen Zinkgraf
CEO Sigma Breakthrough Technologies, Inc.

Preface

There is absolutely no doubt that Lean and Six Sigma as process improvement methodologies deliver results, as proven consistently countless times over literally thousands of projects across hundreds of businesses. What is inconsistent, however, is the efficiency by which the Project Leaders (Belts) and Teams reach the delivered solution and sometimes the effectiveness of the solution itself. Typically, this is considered to be the territory of the Consultant or Master Black Belt (MBB), whose role it is to guide the steps of the Black Belt or Green Belt through the available tools depending on the problem.

Therefore, the best guide requires

- A deep enough experience of how to tackle a specific problem to conclusion with an efficient approach (as a Belt, I want to know exactly what path to follow in my project).
- A broad enough experience to do this across multiple different types of problems that might be addressed in a business (as a Program Leader I need my MBB to know what path to follow for all projects and guide my Belts accordingly).
- Technical skills to be able to guide the Belts in specific tool use (as a Belt, I want to know the practical steps involved in applying each tool).

Interestingly, this expands the common perception of the role of a Master Black Belt as a technical resource and measures that individual in addition by the efficiency and effectiveness of projects that they oversee (i.e., the rate of generation of business value from those projects).

Surprisingly (and fortunately) when asked the route to solution for a particular type of problem, the experienced guides are remarkably consistent in answer—it seems that if you have a specific problem type, then you should follow a specific route to solution. The intent of this book, therefore, is to capture those experiences and for multiple given project types lay down the appropriate routes to solution.

Audiences that find this book valuable are

- Process Improvement Project Leaders (Green Belts and Black Belts), across all industries—Leading projects to improve processes using tools and methodologies that come under the Lean or Six Sigma banners.
- Project Champions or Sponsors—Wondering what questions to ask of their Project Leaders and what they should see in terms of activity, as well as seeking to improve their project selection and scoping skills.
- Technical Mentors (Master Black Belts)—Looking to improve their project and tools mentoring skills and to better select and scope projects.
- Deployment Leaders—Seeking to better select and scope projects to improve the Return on Investment of the Program.
- Consultants—Brushing up on skills as both a Technical Mentor and Deployment Lead.

The book is a little unusual in that it is designed to be a practical tool, used day-to-day by the readers to guide them through how to solve as many different types of business problems as possible using the Lean Sigma methodologies and tools. It is not meant to be a technical reference to take the place of the statistical tomes that are readily available. By analogy, this is how to drive the car, not how the car works.

The book is also unusual in that it is not designed to be read linearly from cover to cover, mainly due to a few simple issues:

- There are a multitude of different problem types
- Each problem type has a different route to solution
- The same tools are used in the solution of multiple problem types
- The application of each tool can vary subtly depending on the problem

The structure is in a form that best helps the reader start with their problem in hand and quickly progress to the solution. To that end, the book has three main parts:

PART I (CHAPTERS 2–5)

Project Roadmaps that describe the route to solution for a wide range of problems. The text lists which tools to use, in which order, and why. To understand application of a particular tool in more detail, the reader should refer to Part III.

PART II (CHAPTER 6)

A Discovery Roadmap used to identify potential projects in a process where there are no obvious targets. This is often useful to businesses that are new to Lean Sigma and are not sure how to identify good projects to work on. To understand application of a particular tool in more detail, the reader should refer to Part III.

PART III (CHAPTER 7)

Individual tools roadmaps explaining in detail how to use each tool.

Throughout this book, I explain which tool to use and why it is used, so that Belts move from blind tool use to truly thinking about what they do and focus on the end goal of improving the process.

Processes and their respective problems are real-world phenomena, requiring practical actions and change. The best Belts I've found were the most practical thinkers, not the theorists, because any tool, even based on the cleverest theory, is only as good as the practical business solution it provides.

Acknowledgments

I'd like to acknowledge Dr. Steve Zinkgraf and fellow Executive Team members at Sigma Breakthrough Technologies, Inc., for giving me the opportunity to capture some of the group's wealth of experience in this book, along with access to the methodologies and materials.

In particular, I'd like to acknowledge Steve himself for leading the way, Daniel Kutz for providing "air cover" during writing, Joe Ficalora for being the technical sounding board, and Joe Costello for his lean insight and more importantly for introducing me to my wife.

Sincere thanks to all SBTI's client leaders with whom I've had the good fortune to learn from over the years: in particular to Dr. Al Landers at Huber Engineered Woods, Paul Fantelli at Lincoln Electric, Antonio Rodriguez at Celanese, Dr. Skipper Yocum, and Dr. John Nimmo at SunChemical, and Tim Tarnowski at Columbus Regional Hospital.

The most important acknowledgement of all has to be to the host of Belts and Project Leaders across hundreds of companies in multiple industries, without whom all of this would be theory—to you we are all truly indebted.

About the Author

Dr. Ian Wedgwood is an Executive Director for SBTI and has over a decade of Lean Sigma Experience. He has both led and facilitated a number of deployments in industries as diverse as electronics, engineered materials, medical devices, chemicals, and healthcare, and has trained and mentored numerous Executives, Champions, and Belts in DFSS, Six Sigma, and Lean. Ian has a strong product development background and co-developed SBTI's Lean Design, Lean Sigma, K-Sigma (accelerated Lean Sigma), and Healthcare methodologies and curricula.

Prior to joining SBTI, Ian worked for the global engineering group Invensys PLC (the merger of BTR and Siebe) with some 1200+ sites worldwide. Starting out as Development Manager for BTR Technology Services, based in the United Kingdom, and then progressing to Technical Manager, he led the development of Six Sigma based software tools and their implementation into BTR's manufacturing sites.

As a Program Manager with BTR C&TG he was involved in key Projects within the BTR Group during strategic acquisitions. One such project, building a new 180,000 square foot manufacturing facility in Tijuana, Mexico, brought Ian to the United States where he still lives with his wife, Veronica, and sons Christian and Sean.

As a Program Manager for Powerware (part of Invensys' Power Systems Division), Ian led the efforts to lean the Divisions' New Product Development processes. Moving back into a Divisional role, he then led Invensys' highly successful Lean Design for Six Sigma deployment. Some 380+ Design Belts within Power Systems alone yielded a 65 times return in less than two years.

He joined SBTI as a Master Consultant in 2001 and SBTI's Leadership Team as its first Executive Director in February 2003.

Ian holds a Ph.D. and a First-Class Honors degree in Applied Mathematics from Scotland's St.Andrew's University.

1 Introduction

Overview

The motivation for writing this book was a disappointing realization over many years of teaching and mentoring Project Leaders that there are plenty of technical texts explaining the painful underlying statistics in Six Sigma and Lean Sigma, but there are hardly any books explaining what to do from a practical standpoint. There are proliferations of books explaining at a high level the overall concept of a project, but next to none that take the Project Leader through a project, step by step. There are a multitude of books explaining just enough on project tools to suck the reader into buying consulting time from the author to apply them, but none that leave the reader in a position of practical self-sufficiency. Most unfortunately of all, there are a whole host of books written by theorists who have never led a project to solve a business problem using the methodologies they espouse, but very few ever written by those who have actually applied this stuff.

The aim here is to be different. The hope is that I provide a book that can be used practically day to day by Process Improvement Leaders (from any industry), Champions, and Consultants to guide them through how to solve as many different types of business problems as possible. It is certainly not meant to be a technical text to take the place of the statistical tomes that are readily available—I'll reference as many as I can of those along the way. By analogy, this is how to drive the car, not how the car works. In a field as passionate as Lean Sigma, I'm sure there will be disagreement at times with the order of tools used, so please remember that this is a guide—not the definitive solution.

I'll also hasten to add at this point that I don't favor Lean over Six Sigma or vice versa. Let's face it—we need them both and, by the end of this book, I probably will have offended both camps equally. The text is most certainly not for purists; it's just about an approach that works.

Intended Audience

The primary audiences for this book are

- The host of Process Improvement Project Leaders (Green Belts and Black Belts), across *all* industries, who are leading projects to improve processes by shortening lead times, increasing capacity, improving yields and accuracy, reducing inventories, and so forth using tools and methodologies that come under the Lean or Six Sigma banners
- Project Champions or Sponsors who are wondering what questions to ask of their Project Leaders and what they should see in terms of activity, as well as seeking to improve their project selection and scoping skills
- Technical Mentors (Master Black Belts) who are looking to improve their project- and tools-mentoring skills and to better select and scope projects
- Deployment Leaders who are seeking to better select and scope projects to improve Return On Investment (ROI) of the Program
- Consultants who are brushing up on skills as both a Technical Mentor and a Deployment Lead

Prerequisites

This book is specifically aimed at a project-based approach to process improvement. In order to ensure a usable text, it will be necessary to make some basic assumptions before leading up to the project—in particular, the existence of the following:

- A clear business reason to do the project.[1]
- A Project Leader (usually referred to as a Black Belt or Green Belt, depending on the level of training) to lead the project. It is usually best to have a Belt that is not from the functional groups impacted by the project if at all possible—that way, the Belt has no preconceived notions of solution and can be relied upon to look at the process with a fresh set of eyes.

---

1 The chapters on Project Identification and Selection in Stephen A. Zinkgraf's book, *Six Sigma—The First 90 Days*, will certainly help set the stage here (Prentice Hall PTR, ISBN: 0131687409).

- A Team comprised of people who live and breathe the process every day. Lean Sigma is certainly a team sport and should not be viewed as a "gladiator" undertaking. There should be no hero mentality to solution of process problems.
- A committed Champion to remove potential roadblocks.[2]
- Time made available for the Team to complete the project, both for the Belt and the Team. If this is not the case, failure is just a few short weeks away.

These elements are absolutely necessary, but in this book I will not spend any more time on them because the focus here will be on the problem-solving roadmap itself and the tools therein.

Another significant assumption here is that the Project Leader will have gone through some basic Lean Sigma or Six Sigma training to at least Green Belt level. It is possible to complete a project just on this text alone, but the intent is for this book to be a *practical* support guide as opposed to a technical teaching guide. I will endeavor to reference key technical texts throughout.

BASICS

In order to better understand the detailed methods of Lean Sigma process improvement, it is important to first have a clear understanding of the basics involved. This begins with simple clarifications of what a process is, how it is defined and then how it is improved.

A PROCESS

The first thing to point out here is that Lean Sigma is a *process* improvement methodology, not a function or an activity improvement methodology. This is a key distinction in framing the project and it is one that Champions frequently get wrong during project identification, scoping, and selection.

A process is a sequence of activities with a definite beginning and end, including defined deliverables. Also, a "something" travels through the sequence (typical examples include a product, an order, a patient, or an invoice). Resource is used to accomplish the activities along the way.

---

[2] The role of the Champion is clearly outlined in Chapter 8, "Defining the Six Sigma Infrastructure" in Stephen A. Zinkgraf's book, *Six Sigma—The First 90 Days*.

If you can't see an obvious, single process in your project, you might have difficulty applying process improvement to it. The start and end points need to be completely agreed upon between the Belt, Champion, and Process Owner (if this is not the Champion). Clearly, if this is not the case, there will be problems later when the end results don't match expectations.

Entities

In the preceding definition of a process, there is a "something" that travels along it. For want of a better name, I'll refer to this as an *entity*. Clearly, this entity can be fundamentally different from process to process, but there seems to be surprisingly few distinct types:

- **Human.** Employees, customers, patients
- **Inanimate.** Documents, parts, units, molecules
- **Abstract.** Email, telephone calls, orders, needs

The trick is to be able to identify the Primary Entity as it flows through the process with value being added to it (for example, a patient or perhaps the physical molecules of a product). There will, of course, be secondary entities moving around the process, but focus should be on identifying the primary.

Belts sometimes find this difficult when the entity changes form, splits, or replicates. For instance, in healthcare (in the ubiquitous medication delivery process), orders are typically written by the physician and so the Primary Entity is the written order. The order can then be faxed to the pharmacy, and is thus replicated and one copy transmitted to the pharmacy fax machine. The faxed order is then fulfilled (meds are picked from an inventory) and effectively the Primary Entity changes to the medication itself, which will be sent back to the point of request.

Similarly, in an industrial setting, we might see the Primary Entity change from customer need to sales order to production order to product.

Deliverables

The last element of the definition of a process is the deliverables. This is often where novice Belts make the biggest mistakes. Simply put, the deliverables are the minimum set of physical entities and outcomes that a process has to yield in order to meet the downstream customers' needs.

The single most common mistake Belts make in process improvement is to improve a process based on what customers say they *want* versus what they truly *need* (more about this in the section, "Customer Interviewing" in Chapter 7).

The deliverables need to be thoroughly understood and agreed upon in the early stages of the project; otherwise later during the analysis of what it is exactly in the process that affects performance, the Belt will have the wrong focus.

If your project doesn't have a start, an end, deliverables, or a Primary Entity, it probably isn't a process and you will struggle to apply Lean Sigma to it. Table 1.1 gives examples of good and poor projects across varying industries.

Table 1.1 Examples of poor versus good projects

Industry	Healthcare	Chemical Manufacturing	Discrete Manufacturing	Service/ Administrative	Transportation and Logistics
Good Projects	Length of stay Emergency department, operating room, care units Accuracy Meds admin/ delivery, charging, billing, patient handoffs Capacity Emergency department, operating room, radiology, lab Lead time Radiology, lab Downtime Equipment, rooms	Accuracy Invoice, yield, assay Capacity Line, product, vessel Lead time Delivery, production, replenishment Downtime Equipment, lines, vessel	Accuracy Invoice, yield Capacity Line, product Lead time Delivery, production, replenishment Downtime Equipment, lines	Accuracy Invoice, delivery, product Capacity Service area, call center, product Lead time Delivery, call hold time Downtime Equipment, servers, lines	Accuracy Invoice, bills of lading Capacity Hump yard, distribution center Lead time Delivery Downtime Locomotive Damage Locomotive, package, radio Inventory Product, packaging

(continues)

Table 1.1 Examples of poor versus good projects (Continued)

Industry	Healthcare	Chemical Manufacturing	Discrete Manufacturing	Service/ Administrative	Transportation and Logistics
Poor Projects	Satisfaction[6] Patient, staff, physician	Reduce healthcare costs			
Poor Projects	Communication[3] Sales and marketing Improve forecast accuracy[4] Cell phone consolidation Improve employee retention Implement XYZ system			Reduce office utility costs Improve quality of master data in SAP/Oracle/etc. File all paper documents electronically Electronic product catalog Reduce DSO from 75 days to 30[5]	

METHODOLOGIES

Six Sigma and Lean are both business improvement methodologies—more specifically, they are business process improvement methodologies. Their end goals are similar—better process performance—but they focus on different elements of a process. Unfortunately, both have been victims of bastardization (primarily out of ignorance of their merits) and often have been positioned as competitors when, in fact, they are wholly complementary.

For the purpose of this practical approach to process improvement

- **Six Sigma** is a systematic methodology to home in on the key factors that drive the performance of a process, set them at the best levels, and hold them there for all time.
- **Lean** is a systematic methodology to reduce the complexity and streamline a process by identifying and eliminating sources of waste in the process—waste that typically causes a lack of flow.

---

[3] Although communication is a process, it is not a fundamental Value Stream in an organization. Instead, look to mending the primary Value Streams first, and then it might even be possible to eliminate the need for person-to-person communication entirely.

[4] It is best to tackle the responsiveness of the process before looking into forecasting (i.e., the more responsive my process, the less I have to worry about forecasting).

[5] Although this is a legitimate project, it is large and difficult for a Green or Black Belt to handle. It usually requires running as a Master Black Belt program of projects.

[6] Satisfaction is a useful metric, but it typically lags in the process and thus becomes difficult to deal with. Also, it is inherently affected by many noises in the process. Try to understand what in the process brings the satisfaction and perhaps target that in the project.

In simple terms, Lean looks at what we *shouldn't* be doing and aims to remove it; Six Sigma looks at what we *should* be doing and aims to get it right the first time and every time, for all time.

Lean Sigma Roadmap

Lean Sigma is all about linkage of tools, not using tools individually. In fact, none of the tools are new—the strength of approach is in the sequence of tools. The ability to understand the theory of tools is important, but this book is about how to apply and sequence the tools.

There are many versions of the Six Sigma Roadmap, but not so many that fully incorporate Lean in a truly integrated Lean Sigma form. Figure 1.1 shows a robust version of a fully integrated approach put together over many years by the thought leaders at Sigma Breakthrough Technologies, Inc. (SBTI).[7] The roadmap follows the basic tried and tested DMAIC (Define, Measure, Analyze, Improve, and Control) approach from Six Sigma, but with Lean flow tools as well as Six Sigma statistical tools threaded seamlessly together throughout. As proven across a diverse range of SBTI clients, the roadmap is equally at home in service industries, manufacturing industries of all types, and healthcare, including sharp-end hospital processes, even though at first glance some tools may lean toward only one of these. For example, despite being considered most at home in manufacturing, the best Pull Systems I've seen were for controlling replenishment in office supplies. Similarly, Workstation Design applies equally to a triage nurse as it does to an assembly worker.

The roadmap is a long way removed from its Six Sigma predecessors and is structured into three layers:

- Major phases
- Minor phases
- Tools and deliverables (how and what)

This is done purposefully to ensure the problem-solving approach isn't just a list of tools in an order. It has meaning inherent to its structure. This is a *crucial* point to practitioners. Throughout this book, I'll explain not only *which* tool to use, but also *why* it is used, so that Belts move from blind tool use to truly thinking about what they are doing and focusing on the end goal of improving the process.

---

[7] Sigma Breakthrough Technologies, Inc. (SBTI) is a professional services firm specializing in Six Sigma and Lean deployments. For more information, see www.sbtionline.com.

Phase	Steps	Tools	Outputs
Define	Initiate the Project	Project Charter; Meeting Effectiveness	✓ Project charter ✓ Project team formed ✓ Clear customer requirements
	Define the Process	SIPOC Map; Value Stream Map	
	Determine Customer Requirements	Brainstorming; Affinity Diagramming; Murphy's Analysis; Interviews; Surveys; Customer Requirements Trees	
	Define Key Process Output Variables	Project Charter; KPOVs	
Measure	Understand the Process	SIPOC/VSM; Input/Output Analysis; C&E Matrix; Detailed Process Maps	✓ Current State Process Maps ✓ Identified and Measured Xs (KPIVs) ✓ Measurement system verified ✓ Current capability of Ys (KPOVs)
	Evaluate Risks on Process Inputs	FMEA	
	Develop and Evaluate Measurement Systems	Data Collection Plans; Data Integrity Audits; Continuous MSA (Gage R&R); Attribute MSA (Kappa Studies)	
	Measure Current Performance	Process Capability; OEE	
Analyze	Analyze Data to Prioritize Key Input Variables	Basic Statistics; Basic Graphs; Statistical Process Control; T-Tests; ANOVA; Non-parametrics; Chi-Square; Regression; Multi-vari Studies	✓ Root causes of defects identified and reduced to vital few ✓ Prioritized list of potential key inputs ✓ Waste identified
	Identify Waste	Spaghetti Diagrams; VA/NVA Analysis; Takt Time; 5S	
Improve	Verify Critical Inputs	Design of Experiments	✓ Finalized List of KPIVs ✓ Action plan for improvement ✓ Future state process maps, FMEA, control plans ✓ New process design/ documentation ✓ Pilot study plan
	Design Improvements	Kanban/Pull; Mistake Proofing; Quick Changeover; Workplace Organization; Process Mapping; Process Documentation	
	Pilot New Process	Training Plans; SPC; FMEA; Control Plans	
Control	Finalize the Control System	Control Plans; Process Documentation; Training Plans; Communication Plans; Statistical Process Control; Documentation	✓ Control system in place ✓ Improvements validated long term ✓ Continuous improvements opportunities identified ✓ New process handed off ✓ Team recognition
	Verify Long-Term Capability	Statistical Process Control; Process Capability	

Figure 1.1 Integrated Lean Sigma Roadmap © SBTI, 2003.[8]

[8] Source: SBTI's Lean Sigma training material.

The best Belts I've found were the most practical thinkers, not the theorists. This is a practical roadmap, and the user should try and focus on the underlying principle of "I'll apply the minimum practical sequence of tools to understand enough about my process to robustly make dramatic improvement for once and for all in my process."

How to Use This Book

The intent for this book is that it be used as a tool to help Project Leaders guide a project, and thus needs to be structured in a form that best helps the reader start with their problem in hand and quickly progress to the solution. I'm sure it is possible to read it from beginning to end; however, it is not designed with that purpose in mind. Its layout probably will be perceived as a little unorthodox, mainly due to a few simple issues:

- There are a multitude of different Problem Categories.
- Each Problem Category has a different route to a solution.
- The same tools are used in the solution of multiple Problem Categories.
- The application of each tool can vary subtly, depending on the problem. This book is structured into three main parts (shown graphically in Figure 1.2):

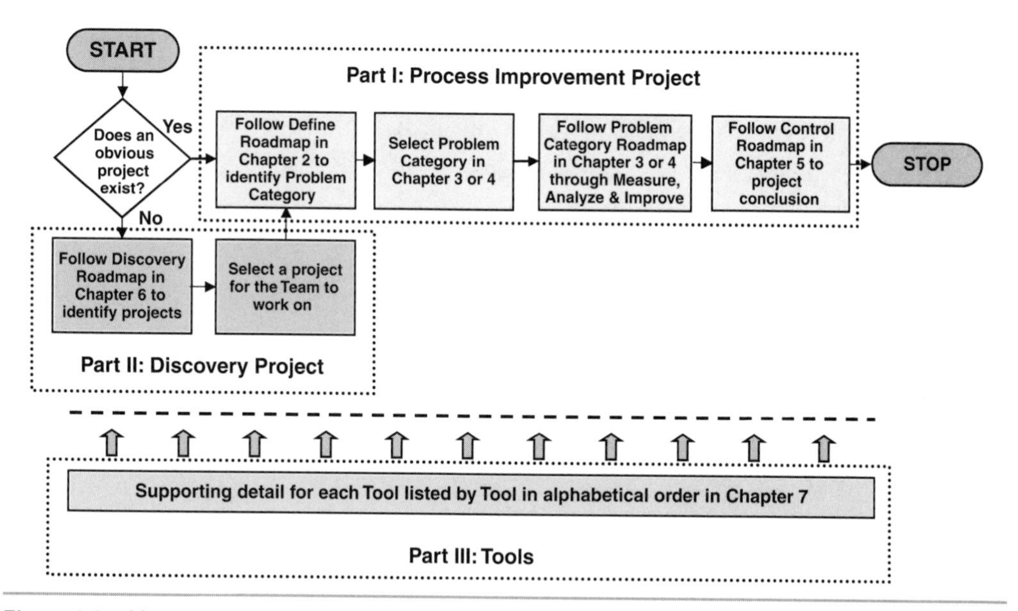

Figure 1.2 How to navigate through this book.

- **Part I (Chapters 2–5).** Project Roadmaps describe the route to a solution for a wide range of problems. The text lists which tools to use (listed in *italics* like this), in which order, why, and in essence forms the detail behind the roadmap shown in Figure 1.1. The Belt/Team should follow the roadmap in this section that best describes the process problem that they are encountering, based on key decision points listed in the text. For more detail on a tool listed, the Belt/Team should refer to the tool detail in Part III, where the tools are listed in alphabetical order.
- **Part II (Chapter 6).** A Discovery Roadmap is used to identify potential projects in a process where there are no obvious targets. This is often useful to businesses that are new to Lean Sigma and are not sure how to identify good projects to work on. The text lists which tools to use (listed in *italics* like this), in which order, and why. For more detail on a tool listed, the Belt/Team should refer to the tool detail in Part III, where the tools are listed in alphabetical order. After the project or multiple projects have been identified in the process using the Discovery Roadmap, one will be selected and the Team will follow the Project Roadmaps described in Part I.
- **Part III (Chapter 7).** Individual tools roadmaps explain in detail how to use each tool.

Thus, in summary:

- If no project is obvious for the process, the Team will follow the Discovery Roadmap in Chapter 6 to its conclusion to identify projects.
- If a project is clear, the Team will follow the Project Roadmap(s) commencing in Chapter 2 to their conclusion to complete the project.
- In both Roadmaps, the text will refer to a sequence of tools and the rationale for the sequence. Details on each tool listed are available in alphabetical order in Chapter 7.

The Project Roadmap in Part II follows this path:

1. A standard set of *Define* tools is applied in sequence at the beginning of any project.
2. At this point, the Belt/Team should have enough understanding of the process problem to select the type of problem that is apparent. The text lists some 25 or so Problem Categories with titles such as "The capacity of the process is too low." Generally speaking, this is at an overall-process level (considering the process as a whole), in which case the categories are listed in Chapter 3. However, there are rare projects in which a significant amount of work has already been done on the process. In this case, the Problem Category might be at a within-process level where a single process step has been identified as being the problem area, in which case the categories are listed in Chapter 4.

3. The Belt/Team selects the Problem Category in Chapter 3 or 4 and follows the *Measure*, *Analyze*, and *Improve* tools roadmaps specific to it.
4. A standard set of *Control* tools is applied in sequence to the end of any project.

Thus, for any project, the user applies the following:

- *Define* tools, standard across all projects
- *Measure*, *Analyze*, and *Improve* tools, pertinent to the specific Problem Category
- *Control* tools, standard across all projects

And, for each tool along the way, practical application detail is available in Chapter 7.

PROBLEM CATEGORIES

To use this book effectively, it will be necessary to identify the Problem Category based on the process issue(s) at hand. This might seem awkward to novice Belts, but it is an important skill to develop. Belts need to be able to step back from the process and see the bigger picture before diving into the detail. Quite often, the inexperienced Champion and Process Owner can be a hindrance at this point by pushing the Belt down a road to solution before truly understanding the underlying problem. The purpose of the Define tools is to provide an understanding of what, from the customer's perspective, the problem truly is and frame it in a measurable form. *Only after the Define tools have been applied can the Belt confidently say which Problem Category he is dealing with.*

AND FINALLY...

Processes and their respective problems are real-world phenomena, requiring practical actions and change. Any tool, even based on the cleverest theory, is only as good as the practical business solution it provides. To reiterate, this is about practical achievement versus theory; thus, at any point in the project, it is important to be able to answer

- What is the question?
- What tool could help answer the question?
- How do I get the necessary data for the tool?
- Based on the data, what is the tool telling me?
- What are the practical implications? (The big "So what?!!" as is it often called)
- What is the next question that arises based on where we've been?

The best Belts maintain this holistic viewpoint; the best Champions and Mentors keep pushing the Belts for the holistic view.

It is probably worthwhile to point out that no project is easy, but hopefully this guide will bring a little clarity and confidence to those who have to navigate through it.

The only thing left to say at this point is "Good Luck!" Even the best Belt needs some of that, too.

2 Define—Tools Roadmap Applied to the Beginning of All Projects

Overview[1]

All projects, no matter what Problem Category they relate to, progress through the same series of tools in the Define phase. The purpose of Define is to ensure that

- The project scope is clear.
- The customer/market value is understood.
- The business value is understood.
- The measures of process performance are agreed upon and baselined.
- Clear breakthrough goals are set for the measures.
- There is business support to do the project in the form of a Project Leader and Team (and all resources are freed up appropriately to do the project).
- There are agreed Champions and Process Owners for the project to ensure barriers are removed.
- This is the right project to be working on.

The essence of all of the preceding is captured in a tool called a *Project Charter*, the completion and signoff of which can be deemed as the exit criteria for Define.

[1] Before reading this chapter, it is important to read and understand Chapter 1, "Introduction," particularly the section, "How to Use This Book."

ROADMAP

In more detail, Belts must apply the following tools to their projects in Define:

To initiate the project, the Belt, Champion, and Process Owner meet to construct the preliminary Charter. They will determine the initial scope, the Team that should be used, the likely metrics, and potential benefits. These items are merely a draft, and the Team will work to better define them during the subsequent tools.

At this point, the Project Team will be mobilized and shortly thereafter the first Team meeting will take place. The Champion and Process Owner more than likely will be there for the first stages of the meeting to validate support for the project and answer any questions that the Team might have with respect to the desired outcomes (but not solutions). Beyond this point, the Champion and Process Owner only join the Team for meetings in an ad hoc manner when requested. Useful tool sets to consider at this time are some of the myriad of Team formation tools that generally fall into two main categories—Team interaction (or interpersonal roles) and meeting effectiveness tools. Some Team interaction or interpersonal roles include

- **Forming, Storming, Norming, and Performing.** The stages that all Teams go through during projects
- **Belbin.**[2] Roles that individual Team members play in project interplay
- **DISC.**[3] Team role characteristics

An example of a meeting effectiveness tool—which aids in defining meeting roles, behavioral expectations, and so forth during Team meetings—would be a SPACER. A SPACER is conducted at the commencement of each meeting to run through Safety, Purpose, Agenda, Code of Conduct, Expectations, and Roles.

There are a whole host of other Team formation tools; however, they will not be covered any further here but are highly recommended to help Team interaction, speed up the project (especially in the initial stages), and ensure a higher likelihood of success.

---

[2] Belbin is the trademark of Belbin and Associates. For more information, see www.belbin.com.

[3] DISC is the trademark of Target Training International. For more information, see www.ttidisc.com.

SIPOC Map
(+ High-Level *Value Stream Map*)

To help define the process, a *SIPOC* (Suppliers, Inputs, Process, Outputs, Customers) is useful, especially in understanding the scope and purpose of the process. Although it seems to be a very straightforward tool, the SIPOC will drive a lot of very important early discussion within the Team about "What really is the question we are trying to answer?"

The central "Process" column of the *SIPOC* can be extracted and rotated 90° to become a High-Level *Value Stream Map*. This requires almost no additional work, but the map is a useful linear representation of the process that can help in communications within the Team and to the rest of the organization. When the appropriate performance metrics are added, it becomes a valuable tool for the business.

After a reasonable scope is determined for the process, the next question is what the Customer Requirements are for the process. In the early Six Sigma days, this was decided quickly between the Process Owner and Champion, and the Team progressed on. More recently, the roadmaps have been strengthened considerably using better Voice of the Customer (VOC) tools, which ensure that after the process has been improved, it genuinely better meets Customer expectations. The tools listed next are in fact a subset of those VOC tools used for developing new products and services in Design For Six Sigma[4,5] (DFSS) and its offspring, Lean Design For Six Sigma[4] (LDFSS), or simply Lean Design. Because we are not striving to completely redesign the process for a new market here (if you are, you might want to look at Process Design For Six Sigma[5]), this subset works well for the majority of projects:

Brainstorming / *Murphy's Analysis*

To help frame what questions to ask the Customers regarding the process, it is useful to spend some time as a Team brainstorming the issues around it. This can be done in the form of traditional Brainstorming, or it is preferable to use the tool known as *Murphy's Analysis*. Note that the output of these tools is absolutely *not* the Voice of the Customer, but rather the Team's internal insights to help guide the structure of the VOC questioning in the subsequent steps.

[4] Learn more about Design For Six Sigma, Lean Design For Six Sigma, and Process Design For Six Sigma at www.sbtionline.com.

[5] See also *Commercializing Great Products with Design for Six Sigma* by Randy Perry and David Bacon (Prentice Hall PTR, ISBN: 0132385996).

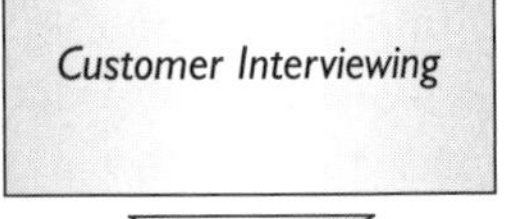

Based on some of the learning from the internal brainstorming, it should be reasonably straightforward to put together an Interview Discussion Guide, Customer Matrix and then go ahead and interview Customers. Richness of Customer data is based on diversity of Customers—not on quantity—and thus, 12–20 interviews should be enough to frame even the most complex process. Interviewing can take 1–2 weeks to complete, so plan the Team's time accordingly.

See "Customer Interviewing" in Chapter 7, "Tools," for more detail.

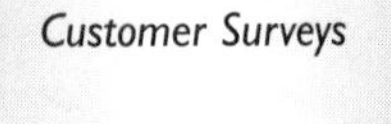

Some processes have a vast multitude of Customers for whom the Team would like some input to ensure that nothing key is missed, but really don't want to spend time interviewing in depth. To gain these Customer inputs, use a survey that is either mailed or handed directly to the Customer or, primarily for internal Customers, posted on the wall, and have them write directly on the surveys. Simply posting the survey on the wall might seem a little hit-and-miss, but it is a surprisingly effective way of reaching respondents that might not take the time to answer in another form.

The survey usually will take one week to get the data back.

Surveys are notoriously error prone, so do not rely on this as your main source of Customer input—use the interviews instead.

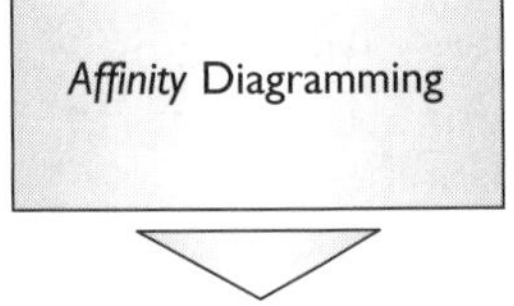

One of the hardest parts of gaining the Voice of the Customer is distilling the copious volume of data down into useful information. This can be done with a simple *Affinity* Diagram or by using a more complex affinity tool such as KJ.[6] For the majority of projects where the number of unique Customer voices is not overwhelmingly large, the simple approach works best.

[6] KJ is known after its inventor, the Japanese anthropologist Jiro Kawakita. For more details, see *A New American TQM: Four Practical Revolutions in Management* by Shoji Shiba, et al. (Productivity Press, ISBN: 1563270323).

Customer Requirements Trees / 5 Whys

After the *Affinity* Diagram is complete, a simple extraction will yield the *Customer Requirements Tree.* This Tree represents a simple hierarchical structure of needs from the highest abstract level down to individual facts and measures. At this point, a Team can say it has captured the Voice of the Customer. To successfully create the tree, it is often best to track the noted metrics back to the set of more essential metrics behind the scenes. A useful tool here is the *5 Whys.*

Use of the *5 Whys* is important throughout the project. Even though it is a very simple tool, it has potent use in the roadmap. I use a slight derivative, which involves asking "Why do I care?" A very recent example in a hospital client demonstrates the importance of using this at this point. The project focus was on the Prepare Visit made by expectant mothers a few weeks prior to the big day. The Team had identified a key Y to be "% of new mothers given access to the Prepare Visit." By asking "Why do I care?" a number of times, the thought flow was as follows:

- More mothers need to go through the Prepare Visit.
- Education given during the Prepare Visit is more readily absorbed than during the Birthing Visit, and valuable doctors' time is wasted reworking the education during the Birthing Visit.
- New mothers need to understand and retain key learning before they return home.
- New mothers are the ones that are required to give correct care to themselves and their newborns after they leave the hospital.
- Sometimes new mothers or babies, after returning home, have to be readmitted to the hospital for what are essentially avoidable reasons.

At this point, the Team realized that the Prepare Visit was in fact a solution to a problem, and the true underlying issue and associated major metric was "Rate of avoidable returns postpartum." This brought a whole new perspective to the problem and hence the project.

Key Process Output Variables (KPOVs)

The Team needs to finalize the primary metrics by which project success will be measured. These are known as the Ys for the process, and are a key focus in the project through the equation $Y=f(X_1, X_2,...X_n)$ where the Xs are all the factors in the process that affect the Ys. See "KPOVs and Data" in Chapter 7 "Tools" for more detail.

Determining the Ys is done by taking the output of the *Customer Requirements Tree* and firming up Operational Definitions of the key metrics. For each metric, a baseline measure is made.

It might be worthwhile to scan through some of the appropriate Problem Categories in Chapter 3, "Global Process Problems," to help identify metrics pertinent to the problem at hand to help complete the KPOVs.

The primary purpose of Define is to make sure this is the right project to be working on. Oftentimes, Belts make the mistake of moving very quickly into Measure and looking at detailed Process Capability studies, when really a simple quick-and-dirty measure of baseline performance would suffice. The purpose of the baseline at this point is just to see if the project is a real project. Later, the Team can look at Measurement Systems and Process Capability and clean up the baseline metric. However, if no baseline data is readily available and the Team has to contrive a sampling regime to get a baseline, it is worth investing time in looking at the validity and reliability of the metric before doing so.

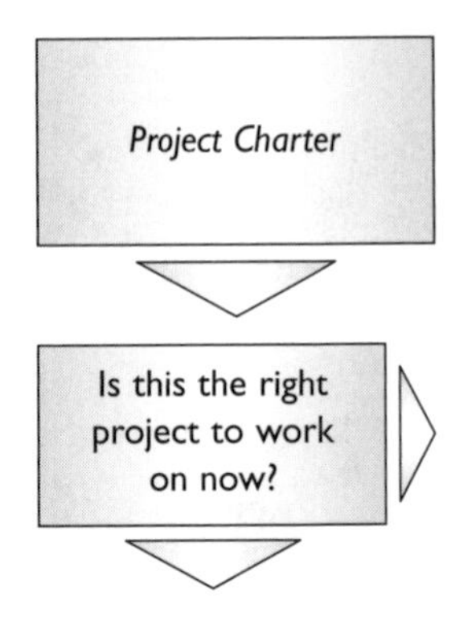

The Belt and Champion return to the *Project Charter* to finalize all the fields based on the Team's work. At this point, the decision is made whether to continue or not.

This is the most important decision point in the whole DMAIC roadmap and is often overlooked. The early commitment to move into Define was not in fact the final commitment to do the project. Only at this point do the Belt and Champion have enough information at hand to make the call.

After the Belt, Champion, and Process Owner sign off on the *Project Charter*, the project can move into the Measure phase. Before doing that, it will be worthwhile to set up some rudimentary Project Management:

Although not covered here in detail, Project Management[7] is a key part of project success. The Belt should look to having at least a:

- Milestone Plan for the project
- Risk Management Plan
- Detailed Work Plan for each Milestone
- Hazard Escalation matrix for those unforeseen circumstances

[7] There are many texts to help projects in general; one method that suits Lean Sigma well is outlined in *Goal Directed Project Management* by Andersen, Grude, and Haug (Kogan Page, ISBN: 0749441860).

At this point, the Team should know enough about the process and the problems it has from insight gained from the VOC tools. The Team does not need to know *why* the problem is occurring (that will come later in the Analyze phase), but the basic understanding of the problem (often captured in a Problem Statement in the *Project Charter*) will allow it to identify the type of problem apparent. A list of generic Problem Categories is shown in Table 2.1. To proceed further in the Lean Sigma roadmap, the Team should select the most appropriate Problem Category from the list in Table 2.1 and go to that section in Chapter 3 (as indicated by the letters A–Y).

Table 2.1 Problem Categories for the process as a whole

Section	Issue	Description
A	On-time delivery problems	There commonly is a failure to have the right entities at the Customer at the right time
B	Process capacity too low	The process cannot generate entities quickly enough to keep up with the pace of demand from the Customer
C	Process defects too high	The process generates too many imperfect entities. Rework or scrap too high. Quality is being "inspected into the process" rather than entities being right the first time
D	% Uptime of process too low	The process is down (not doing value-added work) too much
E	Pace of process too slow	When the process is performing value-added work, it isn't doing so quickly enough
F	Process fails intermittently	The process has enough capacity to keep up with demand, but fails every so often, perhaps missing an entity entirely
G	Process lead time too long	The capacity of the process is sufficient, but lead time through the process is too long
H	Individual steps meet the pace of downstream demand, the process as a whole does not	Despite individual steps within the process all meeting Takt and hence theoretically being able to keep up with the pace of demand, the global process does not meet Takt and therefore cannot keep up
I	Customer demand too variable	Customers demand entities in a highly irregular pattern, making planning and inventory control extremely difficult
J	Too many products	The portfolio of different types of entities that the process can generate is too large, keeping inventory levels high and making planning difficult

(continues)

Table 2.1 Problem Categories for the process as a whole (Continued)

Section	Issue	Description
K	Process deviating from schedule	The process of generating a future schedule is highly inaccurate, thus leading to large variance from plan after the fact (note this is different from Deviation from Forecast—see U)
L	Measurement system is broken	A key measurement system used to judge process performance exhibits high variation and is thus ineffective
M	Performance characteristic not meeting specification	Entity performance metrics (e.g., product strength, purity, accuracy, etc.) are not hitting required target levels
N	Planned maintenance takes too long	When the process is stopped to perform planned maintenance, the associated downtime is too long
O	Setup/changeover takes too long	Process changeover time between individual entities or different entity types takes too long
P	Too much unplanned maintenance	Maintenance conducted is reactive (unplanned) and there is too much of it
Q	Process fails to make product at all	All entities the process generates are defective or the process generates no entities whatsoever
R	Resource usage too high	Headcount used to run the process is too high
S	Process inventory too high	The inventory in the process, whether standard work-in-process or unplanned, is too high
T	Process waste/loss too high	Mass loss (waste) from the process is too high, giving low process yield
U	Process deviating from forecast	The process of generating a future forecast is highly inaccurate, leading to large variance from plan after the fact (note this is different from Deviation from a Schedule—see K)
V	Not enough sales	There are not enough sales of entities from the process
W	Backlog of orders is too high	The process is lagging behind sales, creating a backlog of orders
X	Payments made to suppliers not optimized	Accounts Payable is too high for the business—the timing of payments to suppliers is too variable
Y	Accounts Receivable is too high	The business isn't collecting payments for its products and services quickly enough

Following the roadmaps as indicated in Table 2.1 will home in on the particular problem and identify a means to a solution. This might involve at some point scoping down to focus on one particular process step, identified as the root cause of the issues, rather than treating the process as a whole.

Sometimes we are fortunate enough to have such a clear problem that the individual process step causing the problem is absolutely obvious. In these (albeit rare) instances, it is possible to shortcut the preceding roadmaps and go straight to the Individual Step Roadmaps, as shown in Table 2.2 and described in detail in Chapter 4, "Individual Step Process Problems."

Table 2.2 Problem Categories for a single process step

Section	Issue	Description
1	Process step does not meet Takt	A single process step cannot process quickly enough to keep up with the pace of demand
2	Pace of process step is too low	When a single process step is performing value-added work, it isn't doing so fast enough
3	Cycle time of process step exhibits too much variation	There is too much variability in the time it takes a single step to process an entity

In general (except for the instances listed in Table 2.2), the roadmaps to tackle problems at a single step level are entirely analogous to those for the whole process. If you do not see your Problem Category listed in Table 2.2, refer back to Table 2.1 and select the roadmap from there.

3 Global Process Problems

A: On-Time Delivery Issues

Overview[1]

Virtually all processes are designed to deliver an entity (anything from a single item to a whole shipment) to a Customer, clearly with a Customer's need of having the right entity at the right time (essentially a measure of effectiveness of the process). As the Supplier, we are also interested in the cost to deliver the entity, whether it is in labor or processing costs (measures of efficiency of our process).

Examples

- **Industrial.** Product shipments
- **Healthcare.** Medications delivery/administration, laboratory/radiology turnaround
- **Service/Transactional.** Service delivery, invoicing

Measuring Performance

In considering on-time delivery issues, we are focusing on measures of effectiveness of the process. Thus, to measure delivery performance, typical primary measures would comprise

[1] Before reading this chapter, it is important to read and understand Chapter 1, "Introduction," particularly the section, "How to Use This Book."

- **Accuracy** (are we delivering the right entity, but not necessarily on time) measured as
  - A percentage hit rate—the percentage of the time that we deliver the right thing in the right form in its entirety.

  Or, conversely as
  - A failure rate as a percentage, or in parts per million.
- **Timeliness** (are we delivering an entity on time, not necessarily the right entity) measured as
  - A Replenishment Time from the Customer needing the entity to when it is available to use—this could be the entire Lead Time of our process, or merely a delivery time if we have inventory on hand, ready to ship. It is important to note the difference. For more detail see "Time—Replenishment Time" and "Time—Lead Time" in Chapter 7, "Tools."

TOOL APPROACH

The roadmap commences with

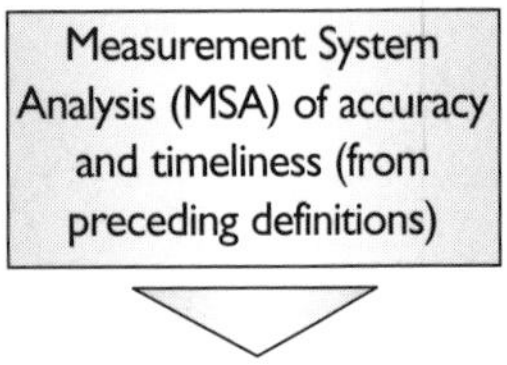

Measurement System Analysis (MSA) of accuracy and timeliness (from preceding definitions)

For accuracy, a full Attribute MSA or Gage R&R might be required. For timeliness, the focus is on validity of the measure and not so much on the reliability of the measure. Thus, just a solid operational definition and validation probably will suffice. This should be conducted for the process as a whole, not for each of the individual steps in the process. For more detail see "MSA—Validity," "MSA—Attribute," and "MSA—Continuous" in Chapter 7.

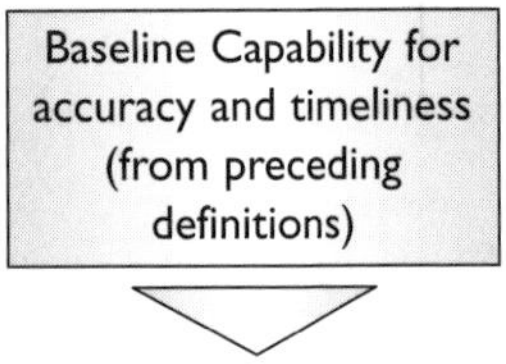

Baseline Capability for accuracy and timeliness (from preceding definitions)

This should be conducted for the process as a whole, not for each of the steps in the process. Later, however, it will be important to capture Capabilities for each step along the process, so if freely available at this point, capture them as well. For more detail see "Capability—Attribute" and "Capability—Continuous" in Chapter 7.

If Accuracy baseline capability is poor

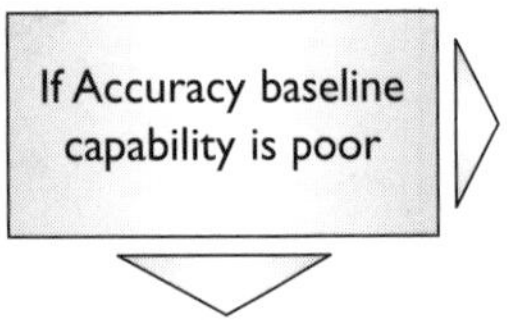

Continue to Section C in this chapter and consider a defect to be "any time that the perfectly correct entity is not delivered."

If Timeliness baseline capability is poor

It will be necessary to understand what is specifically in the process that is affecting the time. Continue down this roadmap.

Calculate Takt Time
Calculate Process Cycle Time
Calculate Process Lead Time

It is important to understand the relationship between the Takt Time (the pace of use of entities), the Process Cycle Time (the actual pace of generation of entities), and the Process Lead Time (the time it takes a single entity to progress the length of the process). For more detail see "Time—Takt Time," "Time—Global Process Cycle Time," and "Time—Process Lead Time" in Chapter 7.

Apply *Demand Segmentation* (if multiple entity types are run through the process)

This will allow us to understand the volume and variation in demand of each of the different entity types that progress through the process. High-volume, low-variation demand entities might have to be dealt with differently than low-volume, high-variation demand entities.

Apply *Demand Profiling* to the highest volume or key entity types

For the higher volume or key entity types that pass through the process, it is important to understand the demand profile over time. Significant spikes in demand or seasonal fluctuations might cause us to set up the process differently.

From applying these tools to timeliness, we should have gained an understanding of whether the issue is either

Capacity of process is too low

On average, the process does not have enough capacity (the Cycle Time is longer than its Takt Time). For more detail see "Time—Takt Time" and "Time—Global Process Cycle Time" in Chapter 7. The process will never generate enough entities to meet demand.

Continue to Section B in this chapter.

Capacity of process is sufficient, but process fails intermittently

On average, the process has enough capacity (the Cycle Time is below the Takt Time), but the process fails intermittently. For more detail see "Time—Takt Time" and "Time—Global Process Cycle Time" in Chapter 7. On average, the process can meet demand (including peaks), but the process is not robust enough to do so on a continuous basis.

Continue to Section F in this chapter.

Capacity of process is sufficient, but Process Lead Time is too long

The Cycle Time is below its Takt Time, but the Process Lead Time is too long. For more detail see "Time—Takt Time", "Time—Global Process Cycle Time" and "Time—Process Lead Time" in Chapter 7. The process can meet customer demand; however, when an entity is required quickly and is not available to ship from inventory, the time taken to generate a whole new entity from scratch is too long.

Continue to Section G in this chapter.

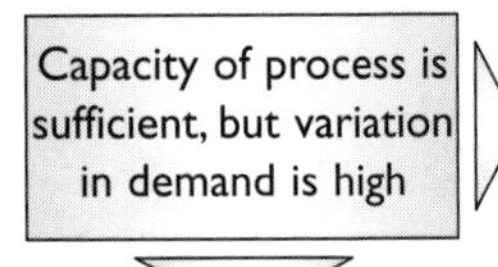

On average, the process has enough capacity (its Cycle Time is below its Takt Time), but variation in demand causes shortages in available entities intermittently. For more detail see "Time—Takt Time" and "Time—Global Process Cycle Time" in Chapter 7. The output of the *Demand Profile* tool (or *Demand Segmentation* tool if many entities are run through the process) will show spikes in demand beyond the current capacity of the process. There are two options in this case: Increase the capacity of the process, or smooth the demand from the Customer/market.

Continue to Section B in this chapter to increase the capacity of the process.

Continue to Section I in this chapter to smooth the demand from the Customer.

An obvious question arises regarding to how to deal with the situation if the problem is due to more than one of the preceding; what would take priority?

Unfortunately, there are no hard-and-fast rules. The approach would be to examine each of the potential options and determine the best approach, perhaps by using an Impact/Effort Graph. In general, it is usually easier to generate more internal capacity rather than change the external reality of customer-demand profiles, but your process/business could be the exception to that.

B: Capacity of Process Is Too Low

Overview

Processes are required to cycle at a fast-enough rate to generate entities at a pace to meet customer (or market) demand. If the process capacity is too low, the process isn't cycling quickly enough and will inevitably fall behind or won't be able to meet spikes in demand when they occur.

It is useful to ask the question before proceeding as to why this low capacity is a problem. If the answer relates to failure to meet delivery to Customers, revert to Section A in this chapter before following the approach outlined here.

Examples

- **Industrial.** Plant/line throughput
- **Healthcare.** Emergency department/surgery throughput
- **Service/Transactional.** Service center capacity

Measuring Performance

In considering capacity of a process, typical primary measures would comprise

- **Throughput** measured as
  - Average number of entities generated by the process over a period of time (typically, a day or month)
- Or **Cycle Time** measured as
  - The time between one entity exiting the process and the subsequent entity exiting the process. This should be faster than the Takt Time. For more detail see "Time—Takt Time" and "Time—Global Process Cycle Time" in Chapter 7, "Tools."

Both metrics effectively achieve the same result; however, it sometimes is useful to describe the problem in both formats to better explain to Team members the implications of the problem.

Tool Approach

If not already done in a previous step, commence with

Measurement Systems Analysis on throughput (or Cycle Time)

Focus should just be on measuring validity (a sound operational definition and consistent measure) of the metric versus a heavy investigation of Gage R&R. For more detail see "MSA—Validity" and "MSA—Continuous" in Chapter 7.

Baseline Capability on throughput (or Cycle Time)

This could be as simple as identifying how many entities the process generated over the period of, say, one month and calculating a throughput rate from that. For more detail see "Capability—Attribute" and "Capability—Continuous" in Chapter 7.

If the throughput of the process is still deemed to be too low, continue with

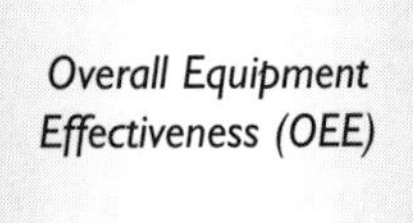

The process would hit maximum capacity if it were running all of the time only doing value-added (VA) activity *and* if it were running as fast as it has ever run *and* if it were generating perfect quality entities.

From the OEE, we should have gained an understanding of whether the issue is either

Rolled Throughput Yield of the process is too low

This is a quality-related problem. Capacity is effectively lost because the process spends time generating defective entities.

Continue to Section C in this chapter.

% Process Uptime is too low

Capacity is effectively lost because the process is spending part of its available time to do VA entity generation doing something else.

Continue to Section D in this chapter.

Pace of process is too slow

Capacity is effectively lost because the process could do the VA work more rapidly.

Continue to Section E in this chapter.

C: RTY, Defects, Accuracy, Quality, Scrap, and Rework Issues

Overview

Processes are required to generate as high a proportion of perfect entities as possible. Any time a nonperfect entity is generated, capacity is lost, and costs are incurred in lost materials, increased labor, and so forth.

Often rework is disguised under other names, so it is useful to spend some time examining the process to understand exactly how cost is being incurred.

Examples

- **Industrial.** Reworking of downgrade product (blending off), downgrading product, line scrap, and material loss
- **Healthcare.** Patient handoff (incompleteness of information), medications delivery/administration accuracy, clinical defects such as Ventilator-Acquired Pneumonia
- **Service/Transactional.** Billing accuracy, order accuracy

Measuring Performance

In considering quality, typical primary measures would comprise

- **Rolled Throughput Yield (RTY)** or **First Time Right** measured as
  - The percentage of entities that make it all the way through the process Right First Time, every step along the way. This is not to be confused with the final Yield of the process, which typically is inflated with rework throughout the process. The RTY is commonly much lower than understood by the business.
- **Primary Performance Metric(s)**
  - The process has a performance characteristic(s) for the entities it generates that is being measured against a specification (e.g., a dimension, strength, or other physical/chemical characteristic).

RTY is a *conformance* metric in that it represents the percentage of occurrences conforming to a specification on a *performance* metric. It is often better to proceed through the roadmap using the performance characteristic(s) itself as the primary metric, rather than a conformance metric such as RTY. If all else fails, use RTY as the primary metric.

TOOL APPROACH

If not already done in a previous step, commence the roadmap with

Measurement Systems Analysis (MSA) on the primary performance metric

More than likely, this will involve a detailed measurement system reliability study (a Gage R&R study for continuous data, or an Attribute MSA/Kappa study for attribute data), rather than just a simple validation. An MSA will be required for each performance characteristic for the entity (e.g., strength and water absorption requires two separate MSAs to be completed). See "MSA—Validity," "MSA—Continuous," and "MSA—Attribute" in Chapter 7, "Tools."

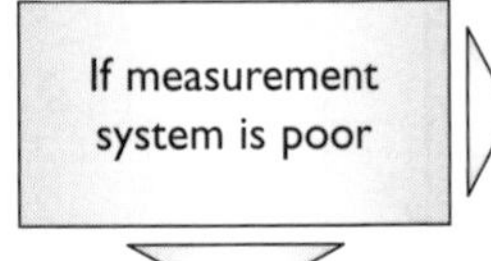

Sometimes the entire problem comes down to an issue with the measurement system. If the MSA conducted previously shows a high % R&R or P/T ratio, move to Section L in this chapter to mend the System.

After the measurement system has been validated, perform

Baseline Capability Study on the primary performance metrice

For attribute data, this will simply be RTY, Defects Per Unit (DPU), or Defects Per Million Opportunities (DPMO). At least 100 data points will be required if the defect rate is high (>5%), and more if the defect rate is lower. For more details see "Capability—Attribute" in Chapter 7.

For continuous data conduct, a full Capability Study to calculate C_p and C_{pk} will be needed. At least 25 data points will be required. For more details see "Capability—Continuous" in Chapter 7.

For both data types, data must be captured over a period long enough for typical process noises to have an effect. For example, if there are monthly fluctuations, data needs to be available across two months or more. Capture one to two weeks' worth of data to get a rough estimate of capability, but continue down the roadmap in parallel while continuing to collect the capability data.

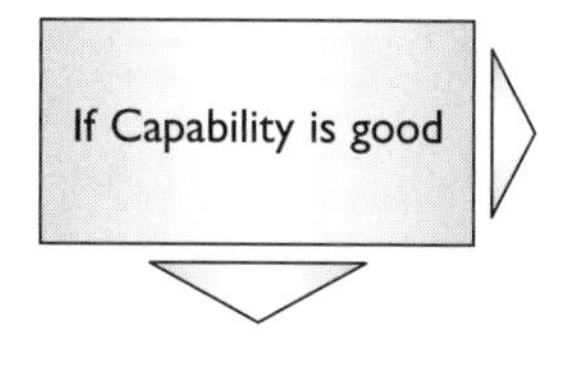

At this point, the problem has been resolved in that the measurement system was the problem (or it's a remote possibility that previous Capability Studies were flawed), in which case see the Control tools in Chapter 5, "Control—Tools Used at the End of All Projects." If not, proceed down the roadmap in this section.

The roadmap to a solution from this point forward relies on the equation $Y=f(X_1, X_2,..., X_n)$, where the Y is the Primary Performance characteristic(s):

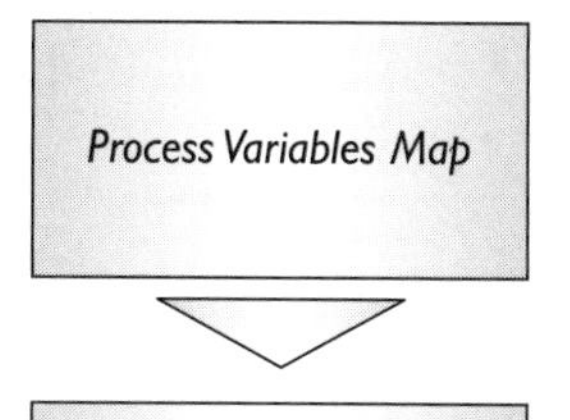

This tool will identify all input variables (Xs) that cause changes in the Primary Performance Metric(s) (Ys), otherwise known as output variables. Any obviously problematic uncontrolled Xs should be added directly to the *Process Failure Mode and Effects Analysis (FMEA)*.

Cause & Effect (C&E) Matrix

The Xs generated by the *Process Variables Map* are transferred directly into the *C&E Matrix*. The Team uses its existing knowledge of the process through the matrix to eliminate the Xs that don't affect the Ys. If the process has many steps, consider a two-phase *C&E Matrix* as follows:

- Phase 1—List the process steps (not the Xs) as the items to be prioritized in the *C&E Matrix*. Reduce the number of steps based on the effect of the steps as a whole on the Ys.
- Phase 2—For the reduced number of steps, enter the Xs for only those steps into a second *C&E Matrix* and use this matrix to reduce the Xs to a manageable number.
- Phase 3—Make a quick check on Xs from the steps eliminated in Phase 1 to ensure that no obviously vital Xs have been eliminated.

Process Failure Mode and Effects Analysis (FMEA)

The reduced set of Xs from the *C&E Matrix* are entered into the FMEA. This tool will narrow them down further, and generate a set of action items to eliminate or reduce high-risk process areas.

This is as far as the Team can proceed without detailed process data on the Xs. The FMEA is the primary tool to manage the obvious Quick Hit changes to the process that will eliminate special causes of variation. At this point, the problem may be reduced enough to proceed to the Control tools in Chapter 5. If not, continue down this roadmap.

Multi-Vari Study

The reduced set of Xs from the *FMEA* is carried over into this array of tools along with the major Ys from the *Process Variables Map*. Statistical tools applied to actual process data will help answer the following questions:

- Which Xs (probably) affect the Ys?
- Which Xs (probably) don't affect the Ys?
- How much variation in the Ys is explained by the Xs identified?

The word *probably* is used because this is statistics, and hence there is a degree of confidence associated with every inference made. This tool will narrow the Xs down to the few key Xs that (probably) drive most of the variation on the Ys.

In service/administrative/transactional processes, it is usually possible to move straight from this point to the Control tools in Chapter 5 without conducting any Designed Experiments unless some kind of (computer) simulation of the process can be made. For more details on Designed Experiments see "DOE—Introduction" in Chapter 7.

Design Of Experiments (DOE)—Screening Design

If there are still a large number of Xs (6 or more) left after the *Multi-Vari Study*, it is best to reduce that set using a Fractional Factorial DOE. This tool will only identify the key Xs and should not be used to investigate interactions or to optimize the process.

DOE—Characterizing Design

The reduced set of Xs from the Screening Design is more deeply understood using a Full Factorial Design. This tool will help us determine the final reduced set of Xs to be controlled in the process, along with any interactions between those Xs. Another output will be the amount of variability explained by these Xs.

It is possible at this point that the process is understood deeply enough and the results are good enough that no further optimization is required. If this is the case, proceed to the Control tools in Chapter 5; if not, continue down the roadmap here.

DOE—Optimization Design

In a small number of instances, it might be appropriate to use Response Surface Methodology (RSM) (or simply *Regression* if we have narrowed down to just one X) to optimize the level of the Xs to maximize performance of the Ys.

After the best levels for the critical Xs have been determined, proceed to the Control tools in Chapter 5.

D: % UPTIME IS TOO LOW

OVERVIEW

Processes are required to be generating entities (doing VA work) for 100% of the shift time. Any time they are not doing VA work can be considered downtime. Note that this is different from the definition typically used by maintenance personnel, which focuses on the *availability* of the process to do any work, rather than the process of actually *doing* work (and that work being all VA work).

This is very similar to the measure known as Utilization, which considers % work time, but does not consider the delineation of value-added (VA) versus non-value-added (NVA) work. For a detailed explanation of % Uptime, refer to "Overall Equipment Effectiveness (OEE)" in Chapter 7, "Tools."

EXAMPLES

- **Industrial.** Uptime of a line, vessel, or plant
- **Healthcare.** Surgery uptime (incision to close)
- **Service/Transactional.** Uptime of a server or carrier

MEASURING PERFORMANCE

In considering uptime, typical primary measures would comprise

- **% VA Uptime** (current shift pattern) measured as the percentage of the existing shift pattern on which the process is performing VA work. Note that VA work does not include
    - Load
    - Unload
    - Rework
    - Inspection

- Planned process maintenance
- Unplanned process maintenance
- Breakdowns
- Scheduled workforce breaks
- Scheduled process downtime of any form
- Changeover time
- Setup time

Often it is useful to also consider the uptime relative to a 24-hour working week, multiple shifts, and so forth, and report these as separate numbers to show the true available capacity of the process, not only that which is limited by current staffing.

TOOL APPROACH

If not already done in a previous step, then apply the sequence of tools as follows:

Measurement Systems Analysis (MSA) on % Uptime

Focus should just be on measuring validity—a sound operational definition and consistent measure of % Uptime versus a detailed investigation of Gage R&R. For more details see "MSA—Validity" in Chapter 7. Determine an exhaustive set of categories of time (e.g., planned maintenance, unplanned maintenance, changeover, etc.). Having a category called "Other" tends to swallow up informative events and consequently any opportunity for improvement, so be as specific as possible.

Baseline Capability Study versus 100% value-added uptime

Use the process operators to capture data for the time categories as defined in the preceding MSA. Ensure they have a clear understanding of the intent of the project to try to reduce the Hawthorne Effect[2] (the Champion might provide valuable input here).

One week's data typically provides a reasonable starting sample for all but the slowest cycling processes, but continue the data capture for long enough that the process time categories are all hit. Consider using historical data to represent large downtimes due to maintenance or forced shutdown.

[2] Individual behaviors might be altered when they know they are being studied. First demonstrated in research at the Hawthorne Plant of the Western Electric Company, Cicero, Illinois (1927–1932).

Pareto total shift time

Categorize the total (shift) time into uptime and all the possible downtime buckets using a *Pareto Chart*. Use the tool to focus the next steps of the project on the biggest sources of lost uptime. There might also be small sources that can be tackled very easily with a quick hit (i.e., negligible cost, dumb stuff). Take these opportunities immediately.

Options will be either

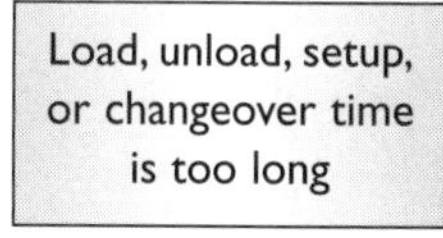

If the changeover for the whole process or for a single step is too long, go to Section O in this chapter.

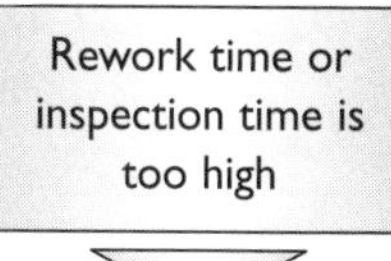

If time is spent reworking, inspecting, scrapping, or reacting in any way to defective entities, it is a quality-related problem and should be addressed using the roadmap in Section C in this chapter to eliminate the source(s) of the defects.

If there is too much unplanned maintenance or the process simply breaks down on a regular basis, go to Section P in this chapter.

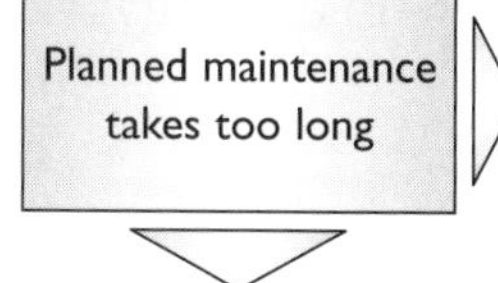

If there is too much planned maintenance or planned maintenance takes too long, go to Section N in this chapter.

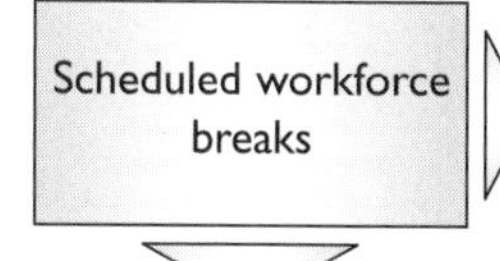

This outcome will not be examined in detail in this book.

Consider using different staffing models, staggered breaks, and so on. Use the operators themselves to come up with the mechanisms to keep the process running during breaks. This helps ensure buy-in, which is typically the most difficult factor in these situations. Consider just keeping the bottleneck (the rate-determining step) in the process running during the breaks—it will require a buffer of inventory ahead of the bottleneck to achieve this.

For any other downtime category, unfortunately, you are on your own. Start with simple tools to map the problem or look to the Problem Categories in this chapter and

Chapter 4, "Individual Step Process Problems," for the smaller problem to see if there are any guiding lights.

E: Pace of Process Too Slow

Overview

Sometimes, even though the process is up, it just isn't cycling fast enough to generate entities to meet downstream demand. *Pace* is the speed of the process when it is up and running doing VA work, which can be a confusing concept for some. Pace only looks at the true VA work that is considered Uptime in the % Uptime calculation. This is not the average process rate that is traditionally captured. For a detailed explanation of pace, refer to "Overall Equipment Effectiveness (OEE)" in Chapter 7, "Tools."

Look to make improvements in pace only after other categories of NVA work have been improved.

Examples

Applicable to all industries is that the hands-on "touch time" in a process is too slow.

- **Industrial.** Vessel (reaction) times too long, assembly time reduction
- **Healthcare.** Procedure time reduction
- **Service/Transactional.** In the transportation industry, the actual transportation time and order processing time at a single step

Measuring Performance

Measures of Pace of a process are usually process-specific, but are generally written in the form of number of entities per unit time. For example:

- Kg product/hr
- Patients/hr
- Widgets/hr
- Invoices (or orders)/hr

Similarly, they can be written as an average time to generate or process an entity:

- Average processing time

The data needs to be captured *only* when the process is running and should not take into account the average slowing of pace due to process stoppages, downtimes, and so forth. Those elements should be accounted for in the % Uptime metric in OEE. If the concern is more one of Uptime, proceed to Section D in this chapter instead.

Tool Approach

First we must get an understanding of the current performance with respect to pace.

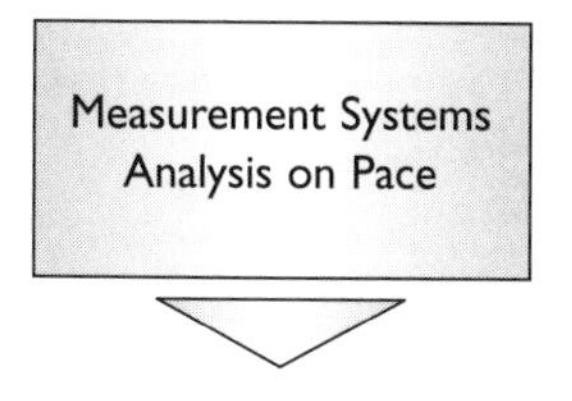

Pace is always a little tricky to determine, so a quick MSA is always worthwhile. Focus should just be on measuring validity—a sound operational definition and a consistent measure of pace while the process is doing VA work versus a detailed investigation of Gage R&R. See also the tool detail for OEE in "Overall Equipment Effectiveness" and "MSA—Validity" in Chapter 7.

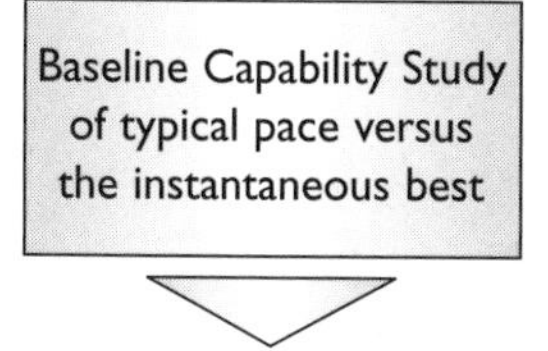

Pace capability is measured as a percentage versus the best the process has ever done (see the tool detail for OEE in Chapter 7). The data should be captured over a period of typically one week to one month (depending on process drumbeat) to get a reasonable estimate of average pace.

The slow pace could be either due to individual steps in the process not cycling quickly enough; or it could be that all the individual steps cycle quickly enough, but something gets lost in the gaps between the process steps. The tools used in the following roadmap aim to understand which case it is

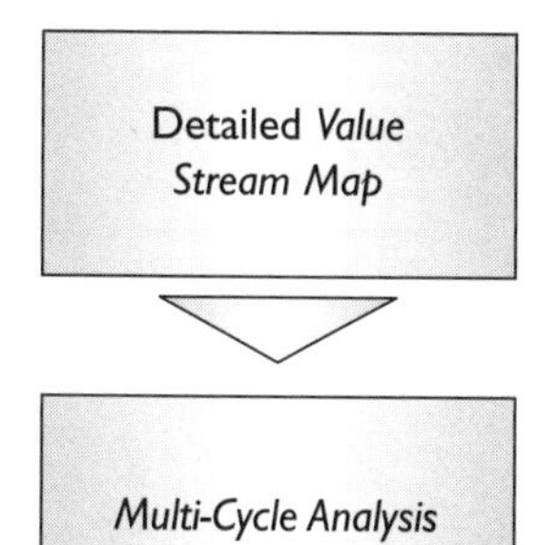

Construct a rigorous *Value Stream Map* including all the detailed steps for the Primary Entity as it progresses through the process. At this point, don't map the substeps of any VA step—that mapping will come later.

Multi-Cycle Analysis

Apply a *Multi-Cycle Analysis* to the VSM to flesh out where the time is being spent along with an indication of variation in times. Usually it is the variability in step times that causes the problems.

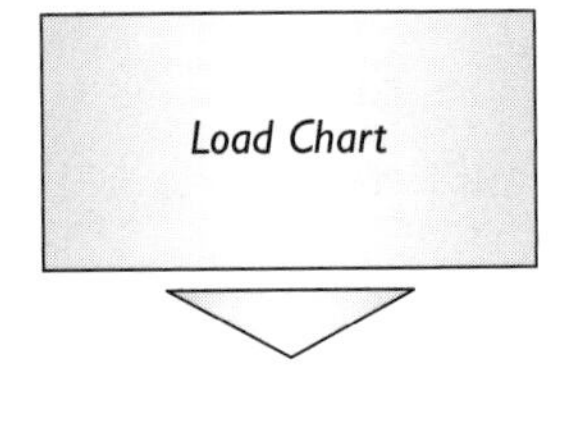

A *Load Chart* for the process will show Takt Time versus Cycle Time. For more details see "Time—Takt Time," "Time—Individual Step Cycle Time," and "Time—Global Process Cycle Time" in Chapter 7. This will highlight individual process steps that aren't cycling quickly enough to meet the pace of demand; or it will show that all steps are meeting the pace of demand, but the process as a whole is failing somehow despite this.

No matter what the *Load Chart* tells us, at this point it is always useful to use the *Value Stream Map* to identify NVA activities. The Team should spend time brainstorming action ideas to eliminate NVA activity in the process, which will in turn increase the pace.

The *Load Chart* will lead us to two possible options:

Global process does not meet Takt but individual steps do

If each individual step along the process is able to perform at a rate fast enough to keep up with the pace of downstream demand, go to Section H in this chapter to determine if either

- The linkage inside the process is failing.
- The delivery to the downstream Customer is failing.
- Supply from the upstream supplier is failing.

Individual process steps do not meet Takt

If, for some steps, the Cycle Time is greater than the Takt Time (the process steps cannot keep up with the pace of demand), then determine if these truly are VA activities. If not, work to eliminate them. If there are VA steps in the process that are too slow, identify the bottlenecks and proceed to the roadmap outlined in "1: A Single Process Step Does Not Meet Takt" in Chapter 4.

F: Process Has Enough Capacity, But Fails Intermittently

Overview

Some processes appear on first sight to have enough capacity to meet downstream demand but for some reason completely fail to deliver at all at times. This is different from accuracy of delivery performance as discussed in Section A in this chapter (and subsequently in Section C), where perhaps a defective or incomplete entity is delivered. Here the issue is more one of process reliability versus process performance: a seemingly good process suddenly misses an entity entirely.

Examples

- **Industrial.** Orders lost in the system, lost shipments
- **Healthcare.** No medication arrives, lost records or charts, missed charge
- **Service/Transactional.** Failed delivery, missed claim

Measuring Performance

This is in essence a process reliability problem, but can be measured using a typical quality approach if the number of failures is reasonably high (say greater than 5%):

- **Rolled Throughput Yield (RTY).** Measured as the percentage of entities that make it all the way through the process every step along the way.

However, if the number of instances of failure is low, a different approach will be needed or otherwise a vast amount of data will be required.

One approach is to use reliability metrics as follows:

- **Mean Time Between Failures (MTBF).** Calculated as the average number of days (or hours or minutes) between instances of failure. Clearly the drive will be to make this number increase.
- **Mean (Normalized) Time Between Failures.** Sometimes the number of entities processed varies over time and hence a normalized version of MTBF is used. For example, in the hospital environment when we consider patients falling (and subsequently injuring themselves), we look at the number of Patient Days Between Falls (how many patients did we have for how many days). If the entity volume is variable, this normalized version is preferable.

Tool Approach

First we must get an understanding of the current performance with respect to failures:

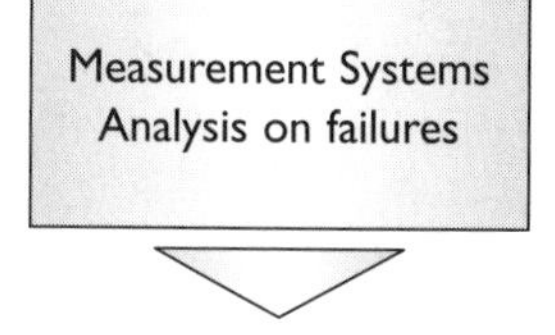

Focus should just be on measuring validity—a sound operational definition and consistent measure of failure versus a detailed investigation of Gage R&R. For more details see "MSA—Validity" in Chapter 7, "Tools." Commonly, these are metrics that have never been tracked before, so initially data collection might have to be done manually until systems can be updated to include them.

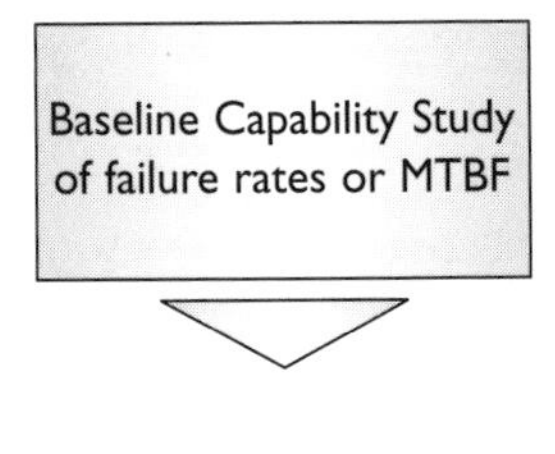

Usually failures are few and far between and hence a longer time interval is required for data collection. The good news is that failures usually hit the business hard and there will likely be reasonably good historical data of the instances of failure. It might take some manual manipulation to get it into the right form, but at least the project shouldn't have to stall waiting on data. Analyze data for a period of typically one month to one year (depending on process drumbeat) to get a reasonable estimate of Baseline Capability. For more details see "Capability—Attribute" and "Capability—Continuous" in Chapter 7.

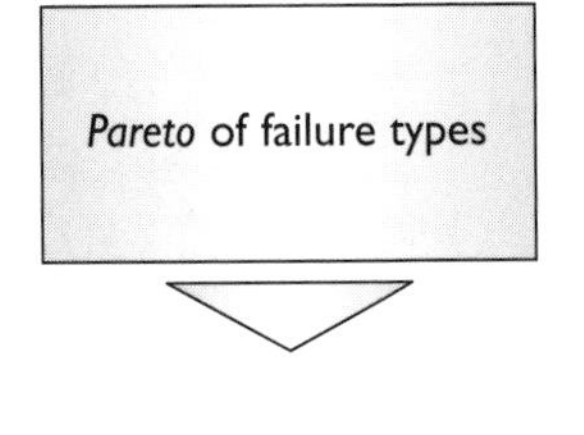

Some causes of failure are so infrequent or unlikely that they shouldn't be the focus of the project. The typical aim of this type of project is to get at least a 50% reduction in failures (or equivalently a doubling of the MTBF). This usually can be achieved by focusing on just a few of the most common or biggest impact failure types. The *Pareto* will highlight these if they aren't already known.

At this point, there might be some obvious solutions that spring to mind, but there really isn't much basis upon which to make a change. Further digging will be required. For novice Belts, the tendency here might be to jump into the $Y=f(X_1, X_2,..., X_n)$ roadmap (as outlined in Section C in this chapter). In fact, a couple of simple mapping tools applied here might help us considerably before taking that route. The best tenet to keep in mind here is "There is only one process that I can consistently do perfectly, every time, for all time and that is no process at all." Seeking to drive out complexity will increase reliability—simple works best!

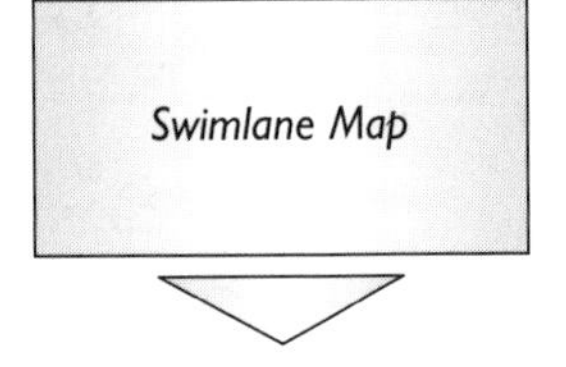

Use the *Swimlane Map* at a high level to identify failures occurring at process handoffs (e.g., missed expectations, poor accountability).

Take the opportunity at this point to remove any obvious NVA activity too.

Some Teams prefer to generate the *Value Stream Map* first and then morph that into the *Swimlane Map*—use whichever method the Team prefers to get to the end result.

If there are many handoffs, the *Handoff Map* will sometimes show a bias in resource involvement and thus show where effort should be focused. At the very least, this map is a great visual tool to explain the issues to the process stakeholders.

Based on the *Swimlane Map* and *Handoff Map*, the Team might have made significant change to the process. If so, work to ensure those changes are well executed with the correct controls in place.

Were changes made significant enough to make dramatic impact on the failure rate?

If there was an absolute, dramatic change made to the process and the Team knows *beyond doubt* that the root cause of failures was reduced, then start to capture new failure Capability data and move to the Control tools in Chapter 5.

However, if the Capability data seems to show no change in failure performance, or if the Team feels that no significant changes were made from the previous tools, then in effect the key Xs that are driving failure are as yet not known. In this case, proceed to Section C in this chapter to narrow down the Xs that cause process failure and consider the Y to be process reliability.

G: Process Has Enough Capacity, But Process Lead Time Is Too Long

Overview

Processes often can generate entities quickly enough if we consider just the time between entities as they exit the process. However, sometimes an entity has to traverse the whole process in a given time to be considered successful. If the Lead Time through the process is too long, the process has failed to meet demand. For more details see "Time—Process Lead Time" in Chapter 7, "Tools."

Examples

- **Industrial.** Long delivery times on special orders
- **Healthcare.** Length of stay in an emergency department/surgery
- **Service/Transactional.** Billing cycle is too long

MEASURING PERFORMANCE

Measuring Process Lead Time is simply a matter of attaching a "runner" document to an entity and time-stamping it as it progresses through the process. The total Process Lead Time is the time from when a single entity enters the process to when that entity exits the process at the other end. The time is measured as an actual clock time versus resource working hours or Total Work Content.

TOOL APPROACH

First we must get an understanding of the current performance with respect to Process Lead Time:

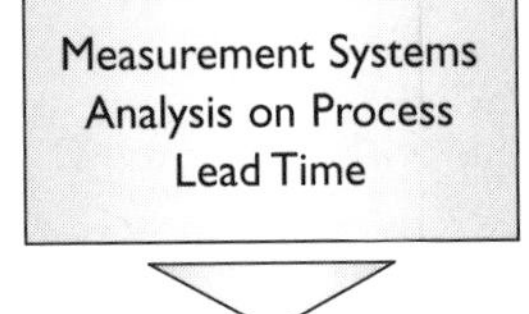

Focus should just be on measuring validity—a sound operational definition and consistent measure of lead time versus a detailed investigation of Gage R&R. For more detail see "MSA—Validity" in Chapter 7. Quite often, the Process Lead Time isn't historically measured from end to end, so be aware of the start and stop points of the metric.

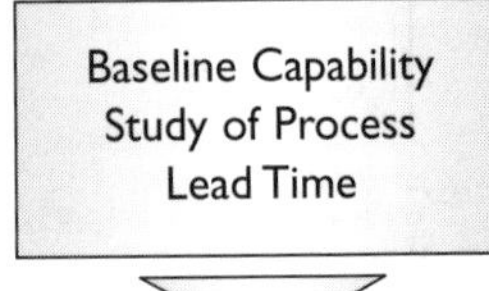

Data capture needn't be too long because it is possible, at least in theory, to get Process Lead Time data on every entity that passes through the process. However, it is important to capture data over a long enough period of time to take into account any external noises that drive the Process Lead Time (volume of demand, different staffing models, etc.).

The data is typically captured over a period of one week to one month (depending on process drumbeat) to get a reasonable estimate.

Process Lead Time can be dramatically reduced by removing NVA activities from the process. A detailed *Value Stream Map* is a tool to identify those activities. Map with a focus on the Primary Entity and particularly the delays it meets. Be sure to consider the whole process from customer-request trigger to customer acceptance. Identify actions to eliminate NVA steps and thus accelerate the Primary Entity.

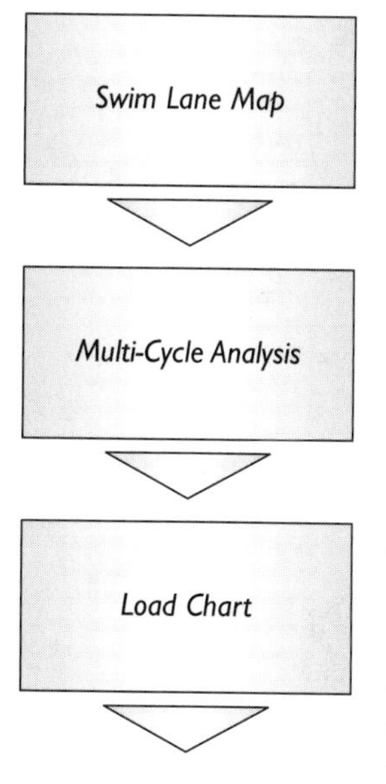

Handoffs through the process can be the cause of significant delay. After the *Value Stream Map* is created, it should be relatively straightforward to add swim lanes to it to identify handoff problems.

Apply a *Multi-Cycle Analysis* to the *Swimlane Map* to flesh out where the time is being spent along with an indication of variation in times. Usually it is the variability in step times that causes the problems.

A *Load Chart* for the process will show Takt Time versus Cycle Time. For more detail see "Time—Takt Time" and "Time—Individual Step Cycle Time" in Chapter 7. All steps should be meeting the pace of demand (or otherwise it wouldn't have enough capacity), but there might be opportunity to redistribute the workload and combine tasks to use fewer operators and hence have fewer handoffs.

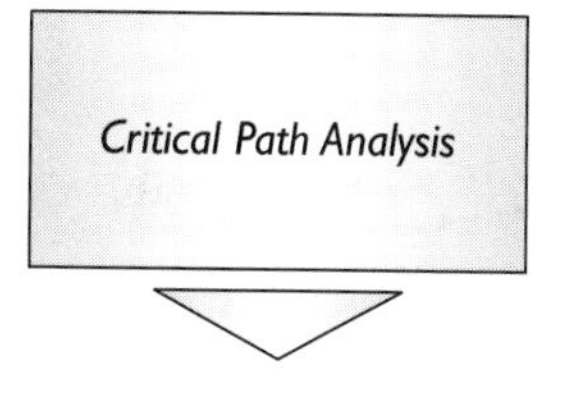

If, after balancing, there still appears to be problems with Lead Time, look to altering the resource structure to allow parallelism of tasks. *Critical Path Analysis* will allow the Team to see the longest chain of activity in the process that is driving the Lead Time. Offloading tasks from the Critical Path will reduce the Lead Time accordingly.

At this point, the Process Lead Time should be significantly shorter. Move to the Control tools as outlined in Chapter 5.

If Lead Time is still too long

If the Process Lead Time is still considered to be too long, there are still a few options available to us:

- Look to accelerating the pace of the individual VA steps (go to "2: The Pace for a Single Process Step Is Too Slow" in Chapter 4).
- Add additional resources to the process at the appropriate places (the load chart will help with this).
- Hold inventory buffers at key locations throughout the process so entities don't have to traverse the whole process to be delivered (only works if the entities aren't unique).

- Seek to anticipate demand from the Customer. (Is there an earlier trigger available to start the entity through the process?)
- Examine options to keep the entity "vanilla" longer. (Is it possible to pre-prepare any parts of the entity so less work is required from the point of request?)
- If all else fails, it might be a case of considering a different process technology.

H: Individual Steps Meet Takt, Global Process Does Not

Overview

If a process as a whole fails to meet the level of downstream demand, its Process Cycle Time is greater than *Takt* Time. For more detail see "Time—Takt Time" and "Time—Global Process Cycle Time" in Chapter 7, "Tools." It is entirely possible that the steps that make up the process are all cycling fast enough, but something falls down between the steps, thus causing the process as a whole to fail.

Examples

Virtually any process in any industry could exhibit this category of problem.

Measuring Performance

The measure of the process meeting Takt is the comparison of the Process Cycle Time against Takt. Process Cycle Time is calculated as

- The average time between entities as they leave the process
- Or equivalently the rate of processing entities

Tool Approach

First we must get an understanding of the current performance with respect to Process Cycle Time:

Measurement Systems Analysis on Process Cycle Time or processing rate

Focus should just be on measuring validity—a sound operational definition and consistent measure of Process Cycle Time versus a detailed investigation of Gage R&R. For more detail see "MSA—Validity" in Chapter 7.

Baseline Capability Study of Process Cycle Time or processing rate

The data is typically captured over a period of one week to one month (depending on process drumbeat) to get a reasonable estimate of Capability. For more detail see "Capability—Continuous" in Chapter 7. Historical data will more than likely be sufficient for the purpose.

High-level *SIPOC* of the Supply Chain

Use *SIPOC* at a 50,000-ft. level to understand supply stream linkage issues with external processes. If there are linkage issues, look to investigating supplier alignment (this could be a whole new project).

In order to enter this particular section, some work must have been done to identify the process steps and the relationship of their respective Cycles Times with Takt. More than likely, this was a *Value Stream Map* and *Multi-Cycle Analysis*. If those tools have already been applied, there is no need to repeat them.

Detailed *Value Stream Map*

Value Stream Map the process with a focus on the Primary Entity and particularly the delays it meets. Be sure to consider the whole process from customer-request trigger to customer acceptance. Identify actions to eliminate NVA steps and thus accelerate the Primary Entity.

Swim Lane Map

Handoffs throughout the process can be the cause of significant delay. After the *Value Stream Map* is created, it should be relatively straightforward to add swim lanes to it to identify changes in responsibility and handoff problems.

Spaghetti Map

Physical layout within the process can cause significant delays and sap capacity. Use the *Spaghetti Map* to identify excess travel distances or NVA flow issues.

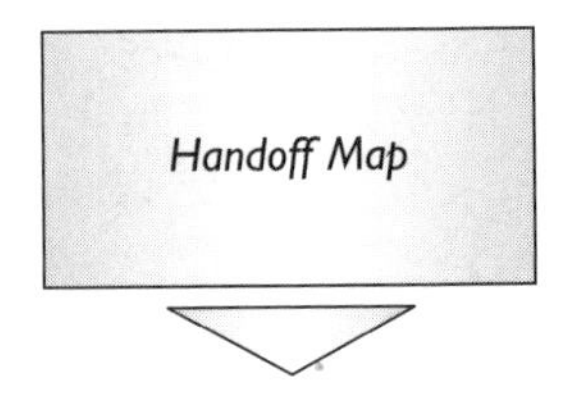

Use the *Handoff Map* if there are apparent multiple revisits to one or more functions. This will identify excess load on a specific function(s) and thus show where effort should be focused.

At the very least, this map is a great visual tool to explain the issues to the process stakeholders.

If from the *Swimlane Map* and *Handoff Map* the process has obvious handoff problems, treat handoff "misses" as a process defect and seek out the key Xs that are driving them. In effect, this is a quality problem using the $Y=f(X_1, X_2,..., X_n)$ roadmap as described in Section C in this chapter. Use accuracy of handoff as the Y in this case. Return to this point after the handoff issues are eliminated.

After handoffs issues are resolved, the process needs to be properly balanced as follows:

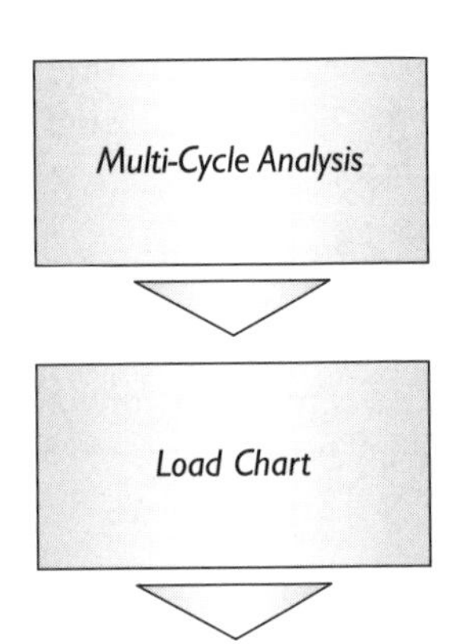

Apply a *Multi-Cycle Analysis* to the *Swimlane Map* to determine where the time is being spent in the process, along with an indication of variation in times. Usually it is the variability in step times that causes the problems.

A *Load Chart* for the process will show Takt Time versus Cycle Time. For more detail see "Time—Takt Time" and "Time—Individual Step Cycle Time" in Chapter 7. All steps should be meeting the pace of demand (or otherwise we wouldn't have come to this particular section), but there might be opportunity to redistribute the workload and combine tasks to use fewer operators and hence have fewer handoffs.

At this point, the process should be performing to Takt and thus the Team can move to the Control tools described in Chapter 5.

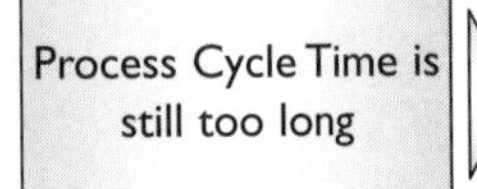

If the Process Cycle Time is still considered to be too long, there are still a few options available to us:

- Look to accelerating the pace of the individual value-added steps (go to "2: The Pace for a Single Process Step Is Too Slow" in Chapter 4).

- Add additional resources to the process at the appropriate places (the load chart will help with this).
- Hold inventory buffers at key locations throughout the process so entities don't have to traverse the whole process to be delivered (only works if the entities aren't unique).
- Seek to anticipate demand from the Customer. (Is there an earlier trigger available to start the entity through the process?)
- Examine options to keep the entity "vanilla" longer (not customized to a particular Customer). Is it possible to pre-prepare any parts of the entity so less work is required from the point of request?
- If worse comes worst, it might be a case of considering a different process technology.

I: Demand from the Customer Is Too Variable

Overview

When a Customer's demand for entities is entirely unpredictable and highly variable, it often becomes very difficult for a process to satisfy the demand. Spikes in demand are usually dealt with using buffers of inventory (stockpiles of entities), time (lengthy delivery times), or capacity (excess resources available). None of these buffers present an ideal solution.

Examples

- **Industrial.** Customer ordering patterns
- **Healthcare.** Outpatient bookings, surgery/emergency department demand
- **Service/Transactional.** Customer ordering patterns, customer usage rates

Measuring Performance

Variation in demand is typically measured as the Coefficient of Variation (COV) of demand, defined as

$$COV = \frac{S}{\overline{X}}$$

where S is the standard deviation of demand (variability) and $\overline{X}$ is the mean of demand (volume). Both are measured in unit numbers of entities, not dollar value.
Entities with a COV less than 1.0 are considered to have smooth demand. Any entities with a COV above 3.0 are considered to be highly variable. For more on COV, see "Demand Segmentation" in Chapter 7, "Tools."

TOOL APPROACH

First we must get an understanding of the current performance with respect to variability in demand:

Measurement Systems Analysis on Coefficient of Variation (COV)

Focus should just be on measuring validity—a sound operational definition and consistent measure of COV versus a detailed investigation of Gage R&R. See "Demand Segmentation" in Chapter 7 for more detail on capturing COV data.

Baseline Capability Study on Coefficient of Variation (COV) of customer demand

The data is typically captured over a period of one month to one year (depending on process drumbeat) to get a reasonable estimate. Historical data will more than likely be sufficient for the purpose.

Caution: COV data is for what a Customer requested (demand) versus what we decided to process (operations planning).

The intervals used should represent typical demand intervals from the downstream Customer. For example, if a Customer orders entities monthly, subdivide the time into monthly buckets. See "Demand Segmentation" in Chapter 7 for more detail on capturing COV data.

High-level *SIPOC* for the Supply Chain

Map the Supply Chain with a high-level *SIPOC* to understand supply stream linkage issues with external processes.

Demand Segmentation

This will allow us to understand the volume and variation in demand of each of the different entity types that progress through the process. High-volume, low-variation demand entities do not present a problem in this category—only the low-volume, high-variation demand entities are the issue.

After the *Demand Segmentation* is complete, it should be easy to see the culprit entity types, specifically in the bottom-right corner of the graph. The first question ought to be "Should these entity types be offered at all?" Entities in this corner cause problems in our process and are very low volume. They should be in our portfolio for a reason: either because they are very high-margin entity types or they are strategic in some way (perhaps they are a new technology that is just being introduced). Any others have obvious cause to be removed using rationalization:

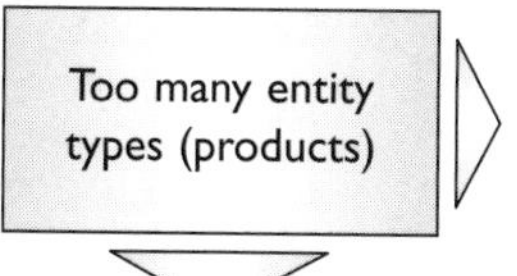

If there are too many entity types, we should consider rationalizing our portfolio somewhat. To do this, go to Section J in this chapter.

If the entity types are valid and cannot be rationalized further, we need a more detailed understanding of the culprits in the bottom-right corner of the Demand Segmentation graph using *Demand Profiling*:

Demand Profiling Customers' demand

The *Demand Profile* plots demand volume over time and gives insight into the pattern of demand as it impacts our process. For the culprit entities, plot a *Demand Profile* for the Customers' ordering (demand) pattern and speak with the sales representative (or equivalent) to understand how the Customers' ordering process works.

Value Stream Map the Customers' ordering process (optional)

As an optional step for large-volume Customers only, it might be worthwhile to add a mapping step at this point for the Customers' ordering process.

For the key culprits, there typically will be a single Customer driving the high variability (multiple Customers tend to smooth demand by the nature of the Central Limit Theorem[3]). There are exceptions to this, such as massive seasonal changes or singular events (the impact of the Super Bowl on beer sales), but these are typically well understood and predictable and the Single Customer Effect is usually the strongest.

[3] Most university-level statistics books give an explanation of the Central Limit Theorem. One text I regularly recommend is *Statistics for Management and Economics* by Keller and Warrack (South-Western College Pub, ISBN: 0534491243), which explains the theory in a practical, readable way.

The Problem Category is aptly titled "Demand from the Customer Is Too Variable." The obvious question at this point (or hopefully earlier!) is "Too variable for what?" Are there missed deliveries? Perhaps the Process Lead Time is too long? Maybe the variability is affecting the ability to forecast? Is it that large levels of inventory have to be kept on hand to buffer the variability? Before continuing, the Team should certainly be able to articulate this because the next steps will probably involve interaction with Customers, and they'll need a good story!

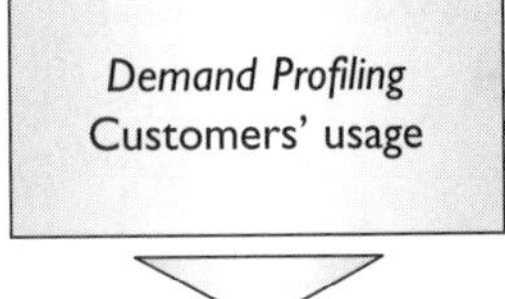

Use *Demand Profiling* again, but this time on the customer usage pattern for the entity, as opposed to ordering pattern. Quite often, customer usage is smooth, but the Customer will make large batch requests of us, the Supplier. Obviously, this will take some interaction with the Customer to achieve.

Customer usage is smooth

This is best-case scenario, but only if we can take advantage of that smooth usage in our process. Options include

- Altering the demand quantities and frequency
- Installing "blanket orders" and "call off" whereby Customers agree to order a larger amount over a long period of time, but take delivery in regular, smaller quantities, paying as they go
- Investigating the triggers in the process to find options of earlier warnings of demand
- Utilizing a Pull System with the Customer. For more detail see "Pull Systems & Kanban" in Chapter 7.

For whichever solution is selected, implement the solution and move to the Control tools in Chapter 5.

Customer usage is irregular

This is tougher. Options include

- Reducing the Process Lead Time, in which case go to Section G in this chapter.
- Working to increase warning time or trigger from the downstream Customer (can we know any earlier?).
- Looking at the demand profile of the Customer's Customer and identifying opportunity (this will only work if you are considered to be a Strategic Supplier—that is, the Customer wants to listen to you).

- Consolidating the entities across a number of sister sites in the company, so only one site deals with this specialty request and thus sees higher volume (and typically less variability, accordingly).
- Constructing a line (process area) that specifically deals with the low-volume, high-variability entities. As such, this area would need rapid changeover ability, a responsive custom approach, and so on.
- Holding inventory buffers at key locations throughout the process, so entities don't have to traverse the whole process to be delivered (only works if the entities aren't unique).
- Examining options to keep the entity "vanilla" longer (not customized to a particular Customer). Is it possible to pre-prepare any parts of the entity so less work is required from the point of request?
- Moving to a Platform Technology approach or modular entity.
- Considering a different process technology.

For whichever solution is selected, implement the solution and move to the Control tools in Chapter 5.

J: Too Many Entity Types (Products)

Overview

Due to many reasons, we might come to the conclusion that we have too many products (different entity types) in our portfolio. Although this is more of a product or service design-related problem, there are some simple approaches that yield beneficial results.

Examples

- **Industrial.** Too many Stock Keeping Units (SKUs), product codes, or manufacturing codes
- **Healthcare.** Too many surgery offerings, lab tests
- **Service/Transactional.** Too many service combinations, possible transaction types

MEASURING PERFORMANCE

Measuring can be considered very easy—just count the number of entity types. Conversely, we might be more interested in margins by entity type, which can involve some convoluted financial calculations. This is probably one for the finance leaders to decide what benefit they are looking for in reducing the number of entity types.

TOOL APPROACH

First we must get an understanding of the current performance:

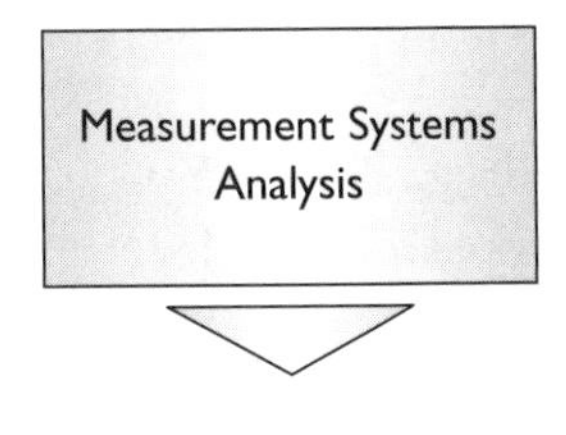

First the metric itself needs to be agreed upon by the Champion, Process Owner, and Belt. Next look at validity of the metric—a sound operational definition and consistent measure versus a detailed investigation of Gage R&R. For more detail see "MSA—Validity" in Chapter 7, "Tools." Even something as simple as counting the number of different entity types can lead to disagreement—don't assume anything!

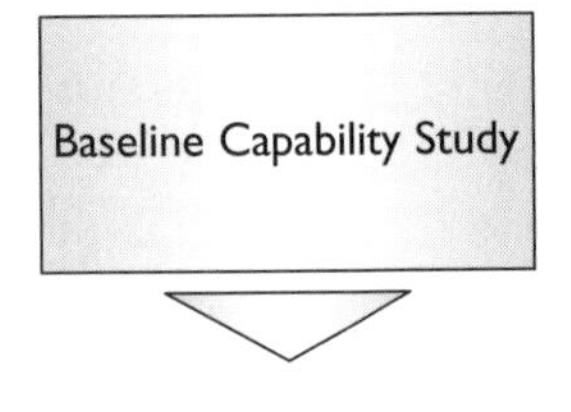

Make a current count based on the chosen metric. There will always be somewhat of a moving target, so draw a line chronologically and make the count at that point. There will be some disagreement on what constitutes a separate product or entity type. Revert back to the clear operational definition from the MSA on a regular basis.

If not already done, conduct a *Demand Segmentation* on the entity types:

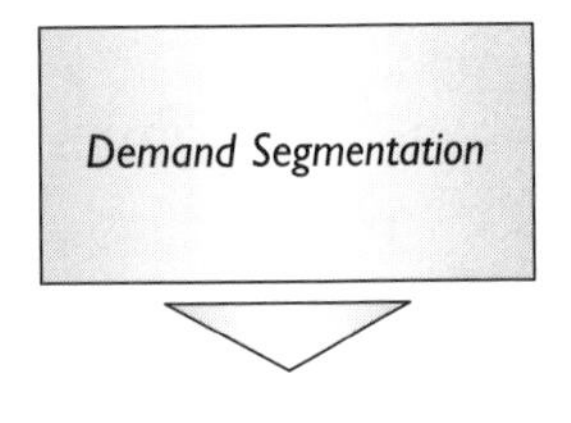

Demand Segmentation will allow us to understand the volume and variation in demand of each of the different entity types that progress through the process. It will highlight those entity types that are highly variable in demand with low volume. It is this set of entity types that will be the focus rather than the high volumes.

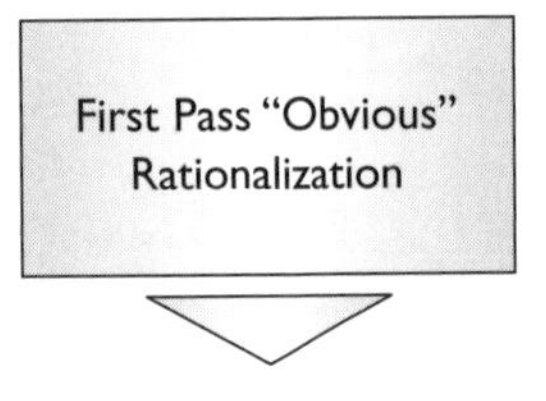

The first examination of the entity types will be fairly low tech, but could be controversial depending on the businesses ability to "let go" of types.

The Team and any necessary internal consultants examine the list of entity types looking for exact matches and therefore obvious redundancy. This sounds trivial, but can account for as much as a 50% reduction in types.[4]

Second Pass *Cause & Effect Matrix* Rationalization

Construct a *Cause & Effect Matrix* on the entity types. List the types as the Xs and list the following Ys, noting that the value of each might have changed after the first pass reduction:

- Profitability
- Volume
- Number of Customers
- Strategic importance (e.g., new technology)
- Uses standard raw materials (in the C&E, use the relationship scoring as follows: 0–all standard to 9–problematic)

The bottom end of the C&E lists prime targets for removal from the portfolio. Look to

- Discussing the entity type with the Customer to see if anything else is suitable (VOC)
- Educating the Customer on substitute entities

There will inevitably be low-volume, high-variability types that we have to keep in our portfolio. For these consider

- Outsourcing the entity type
- Consolidating the entity type across a number of sister sites in the company, so only one site deals with this specialty request and thus sees higher volume (and typically less variability accordingly)
- Moving to a Platform Technology that allows a modular entity
- Developing a whole new replacement entity that is generic enough to replace multiple entity types (Lean Design For Six Sigma would pay dividends in this case)[5]

---

[4] One SBTI ink-manufacturing client applied this simple sweep to its portfolio of inks and discovered an array of 59 different white products. The quote "A white is a white is a white" pretty much sums up the impact of the reduction.

[5] See *Commercializing Great Products with Design for Six Sigma* by Randy Perry and David Bacon (Prentice Hall PTR, ISBN: 0132385996).

- Constructing a line (process area) that specifically deals with the low-volume, high-variability entities. As such this area would need rapid changeover ability and a responsive custom approach
- Examining options to keep the entity "vanilla" longer—is it possible to pre-prepare any parts of the entity so less work is required from the point of request?

For all of the solutions outlined, Control is the hardest piece. The portfolio of entity types came to be in this proliferation for a reason. Control will involve looking at the Entity Change process and ensuring that new entity types cannot be randomly added without first considering the use, or removal, of existing types.

Proceed to the Control tools described in Chapter 5 and sincerely the best of luck on this one!

K: HIGH SCHEDULE VARIATION

OVERVIEW

Scheduling, from a process perspective, is effectively a guess as to what the process should be doing at a certain point in time in the future. Interestingly enough, this Problem Category (and the closely related Forecasting equivalent in Section U in this chapter) seems to be one of the first problems requested early in any deployment of Lean Sigma. This is usually due to a misunderstanding of the root cause of the problem. Often (mistakenly), the belief is that if a perfect schedule could be generated, then running the process would be straightforward. In fact, the reality is that it is a more responsive process that creates better scheduling, rather than the other way around.

Hence, the initial focus should be on reducing Process Cycle Time and more importantly Process Lead Time to make the durations more predictable and thus be able to generate a better schedule. For more detail see "Time—Process Lead Time" and "Time—Global Process Cycle Time" in Chapter 7, "Tools." In parallel with efforts to make the process more responsive, look at the interaction with downstream Customers to smooth variability in demand (understand causes of cancellations, for example). Finally, after all opportunity has been captured, look at the scheduling approach (usually an algorithm) itself.

EXAMPLES

- **Industrial.** Not a common problem, perhaps delivery scheduling
- **Healthcare.** Operating room scheduling, outpatient scheduling
- **Service/Transactional.** Delivery scheduling

Measuring Performance

The most common way to measure performance of a schedule is to use variance to schedule. Note that the term *variance* here is not the statistical term *variance*; it just means the difference between actual and planned.

This is measured as follows:

- Consider the scheduled start time for each scheduled entity to be zero. If an entity actually starts early versus schedule, record a negative offset in time (5 minutes early would read –5 minutes) and conversely a positive offset if it actually starts late. The rollup would be by the mean and standard deviation of this column of offsets, with a goal of zero mean and minimal variation (standard deviation).

Tool Approach

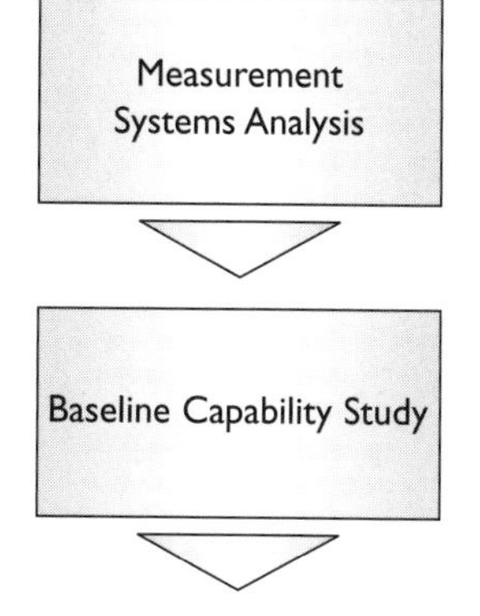

First the variance metric itself needs to be agreed upon by the Champion, Process Owner, and Belt. Next, look at validity of the metric—a sound operational definition and consistent measure versus a detailed investigation of Gage R&R will suffice. For more details see "MSA—Validity" in Chapter 7.

Take a baseline measure of variance to schedule. This will always be somewhat of a moving target, so pick a point in time and stick to it. For more details see "Capability—Continuous" in Chapter 7.

There might be a need to stratify the metric by entity type (e.g., variance by procedure type in a surgery). This is not absolutely necessary at this stage, but will need to be done later anyway, so it's usually best to just go ahead and stratify as well as get a total variance across all entity types.

The vast majority of improvements to scheduling doesn't come from improving the scheduling processes directly, but by eliminating noise from the operations process itself (i.e., the process that is being scheduled). Eliminating noise in the operations process will dramatically improve our ability to schedule accurately, but after the noise in the process is reduced, there might still be genuine reasons to look at scheduling.

In effect, we have simplified the problem into two phases as follows:

1. Improve the operations process.
2. Improve the scheduling process.

For Phase 1, there are a few Problem Categories to resolve in order of importance:

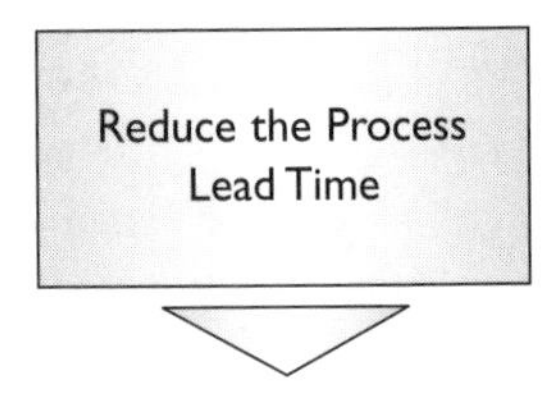

The tools approach that enables a shortening of Process Lead Time will help make the process duration more predictable. For more details see "Time—Global Process Lead Time" in Chapter 7. This is usually a good candidate for a kaizen[6] event in that typically no science is required; it's really just a case of removing NVA activity and streamlining the process.

Go to Section G in this chapter and then return to this point.

Interestingly, the majority of variability in process duration doesn't appear in the VA work done on the entity, but rather in the NVA work done in changeover between entities or entity types.

Again, this is good candidate for a kaizen.

Go to Section O in this chapter to resolve this and then return to this point.

During the work in reducing the Process Lead Time and the Changeover Times, there will have been baseline data captured around the performance of the process. Examine this data to ensure that the following don't apply (if they do, it might be necessary to resolve them first):

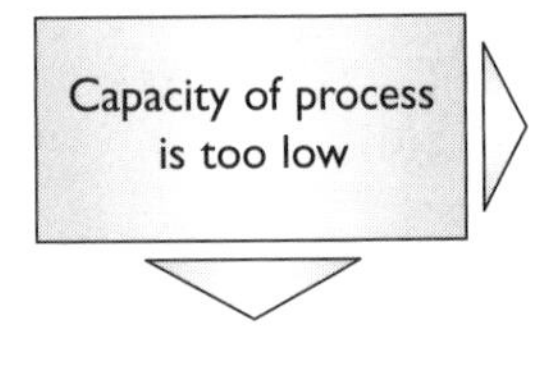

On average, the process does not have enough capacity (the Cycle Time is longer than its Takt Time). For more details see "Time—Takt Time" in Chapter 7. The process will never generate enough entities to meet demand.

Go to Section B in this chapter to resolve this and then return to this point.

---

[6] Kaizen is a 2–5 day rapid change event focused on streamlining a process by removing NVA activities and improving flow. The event gathers a team of the right people together, those that live and breathe the process every day, along with an objective facilitator. The team is charged with understanding and resolving the process issues during the event itself using Lean tools and techniques.

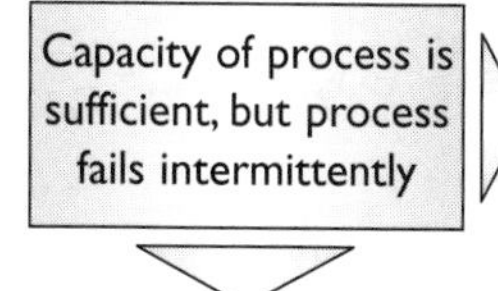

On average, the process has enough capacity (the Cycle Time is shorter than the Takt Time), but the process fails intermittently. On average, the process can meet demand (including peaks), but the process is not robust enough to do so on a continuous basis.

Go to Section F in this chapter to resolve this and then return to this point.

After the Phase 1 work, the operations process should be a lot more consistent and thus predictable. It is often the case that the scheduling problem has been resolved at this point. To that end, it is important to revisit Capability at this point.

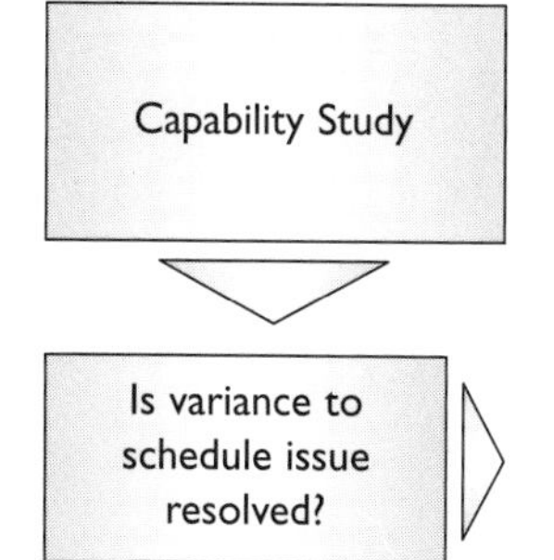

Take a measure of variance to schedule. If stratification was used in the baseline performance, use it again here.

From the Capability data, judge whether the variance to schedule meets business requirements. If it does, proceed to the Control tools in Chapter 5. If not, continue to Phase 2 (next).

For Phase 2, the Team needs to look to the processes that bring bookings to the schedule and what drives variability in the durations of those entities. This will come from examination of external factors as well as internal (operations) factors.

Understand the Input: Stratification by Variability in Duration

Some entity types have consistent process duration (and hence are easy to schedule precisely), whereas others are highly variable (and thus difficult to schedule precisely). It is beneficial to consider each separately.

List all the entity types and for each collect data for their duration for 25 to 100 data points. Calculate the standard deviation for each. Sort the types by standard deviation. If fewer than 25 data points are available, consider using the range instead of the standard deviation.

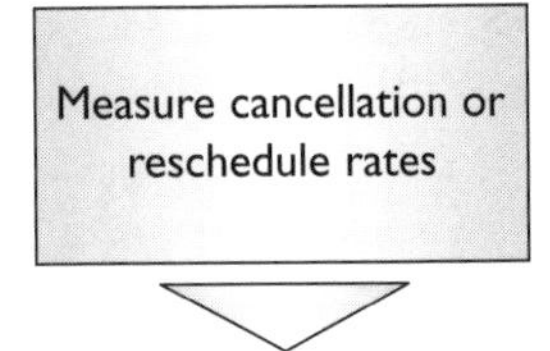

Cancellations and rescheduling often drive significant scheduling problems. For each of the stratified entity types, gather cancellation and rebooking rates (simply as number canceled/rebooked versus number booked) for a period of one month or more.

These two steps will allow the Team to segregate the entity types into three populations:

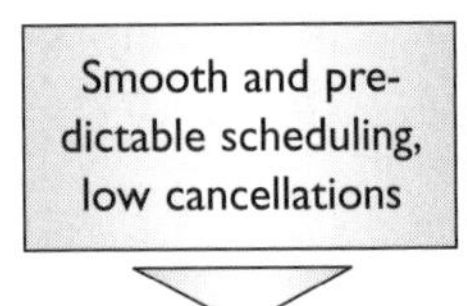

Leave as is.

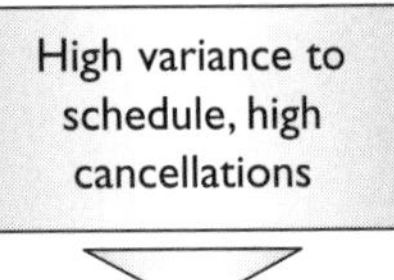

Seek to understand the cancellation process; look at the process of how the Customer books (from their perspective) and what impacts them showing up. Go to Section C in this chapter to identify Xs that drive cancellation and rebooking, and determine how to reduce them or get more advanced warning.

Return here to continue down this roadmap.

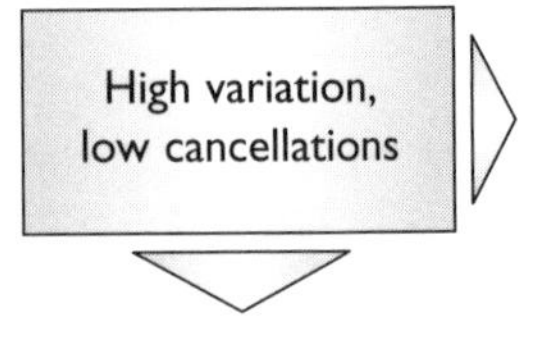

The factors driving variance to schedule are not cancellations and rebooking. The scheduling algorithm used is not representative of the Xs that drive variation in duration of the operations process. Continue in this roadmap.

The roadmap from this point forward is one of the most unusual in Lean Sigma in that the Measure/Analyze will be done on the operations process to identify the Xs that drive the duration, and the Improve/Control will be done on the scheduling process to utilize the identified Xs to better approximate the duration during scheduling.

Process Variables Map on the operations process

This tool will identify all input variables (Xs) that drive duration in the operations process (not the scheduling process). It might be worthwhile to also use a *Fishbone Diagram* at this point to ensure no Xs are overlooked.

Cause & Effect (C&E) Matrix on the operations process

The Xs generated by the *Process Variable Map* are transferred directly into the *C&E Matrix*. The Team uses its existing knowledge of the process through the matrix to eliminate the Xs that don't affect duration. If the process has many steps, consider a three-phase *C&E Matrix* as follows:

- Phase 1—List the process steps (not the Xs) as the items to be prioritized in the *C&E Matrix*. Reduce the number of steps based on the effect of the steps as a whole on duration.
- Phase 2—For the reduced number of steps, enter the Xs for only those steps into a second *C&E Matrix* and use this matrix to reduce the Xs to a manageable number.
- Phase 3—Make a quick check on Xs from the steps eliminated in Phase 1 to ensure that no obviously vital Xs have been eliminated.

Failure Mode and Effects Analysis (FMEA) on the operations process

The reduced set of Xs from the *C&E Matrix* are entered into the *Process Failure Mode and Effects Analysis*. This tool will narrow them down further, along with the useful byproduct of generating a set of action items to eliminate or reduce variation in process duration.

Multi-Vari Study

The reduced set of Xs from the FMEA is carried over into this array of tools with the Y being operations process duration. Statistical tools applied to actual process data will help answer the questions:

- Which Xs (probably) affect the duration?
- Which Xs (probably) don't affect the duration?
- How much variation in the duration is explained by the Xs investigated?

The word *probably* is used because this is statistics and hence there is a degree of confidence associated with every inference made. This tool will narrow the Xs down to the few key Xs that (probably) drive most of the variation in duration. In effect, a simple model can be generated that derives the duration based on the levels of the Xs for a particular entity.

When the *Multi-Vari Study* is complete, there should be a reasonably sound understanding of which Xs drive the duration and how. There will inevitably be some unexplained variation (noise) too. Often at this point the desire is to optimize to the nth degree, but generally, due to the noise in any operations process, a good approximation is as good as it gets.

Sometimes it is possible to take this roadmap further by looking at Designed Experiments, but in general there should be enough understanding at this juncture to resolve the lion's share of the problem.

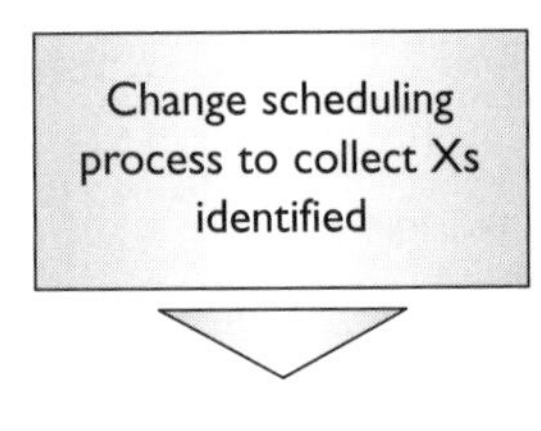

In order to get a better schedule, the scheduling processes and procedures need to be updated to collect the key X data (identified in the *Multi-Vari*) for all entities. For instance, in the surgery scheduling example, if Body Mass Index (BMI) is identified as a key driver for surgery duration, it needs to be asked for, or calculated, during the scheduling process.

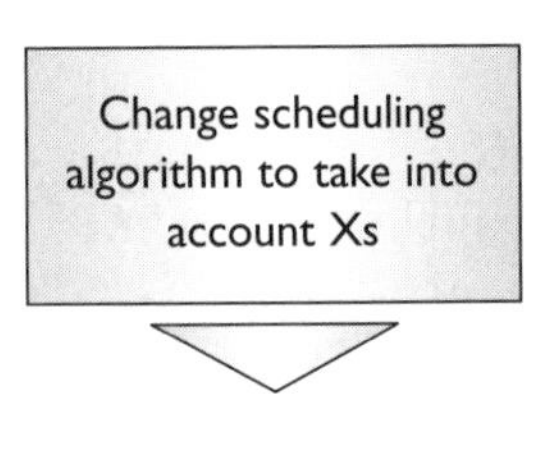

The scheduling algorithm needs to be changed to reflect the new understanding of which Xs drive duration and how. There is a vast array of scheduling software on the market, so there is no single answer here. Quite often, businesses resort to manipulating the schedule by hand or use look-up tables for an interim period until a software update is made.

After the improvements have been made to the scheduling system, move to the Control tools in Chapter 5.

L: MEASUREMENT SYSTEM BROKEN

OVERVIEW

It is a regular occurrence in businesses (and especially true in Service/Transactional processes) to find that a key measurement system that is relied upon to judge whether an entity is within specification or is just not up to par. The impact of this can be huge because the process might be delivering defective entities to the Customer and reworking good ones based on an unsound measurement.

Usually this is a subproject discovered by a larger project looking at the operations process itself. Nevertheless, it can yield significant results in its own right.

Examples

- **Industrial.** Production gages used to judge final quality
- **Healthcare.** Charging, triage
- **Service/Transactional.** Quoting accurately, weighing product

Measuring Performance

Performance of a measurement system is built around the metrics used in Measurement Systems Analysis, namely % R&R, P/T Ratio, Repeatability, Reproducibility, and Distinct Categories. Notice that this list only contains the *MSA* metrics for continuous metrics (see "MSA—Continuous" and "KPOVs and Data" in Chapter 7, "Tools").

There are attribute metric equivalents, but in general attribute measurement systems are usually limited at best for this type of use. Strive very hard in the early stages of the project to identify a related continuous metric that can replace the attribute metric. This will pay off in dividends later. If there seems to be no continuous metrics available (despite considerable effort by the Team looking for them—hopefully, you are catching on to the hints here because they are being laid on quite thick), all is not lost. Proceed using this roadmap, but replace the Gage R&R analysis used with a Kappa study or Attribute MSA (see "MSA—Attribute" in Chapter 7).

Tool Approach

In most roadmaps, in order to judge process performance (capability), an MSA is conducted first. Here the MSA is the Capability Study:

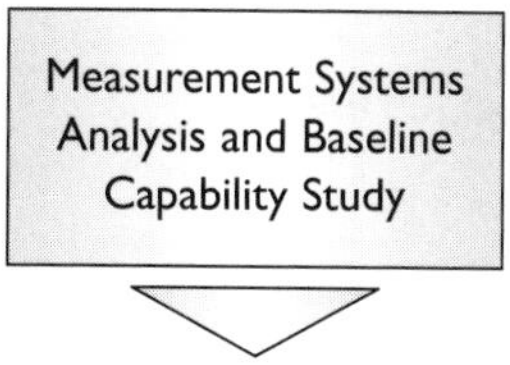

Measurement Systems Analysis and Baseline Capability Study

For continuous data, the Capability Study is the Gage R&R and the measures of capability will be % R&R, P/T Ratio, and Number of Distinct Categories. Use the method described in "MSA—Continuous" in Chapter 7.

For attribute data, the Capability Study is a Kappa Study (for classification) or Attribute MSA (for measurement), and the measures of capability will be Kappa or Percentage Agreement (within and between appraisers). See "MSA—Attribute" in Chapter 7.

Measurement system is poor

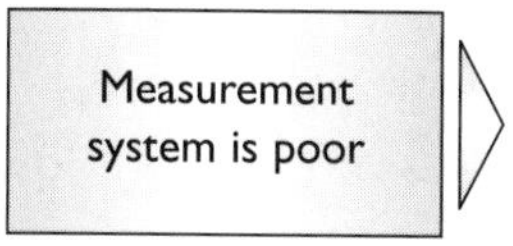

Remember that a measurement system isn't just the gage that's used within it. This problem can be treated as its own process improvement project, where the measurement process is the focus. The defect reduction roadmap using $Y=f(X_1,..., X_n)$ works well.

Consider the Ys to be Repeatability, Reproducibility, and Linearity (and possibly Discrimination). For reproducibility and repeatability, allocate the weighting in the C&E Matrix based on the % contribution from the Gage R&R Study (if high reproducibility, weight that higher). If the Gage (or metric) needs to perform over a broad range of values, give Linearity a high weighting.

Go to Section C in this chapter.

OTHER CONSIDERATIONS

Sometimes there are limits to the measurement system that prevent it from having the precision required. Following the preceding roadmap might only get the system to be borderline acceptable. In this instance, it is possible to use a workaround known as a D-Study, which involves taking multiple readings and averaging them. Clearly this isn't the best scenario, but it could be the only practical path available short of investing in new measurement system technology.

Improving measurement systems is a whole area of study in itself and involves complexities such as

- **Destructive testing.** The same entity cannot be measured twice (e.g., blood sample lab testing, chemical testing, test to failure, etc.). This makes the MSA very difficult because it requires a measure of multiple readings being made for the same entity.[6,7]
- **In-line testing.** The test is done automatically within the process itself as the process proceeds. Therefore, there is no reproducibility element and, in fact, the test is akin to destructive testing because it is impossible to measure twice under exactly the same conditions.

There are solutions to examining these measurement systems such as

---

7 See *Measurement Systems Analysis (3rd Edition)* developed by the American Society for Quality and the Automotive Industry Action Group. See also *Concepts for R&R Studies (2nd Edition)* by Larry Barrentine (ASQ Quality Press, ISBN: 0873895576).

- **Process variation studies.** Using Nested ANOVA to understand the relative variation in test, sample, operator, batch, process, time, and so on.[8]
- **Reference materials.** When using a gage to measure something highly variable (again, difficult to replicate a test), use a more consistent material with similar properties as a reference material to validate the gage.[9]

Each of these areas requires significant explanation and will not be discussed further here.

M: Performance Characteristic Not Good Enough

Overview

Many processes, particularly in manufacturing industry, are judged heavily on a performance characteristic such as strength, assay, thickness, viscosity, color, flatness, and so forth.

When these characteristics are judged by the Customer to be not good enough, in effect we are creating defective entities. This is essentially the same problem as described in Section C in this chapter for defect reduction and thus will not be considered as a separate Problem Category.

Examples

- **Industrial.** Chemical production (assay, strength), materials production (strength, thickness, flatness), widget production (gain, strength, flatness, speed)
- **Service/Transactional.** Call centers (quality and accuracy of answers, ability to fix problem on the first pass, etc.)

Measuring Performance

Use the performance characteristic itself as the measure. These metrics tend to be continuous in nature and make excellent Ys for use in Section C in this chapter. Statistical power will be lost if they are converted to an attribute measure of defectiveness (meeting

---

[8] For an example, see Statistics for Experimenters: Design, Innovation, and Discovery, 2nd Edition (Wiley-Interscience, ISBN: 0471718130), pp. 571-583.

[9] See the shingle testing study presented in *Quality Engineering Magazine*, 1998; article: "Using Repeatability and Reproducibility Studies to Evaluate a Destructive Measurement Test Method," Quality Engineering, Vol. 10, No.2, December 1997, pp. 283-290 by Phillips, Aaron R; Jeffries, Rella; Schneider, Jan; Frankoski, Stanley P.

specification or not). In Lean Sigma terms, we have taken a useful *performance* metric and converted it into a not-so-useful *conformance* metric.

TOOL APPROACH

See Section C in this chapter.

Special attention should be paid in particular to the *Measurement Systems Analysis* because oftentimes this in itself can resolve the problem.

N: PLANNED MAINTENANCE TAKES TOO LONG

OVERVIEW

Regularly (particularly in the manufacturing industry) we purposefully shut down the process to conduct needed repair or upgrades to it. In the case of a sold-out process, every second lost represents lost revenue from the process. Even if the process is not sold out, the need to maintain delivery to Customers will invoke one or more of the standard buffers of

- **Time.** Promise longer delivery times.
- **Inventory.** Build stock prior to the shutdown to cover the downtime.
- **Capacity.** Keep extra (usually costly) equipment on hand to substitute for the lost capacity during the shutdown.

The aim with this type of project is to ensure that when the process is shut down, it is brought back online as quickly as possible.

EXAMPLES

- **Industrial.** Vessel cleaning, line maintenance
- **Healthcare.** Operating room sterilization, renovation, and upgrades
- **Service/Transactional.** Haulage equipment maintenance (locomotives, tractor units)

MEASURING PERFORMANCE

Measurement in this case is for the clock time from when the process stops to when the process starts again. During the downtime, we might also consider the Total Work

Content (the key driver for labor cost), but this is basically a secondary metric and could be considered to be an X.

Tool Approach

If we define the problem as the shortening of downtime, it is entirely analogous to rapid changeover/setup and thus we can proceed directly to Section O in this chapter.

O: Setup/Changeover Takes Too Long

Overview

The majority of processes don't have the luxury of processing the same entity type continuously for all time; thus, at some point (or regular points), they have to switch over to another entity type. This causes a downtime and hence lost capacity and responsiveness. Sometimes even the time between entities of the same type has the same impact. The former problem is often countered with the very poor solution of running long campaigns of the same entity type ("we can't afford the time to stop and change"), which causes lower delivery performance and increased inventory levels. It is much better to tackle the root-cause problem of the long changeover time instead.

Examples

- **Industrial.** Line changeover, line setup, vessel cleaning
- **Healthcare.** Room setup, operating room turnover
- **Service/Transactional.** Freight vehicle change (e.g., road to rail)

Measuring Performance

Changeover time is measured as actual clock time from the end of the last *value-added* step to the start of the first *value-added* step for the next entity (type). The words *value-added* here are the key to really understanding and capturing potential opportunity.

Tool Approach

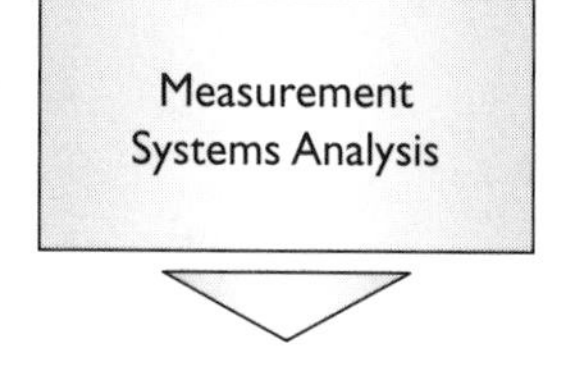

Look at validity of the time metric; a sound operational definition and consistent measure versus a detailed investigation of Gage R&R will suffice. For more details see "MSA—Validity" in Chapter 7, "Tools."

This whole topic can be grouped under one tool called Setup Reduction or otherwise know as *Rapid Changeover* or Single Minute Exchange of Dies (SMED)[10]. Most processes have several different types of setup or changeover depending on the entities being run pre- and post-changeover. These will vary in complexity and frequency. The best way to start is to select the most frequent and most complex changeover. Often this is the superset of all activities that make up other changeovers (i.e., the worst-case scenario that involves all aspects of a change).

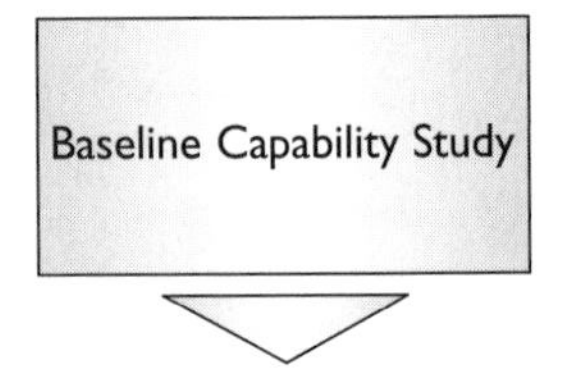

Take a baseline measure of changeover time for the chosen changeover. This might be calculable from historical data or could be tracked as a matter of course. If not, capture 10 or more data points to get a reasonable average of time.

The roadmap to reduce the time is then as follows (see also "Rapid Changeover (SMED)" in Chapter 7 for more detail):

A full *5S* will be conducted later, but it is useful to quickly take a baseline at this point.

Identify all the roles involved in the changeover and video each role (multiple video recorders are required here) from the identified start of the changeover to its end.

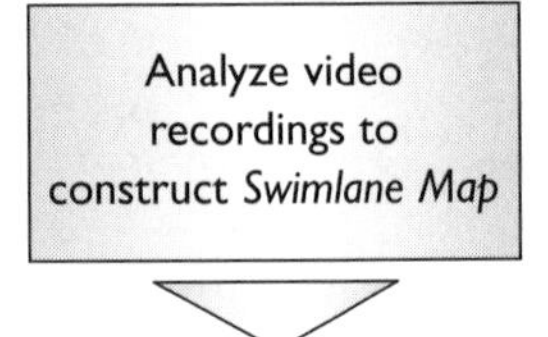

Taking each video in turn (or preferably in parallel using subteams), identify the activities and steps for each role. Lay them out on a *Swimlane Map*. Timings should be included from the video for all steps.

[10] See *A Revolution in Manufacturing: The SMED System* by Shigeo Shingo (Productivity Press, ISBN: 0915299038).

Spaghetti Map	From the videos, construct a *Spaghetti Map* showing movement for all roles and other key entities (material, supplies, information, Primary Entity, etc.). Calculate the travel distances for each of the roles.
Setup Reduction Analysis (Internal / External)	Use the internal/external analysis from the Setup reduction tool to identify which activities should be eliminated (as NVA) and which could be done pre- or post-setup.
Conduct *5S* in changeover area	The vast majority of time reduction will come from better organization and layout. A *5S* undertaking in the area will ensure that the workstations make sense and the materials and supplies are stored close to point of use (POU).
Critical Path Analysis to construct new *Swimlane Map*	After the changeover is stripped to its bare minimum, structure the remaining tasks into a sensible order, taking into account skills of roles (does an activity require a "license" to do it?). *Critical Path Analysis* helps reduce the overall time by identifying the longest chain of activity.
Construct Roles and Metrics	From the *Critical Path Analysis/Swimlane Map*, construct role checklist for the changeover and move to implement the new structure. Test out the new method and tweak as necessary.
5S remeasure	Remeasure the process using the *5S* measurement tool.

After the changeover looks reasonable (the tendency is usually to try and "over-think" it), move to the Control tools in Chapter 5.

The preceding work represents improvement to only one changeover type, but hopefully encompasses learning that could be implemented across the board.

Create a rollout plan and Teams to take the approach to other changeovers by order of volume and complexity.

P: Too Much Unplanned Maintenance

Overview

Sometimes a process breaks down at an unacceptably high rate, causing lost capacity, impacted delivery times, or just plain expensive maintenance. Clearly there are two elements here:

- Frequency of failure
- Duration of failure

They combine to form a total cost of failure made up of (at least)

- Missed deliveries (lost revenue and possible loss of future revenue)
- High inventory
- Direct maintenance costs (parts, labor, etc.)
- Additional equipment kept to alleviate the problem (extra capacity or parts)
- Production defects (scrap and rework) attributable to breakdowns
- Lower reliability product

Examples

- **Industrial.** Plant/equipment breakdown, line breakdown
- **Healthcare.** Information systems equipment failure (servers, systems), medical equipment breakdown (scanners, test equipment)
- **Service/Transactional.** Server breakdown, equipment breakdown (locomotives, tractor units, tracking systems, etc.)

Measuring Performance

If the failure rate were very high, it would be possible to count actual failures in a given period (breakdowns per month). More than likely, the failures per month is small and thus a better metric is the Mean Time Between Failures (MTBF).

If duration of downtime were the key issue, measuring Total Downtime Due to Unplanned Maintenance would be a good metric. It will also be a good idea to roll this all back up into one metric of lost monies (revenue or profit dollars).

TOOL APPROACH

The approach here would be to quickly determine what the best metric to drive the project would be:

- Total Downtime due to Unplanned Maintenance

or

- Total Number of Incidents of Unplanned Maintenance

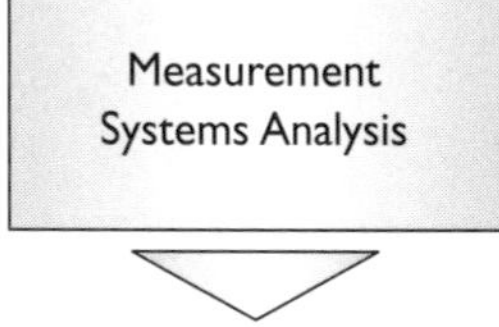

For both measures, we will need to look at the validity of the metric; a sound operational definition and consistent measure versus a detailed investigation of Gage R&R will suffice. For more details see "MSA—Validity" in Chapter 7, "Tools."

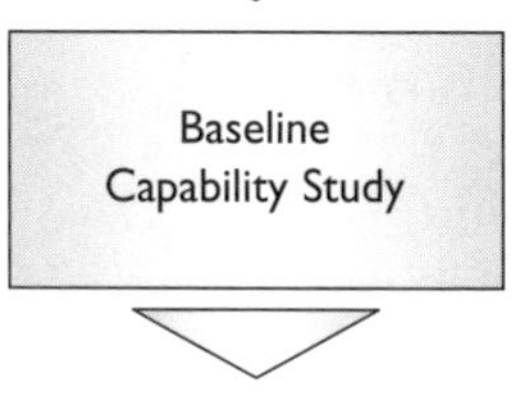

Take a baseline measure of both metrics for a one-month period. Historical data will probably be good here for number of incidents, but might be a bit sketchy for duration of downtime. If it is not available, use whatever data is available on hand to make the call between the two metrics and set up a data collection to confirm the choice (i.e., continue on in parallel).

If plenty of sound data is available, using a year's worth will be more than enough.

It will also be necessary for each breakdown to get a cause of failure. This will be used later in the roadmap.

From the Capability data, it should be apparent whether the issue is one of the downtime being the major issue, the number of incidents, or both.

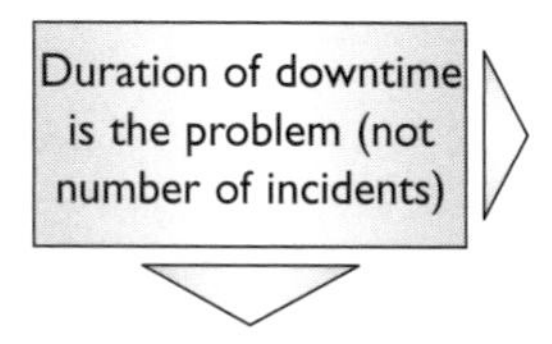

If duration of downtime due to unplanned maintenance is the problem, the objective would be to reduce the time taken to bring the process back online after it is down. This is one of the tools in *Total Productive Maintenance (TPM)* and is known as Breakdown Maintenance.

Go to Section N in this chapter to resolve this problem.

Duration of downtime and number of incidents are both a problem

If both issues prevail, the first thing would be to reduce the time taken to bring the process back online after it is down. This is one of the tools in *Total Productive Maintenance (TPM)* and is known as Breakdown Maintenance.

Go to Section N in this chapter to resolve this problem and then return to this roadmap to reduce the number of incidents.

At this point, either the time taken to bring the process back online has been minimized, or it was never a problem in the first place. The focus from this point forward is on what is causing it to go down in the first place. We would still work by our original metric (either time or number of incidents):

Examine failure information to identify major categories of root cause

Processes tend to break down based on a limited number of reasons. Simple examination of historical failure data, coupled with brainstorming with the maintenance group, will identify the finite number of clusters of failure types.

Pareto by time or number of incidents

Take the data used in the baseline Capability Study and determine the primary causes of failure using a *Pareto Diagram*. It is common for a few root causes to be generating the vast majority of breakdowns.

From this point forward, the focus would be on the top 70%–80% of failures (perhaps two to three reasons).

The roadmap from here is based on the equation $Y=f(X_1, X_2,..., X_n)$, where Y is either the Number of Incidents of Unplanned Maintenance or the Total Time Lost Due to Unplanned Maintenance. Throughout the following roadmap, the focus will only be on the process steps that break down, not the whole process. The reason for this is twofold:

- We don't want to spend time examining steps that aren't breaking down.
- The effect of earlier steps can be listed as an X on the steps we are examining.

Process Variables Map and *Fishbone Diagram*

A combined use of tools tends to work well here. Both tools are used to identify all the input variables (Xs) that could cause breakdown. Use the *Process Variables Map* first and then use a second pass of building a *Fishbone Diagram* for each process step to ensure absolutely all the Xs have been identified. Remember: Only list Xs for steps where the breakdown occurs.

Any obviously problematic uncontrolled Xs should be added directly to the *Process Failure Mode and Effects Analysis (FMEA)*.

Cause & Effect (C&E) Matrix

The Xs generated by the *Process Variables Map/Fishbone Diagram* combination are transferred directly into the *C&E Matrix*. The Team uses its existing knowledge of the process through the matrix to eliminate the Xs that probably don't cause breakdown. At this point, there are usually just a few process steps being examined, so a single Phase *C&E Matrix* will suffice, but if for some reason many steps are involved, consider a three-phase *C&E Matrix* as follows:

- Phase 1—List the process steps (not the Xs) as the items to be prioritized in the *C&E Matrix*. Reduce the number of steps based on the effect of the steps as a whole on breakdowns.
- Phase 2—For the reduced number of steps, enter the Xs for only those steps into a second *C&E Matrix* and use this matrix to reduce the Xs to a manageable number.
- Phase 3—Make a quick check on Xs from the steps eliminated in Phase 1 to ensure that no obviously vital Xs have been eliminated.

The Ys used for the *C&E Matrix* would be the primary failure types with the importance rating relating to the frequency of occurrence of that type.

Process Failure Mode and Effects Analysis (FMEA)

The reduced set of Xs from the *C&E Matrix* is entered into the FMEA. This tool will narrow them down further, along with generating a set of action items to eliminate or reduce high-risk process areas.

This is as far as the Team can proceed without detailed process data on the Xs. The *FMEA* is the primary tool to manage the obvious Quick Hit changes to the process that will eliminate special causes of breakdown. At this point, the problem might be reduced enough to proceed to the Control tools in Chapter 5. If not, continue down this roadmap.

The project is now in the interesting position that the Belt should set up a longer-term data collection in the form of a *Multi-Vari Study*. During this study, the Belt should monitor data collection but is considered to be freed up enough to work on other things/project work. The *Multi-Vari Study* will be conducted on as many parallel lines as is deemed sensible to gather enough information. It would run for at least four months, although it would be useful to continue even longer to gain even more insight, especially if the MTBF is greater than seven days.

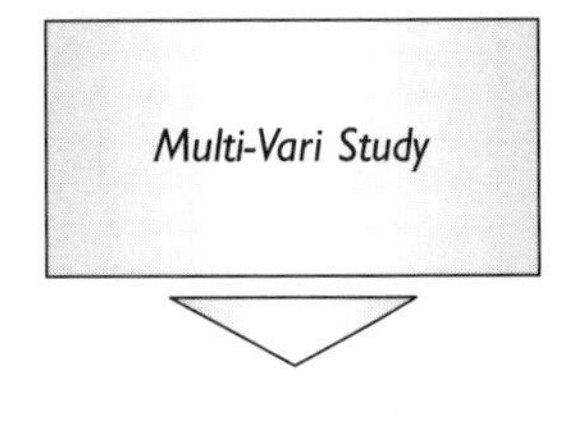

The reduced set of Xs from the FMEA is carried over into this array of tools along with the Y being MTBF. Statistical tools applied to actual process data will help answer the questions:

- Which Xs (probably) affect the MTBF?
- Which Xs (probably) don't affect the MTBF?
- How much variation in the MTBF is explained by the Xs investigated?

The word *probably* is used because this is statistics and hence there is a degree of confidence associated with every inference made. This tool will narrow the Xs down to the few key Xs that (probably) drive most of the variation in MTBF.

At this point it is usually best to move straight to the Control tools in Chapter 5. It is uncommon in process reliability (versus product reliability) projects to undertake Designed Experiments, but is possible.

To do this, the Xs from the *Multi-Vari Study* would be taken through the roadmap of Screening DOE, Characterizing DOE, and Optimization DOE. This is an interesting philosophical area for debate, but will not be addressed further here.

Q: Process Can't Make Product at All

Overview

Sometimes (thankfully, very rarely) processes lose the capability to make product (in specification) at all. Every entity they generate is a defect or the process just cannot generate any entities. This might seem impossible to some but, working in multiple industries, I've seen more than a handful of these cases.

This Problem Category is probably least suited to Lean Sigma, but hopefully the roadmap shown might give some pointers to strengthen a better one out there.

EXAMPLES

- **Industrial.** Chemical plant yields fall to zero; all products generated are defective[11]

MEASURING PERFORMANCE

The performance characteristics on which the product specifications are based are the best measures here (e.g., if a specification for viscosity is being used to determine that the product is defective, use viscosity as the metric). These metrics tend to be continuous in nature and make excellent Ys for tool use. Statistical power will be lost if they are converted to an attribute measure of defectiveness (meeting specification or not). In Lean Sigma terms, we would have taken a useful *performance* metric and converted it into a not-so-useful *conformance* metric.

TOOL APPROACH

The roadmap here is debatable. Time obviously plays a critical factor in any discussion here. This is usually an "all hands on deck" situation. The tools listed are rigorous, but might be perceived as not quick enough—I'm in complete agreement, but rushing to a solution often takes longer (more haste, less speed). The general path would be to assume that at some point in time the process was working and therefore something has changed—in which case, the focus is to determine what *did* change.

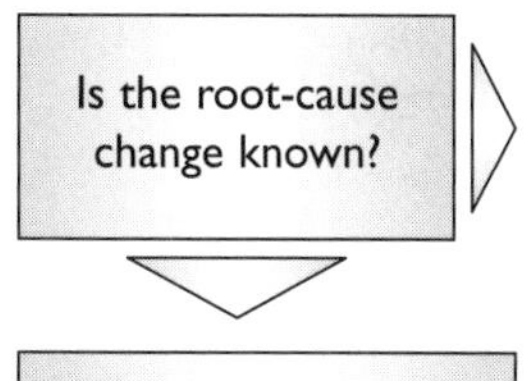

If it is obvious what changed to cause the issue, skip the following steps and proceed to the * step later in this section. Otherwise, continue down this roadmap.

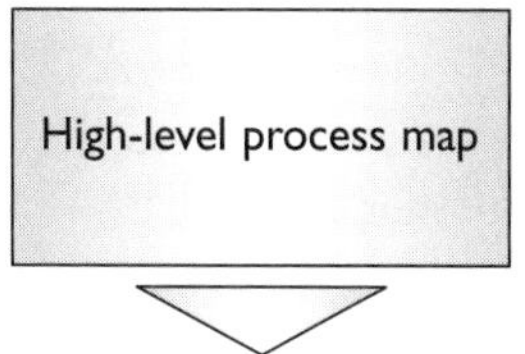

If one doesn't exist, quickly lay down a map of the process. This is just for the major steps in the process at this point. Don't detail out all the substeps.

[11] At one UK rubber company in 1995, I witnessed the making of 500 units to get enough good ones to meet an order for 30 units.

Identify the point of process failure

Narrow down the point of failure in the process. This is done by tracking step by step through the process and determining whether the output at each step meets known requirements. The exact point might not be determined, but at least the section of the process in which the failure is occurring can be identified.

Is the root-cause change known?

If it is now obvious what changed to cause the issue, skip the following steps and proceed to the * step later in this section. Otherwise, continue down this roadmap.

For the following steps, it is usually worthwhile having someone onsite at the process, checking on possibilities as they arise from the tools. Parallelism is good here.

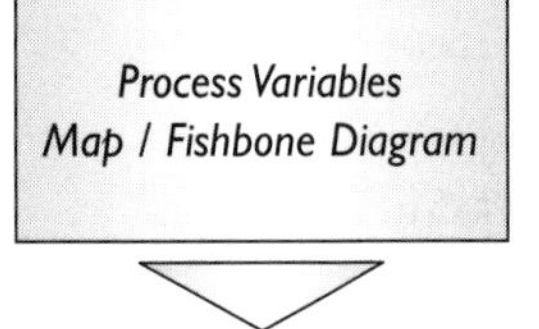

Use both the *Process Variables Map* and *Fishbone Diagram* (to ensure nothing is missed) to identify all input variables (Xs) that cause changes in the Primary Performance Metric(s) (Ys) that are failing. Only map the section of process where the failures are occurring. Any obviously problematic uncontrolled Xs should be examined at the process site immediately.

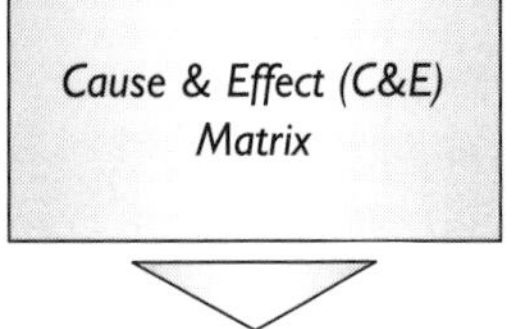

The Xs generated by the *Process Variable Map* are transferred directly into the *C&E Matrix*. The Team uses its existing knowledge of the process through the matrix to eliminate the Xs that don't affect the Ys. There should not be many process steps involved, so a single-phase C&E should suffice.

Feed the results (narrowed-down list of Xs) of the *C&E matrix* to the onsite personnel to have them examined.

Process Failure Mode and Effects Analysis (FMEA)

The reduced set of Xs from the *C&E Matrix* is entered into the FMEA. This tool will narrow them down further, along with generating a set of action items to eliminate or reduce high-risk process areas.

* At this point, the root cause of failure could be known or narrowed to a small number of possibilities that should be investigated quickly to determine which cause is most likely. After the root cause is known, the question would be "Can it be reversed?" with the following answers:

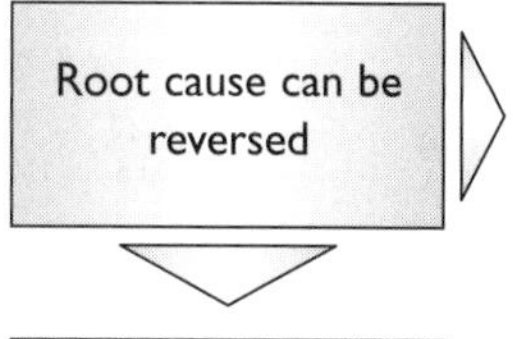

Change back immediately and then proceed to Control tools in Chapter 5 to prevent the change occurring in the future.

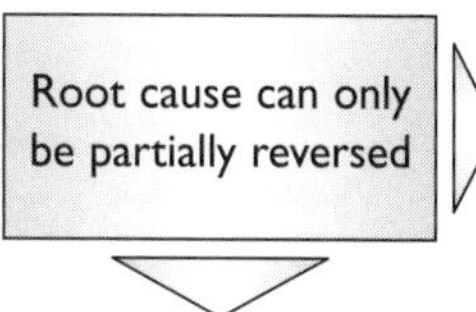

Change back as far as possible immediately. From this point forward, it might be possible to use the roadmap to enhance the performance characteristic shown in Section M in this chapter.

If you've come to this point, things don't look good. (Sorry!) There are limited options available, such as seeking a secondary supply, but really this will invoke a "back to the drawing board" approach. In this case, Design For Six Sigma[12] or Six Sigma Process Design[13] could play a significant role. Sincere good luck to you!

R: Resource Usage Is Too High (Headcount Reduction)

Overview

Unfortunately, this project gets identified on a regular basis early in a deployment. Fortunately (in most cases), a quick rethink shows that the issue is not one of too many heads to support the revenue generated, but too little revenue generated for the headcount. It is always better to grow out of this problem than to shrink headcount. Often an investigation of delivery performance or capacity can identify a better road. Traditional thinking is often based on a shortsighted slash on the one resource that might give a significant return.

Nevertheless, there are legitimate reasons for doing this type of project, especially in industries such as healthcare where there is a shortage of key skills and we would like to reduce in one area to reuse/expand in another. Finally, there are those (albeit rare) cases where there truly is a burden of a headcount that is too high.

Examples

Any process where there is desire to reduce headcount, whether it be because of low availability of certain skills, high turnover problems, or just a desire to cut high labor costs.

---

[12] See *Commercializing Great Products with Design for Six Sigma* by Randy Perry and David Bacon (Prentice Hall PTR, ISBN: 0132385996).

[13] No known published literature exists at the time of writing. See www.sbtionline.com for more details.

MEASURING PERFORMANCE

The usual headcount costs are the best measure to use in this type of project (i.e., labor costs, salaries, benefits, etc.). It is also often worth considering the Total Work Content as a measure of performance because headcount is based directly on this.

TOOL APPROACH

The roadmap approach in this problem is primarily based on examining resource activity to reduce work content of NVA activities.

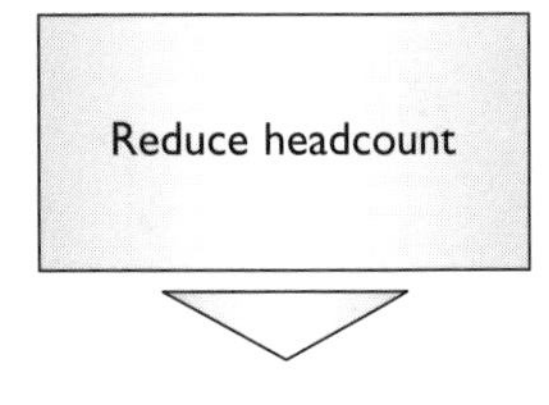

It might at first glance seem that cutting heads should be the last step in the process. In fact, it would be if the resource that was being cut were not people. If management is absolutely set on cutting resources it is always better to cut first and then recover with a project, rather than vice versa. If the cut were made after the project, it would effectively sound the death knell for the program as a whole—Lean Sigma would be linked forever with headcount reduction and no one would want to be involved in projects thenceforth. It is far better for the project to be seen as a huge help after a headcount reduction to make things workable.

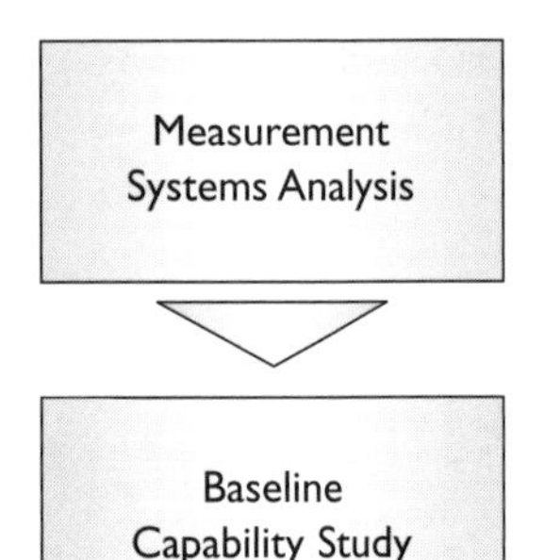

For our headcount measures (whatever we choose), we will need to look at the validity of the metrics; a sound operational definition and consistent measure versus a detailed investigation of Gage R&R will suffice. For more details see "MSA—Validity" in Chapter 7, "Tools."

Take a baseline measure of costs. It is highly likely that historical data is available and is fit for the purpose.

After we have a solid baseline, the project can move relatively quickly and is a good kaizen opportunity.

Detailed *Value Stream Map*

Construct a rigorous *Value Stream Map* including all the detailed steps for the Primary Entity as it progresses through the process. Ensure that time is spent mapping the (secondary) resources (i.e., the people, as they are involved in the process).

Apply a *Multi-Cycle Analysis* to the *Value Stream Map* to identify where the resource time is being spent along with an indication of variation in times.

From the data also identify the Total Work Content in the process.

NVA / VA Analysis

Use the *Value Stream Map* and data from the *Multi-Cycle Analysis* to identify NVA activities. The Team should spend time brainstorming action ideas to eliminate NVA activity in the process, which will in turn decrease the Total Work Content.

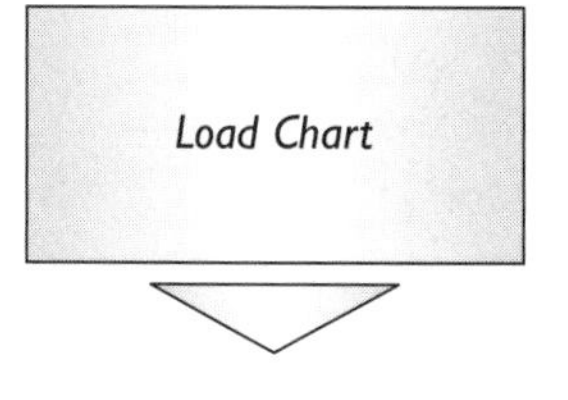

A *Load Chart* for the process will show Takt Time versus Cycle Time. For more details see "Time—Takt Time" and "Time—Individual Step Cycle Time" in Chapter 7. This is a crucial element in the picture. It will only be possible to cut headcount down to the point that the number of resources meets the Total Work Content for the level of demand. Any lower and the process will not meet the pace of customer demand.

At this point, the Team should have a reasonable understanding of where the resource time is spent and the process shouldn't have operators spending time on too many NVA activities. To take this further, the Team should look to other Problem Categories for solution. Examine resource time to further identify

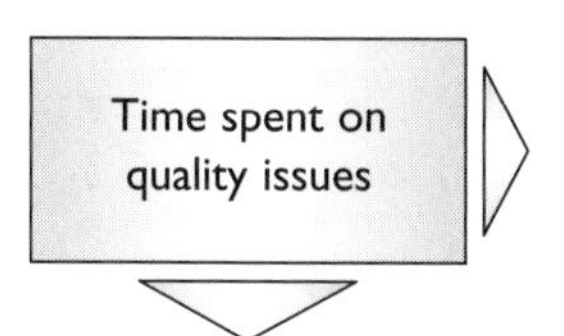

By increasing the inherent quality level of the process, resource time could be freed up from scrapping, reworking, inspection and so forth.

Go to Section C in this chapter to address this problem.

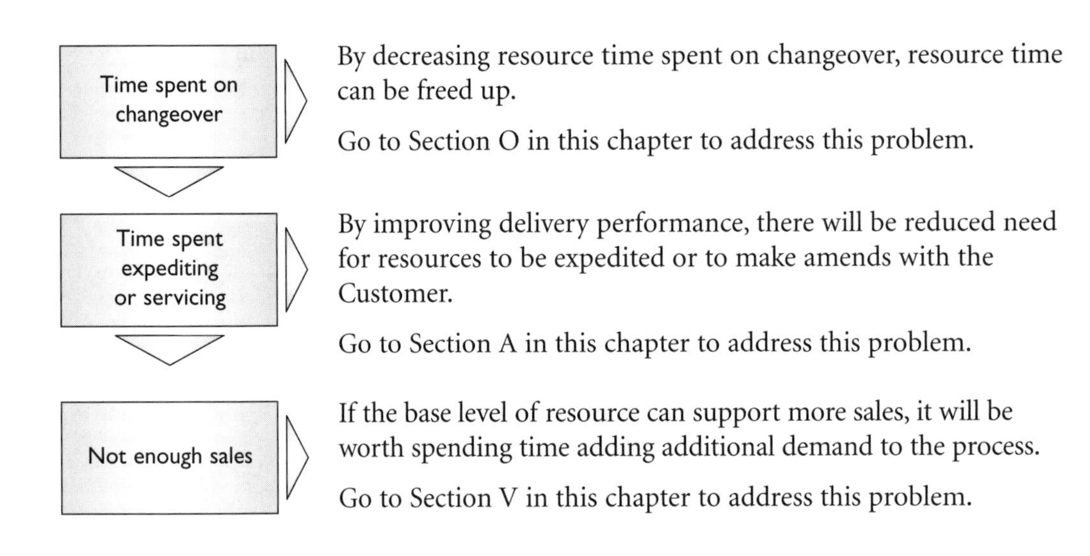

By decreasing resource time spent on changeover, resource time can be freed up.

Go to Section O in this chapter to address this problem.

By improving delivery performance, there will be reduced need for resources to be expedited or to make amends with the Customer.

Go to Section A in this chapter to address this problem.

If the base level of resource can support more sales, it will be worth spending time adding additional demand to the process.

Go to Section V in this chapter to address this problem.

S: INVENTORY IS TOO HIGH

OVERVIEW

Virtually all processes need standard levels of inventory of some sort along the process so that they can meet the delivery requirements of the downstream Customer. This is usually because the Process Lead Time is too long and thus when a trigger is received from the Customer to deliver an entity, there isn't enough time for the entity to traverse the whole process (see Figure 3.1). For more details see "Time—Process Lead Time" and "Time—Replenishment Time" in Chapter 7, "Tools."

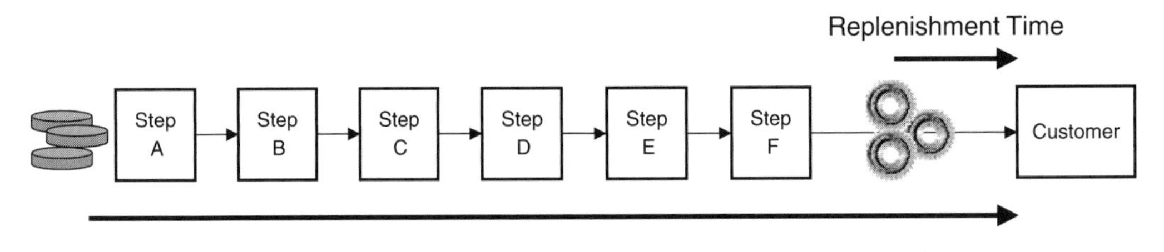

Figure 3.1 Replenishment Time versus Process Lead Time.

Other reasons for inventory might include the use of inventory in front of a bottleneck process step to ensure that the bottleneck is never starved of something to work on (time lost on the bottleneck is time lost forever for the whole process).

- Also, for "supply chain" reasons, we keep a stock of inventory at the beginning and end of the process (often known as Raw Materials and Finished Goods). We must look to understanding the Supply Chain to set the appropriate inventory levels.

Projects of this type usually take on three major forms:

- Rightsizing the inventory based on the existing process and replenishment requirements
- Understanding demand to reduce the need for inventory
- Making the process more responsive (shorter Process Lead Times) to reduce the need for elevated levels of inventory

Examples

- **Industrial.** Raw material, work in progress, finished goods inventory reduction
- **Healthcare.** Materials and supplies inventory
- **Service/Transactional.** Materials and supplies inventory, distribution center inventory levels

Measuring Performance

Inventory can be measured in some interesting ways:

- **Days On Hand (DOH).** If my process uses 10Kg of raw materials per day and I have 80Kg, then I have 80Kg/10Kg=8 days of raw materials on hand.
- **Inventory turns.** If the cost of raw materials for sales made in one year is $1,000 and there is $200 of raw material inventory, that inventory will turn $1,000/$200=5 times in the year.
- **Inventory dollars.** The sum of the dollar cost of the inventory.

Although the reduction of the latter is most likely to be the end goal of a project of this sort, often DOH is a more enlightening metric because it gives a direct indication of validity of having the inventory.

TOOL APPROACH

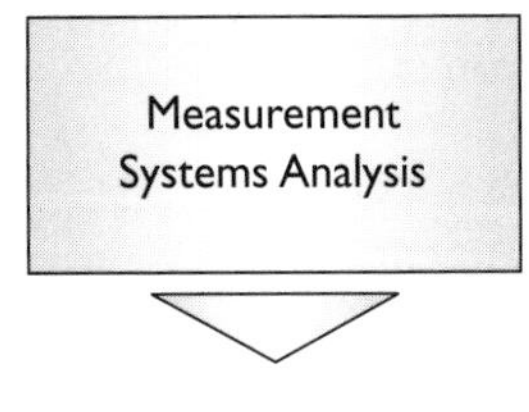

For Days On Hand and Total Inventory Dollars, we will need to look at the validity of the metrics; a sound operational definition and consistent measure will be required. We will also need a Gage R&R on the metric if extensive gage-based counting or weigh counting is involved. For more details see "MSA—Validity" and "MSA—Continuous" in Chapter 7.

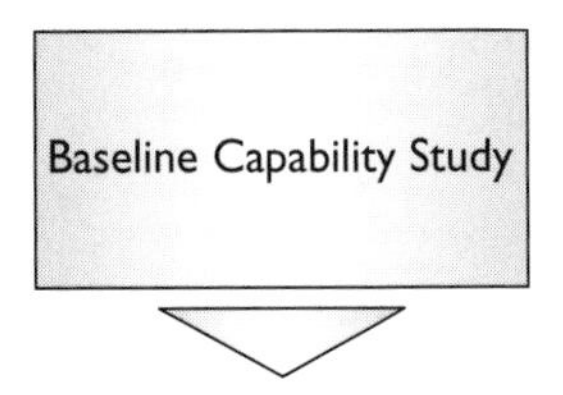

Take a baseline measure of Days On Hand and Total Inventory Costs. It is highly likely that data is directly available and is fit for the purpose, unless our Gage R&R highlighted problems in the measurement system.

As explained in the Overview, the solution can be based on three major areas:

- **Rightsizing the inventory** based on the existing process and replenishment requirements (internal adjustments to inventory management processes)
- **Understanding demand** in order to reduce the need for inventory (increased understanding of external realities)
- **Making the process more responsive** (shorter Process Lead Times) to reduce the need for elevated levels of inventory (changing internal operations processes in order to need less inventory)

Taking the first of these, rightsizing the inventory, can yield significant reductions just by setting the levels of inventory to the correct amount needed to run the process as follows:

Pareto the levels inventory by entity type

Determine as a dollar value the levels of inventory for each entity type. Clearly the focus in the early stages should be on reducing the highest dollar value types.

Calculate Days On Hand for each entity type

Starting at the top of the *Pareto*, calculate from the average daily usage of an entity type and the volume on hand of each entity type the Days On Hand of the inventory per item.

At this point, there can be an immediate improvement made by simply setting a rational limit (beyond which it makes no sense to carry the inventory, for example, 90 days). Inventories beyond this level should be worked down immediately or dealt with appropriately.

Demand Segmentation

Use *Demand Segmentation* to separate the high-volume, low-variation entity types from the low-volume, high-variation entity types. These can be dealt with differently.

High-volume, low-variation entity types

High-volume, low-variation entity types are effectively the low-risk inventory. By examining the Replenishment Time from the supplying processes, it will be possible to reduce the Days On Hand of these types even further.

Set Reorder Points and Economic Order Quantities[14] based on daily usage and Replenishment Time for these types.

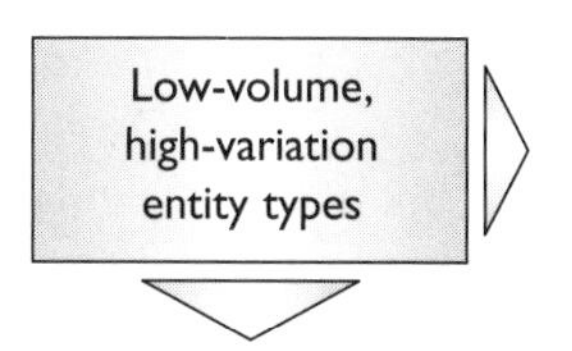

Low-volume, high-variation entity types are effectively higher risk inventory. They will need further examination to determine the correct Reorder Points and Economic Order Quantities.

Starting with the highest-value items from the *Pareto*, work through the low-volume, high-variability types and plot a *Demand Profile* for each.

The *Demand Profile* will show the minimum inventory required on hand at any point. From this, the Replenishment Time and the *peak* daily usage, calculate the Reorder Point and Economic Order Quantities.

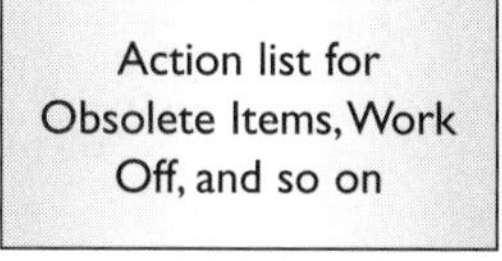

The preceding simple mechanisms will have identified significant levels of inventory needing to be worked off, made obsolete, or dealt with appropriately.

Create an Action List to formalize this and begin to implement.

[14] For details on Reorder Points and Economic Order Quantities, see for example *Manufacturing Planning And Control Systems For Supply Chain Management*, ISBN: 007144033X Publisher: McGraw-Hill Professional, Thomas E. Vollmann, William Lee Berry, David Clay Whybark, F. Robert Jacobs.

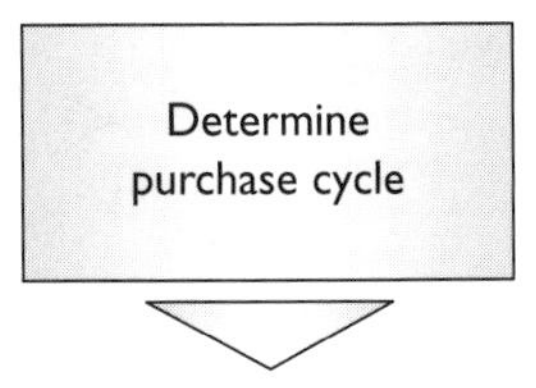

The controls for inventory will be in Purchasing or Operations Planning, depending on where the inventory was located.

Move to the Control tools in Chapter 5 with focus on preventing inventory levels moving from their prescribed levels.

The preceding roadmap represents just the setting (usually reduction) of inventory to the correct levels. In fact, the true levels of inventory needed are determined by

- The ordering pattern of the entities
- The replenishment rate of the entities
- The time to replenish entities

These are driven entirely by the (internal) process generating the entities and the (external) process using the entities. In order to reduce inventories, we must tackle these processes, rather than the inventory management process.

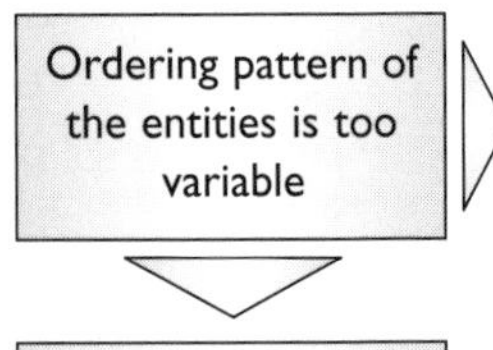

The inventory level is too high because the variation in demand from the Customer is too high.

In this case, go to Section I in this chapter.

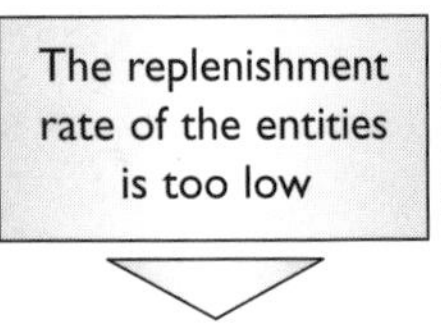

The inventory level is too high because, on average, the process can keep up, but the capacity is too low if a surge in demand appears.

In this case, go to Section B in this chapter.

The time to replenish
entities is too long

The inventory level is too high because the time to replenish an entity (the Process Lead Time) is too long.

In this case, go to Section G in this chapter.

T: Waste/Process Loss Too High

Overview

Common to most industrial processes, where materials are processed in multiple steps from raw materials through intermediates to finished goods, there are opportunities to lose molecules of product throughout the process.[15] The objective of this type of project is to minimize the loss from the process.

EXAMPLES

- **Industrial.** Chemical processing (e.g., ink and pigment production)

MEASURING PERFORMANCE

Process loss is measured using a mass balance equation, such as

$$\text{(I)}\quad \text{Manufacturing Yield(\%)}=100\times\frac{\sum_{i=1}^{n}\text{Outputs (Kg)}}{\sum_{i=1}^{n}\text{Inputs (Kg)}}$$

$$\text{(II)}\quad \text{Raw Material Losses (\%)}=100\times\left[\frac{\sum_{i=1}^{n}\text{Inputs (Kg)}-\sum_{i=1}^{n}\text{Outputs (Kg)}\pm\Delta\text{Stock (Kg)}}{\sum_{i=1}^{n}\text{Inputs (Kg)}}\right]$$

where

- Outputs are the finished product exiting the process.
- Inputs are the components and raw materials entering the process.
- Stock is the inventory of materials through the process.

TOOL APPROACH

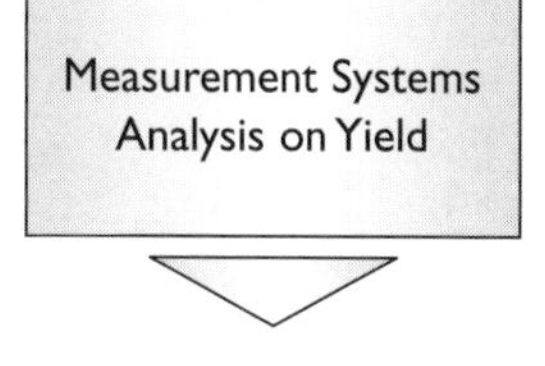

Initially, the focus is on measuring Yield (from equation I), so the ability to identify Mass In and Mass Out will be important. At this stage, the interest is only at the high level in the ability to identify what the business thinks is the mass of raw materials purchased and the mass of finished products shipped. This will just be a validation at this point, and not a Gage R&R Study on the weigh-scales, for example. For more details see "MSA—Validity" in Chapter 7, "Tools."

[15] This is especially common in chemical and pharmaceutical processes.

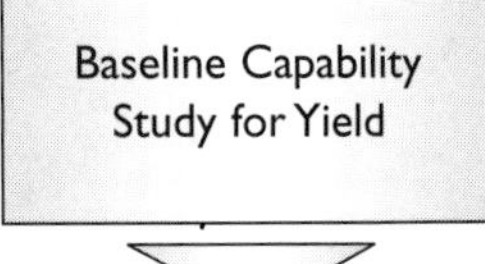

For the Yield as validated in the preceding MSA, take a baseline measure for a period of one month or more. This data will almost certainly be available historically.

High-level *Value Stream Map* of physical transfers

Construct a high-level *Value Stream Map* highlighting in particular any physical transfers of material, or points where material loss could occur. Any low-cost, low-risk, quick-hit opportunities should be taken at this point.

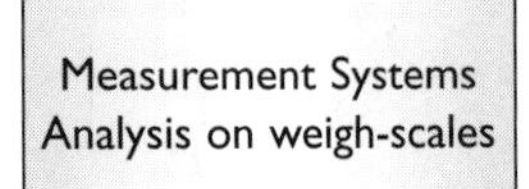

The Team will need to collect loss data throughout the process, so for each measuring device (scale, weigh-counter, etc.) there should be an up-to-date MSA. This should be a full Gage R&R Study. For more details see "MSA—Continuous" in Chapter 7.

The majority of unaccounted loss is typically at the beginning and end of a process, so the focus should be here first.

Pareto on Raw Materials Usage (by Dollars)

Taking Raw Materials and components first, the approach will be to identify process loss using a Capability Study. However, given that there are probably large numbers of Raw Materials, a *Pareto* will provide a prioritized approach. A *Pareto* by dollar value for usage rates will be best because the focus is on cost reduction.

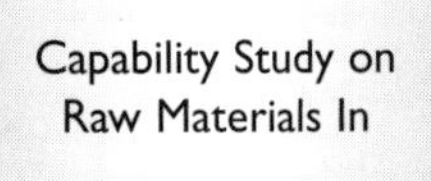

Stated and actual content of purchased containers invariably differ significantly. Even though it might state 2.5Kg on the container from a supplier, it is highly likely that the material content will differ from this amount, often representing significant loss (and less frequently gain) in the process. Using the prioritized order from the *Pareto*, examine the capability of actual versus stated contents of containers. Use a normalized scale with zero being the correct content, underfill the negative content, and overfill the positive content. For more details see "Capability—Continuous" in Chapter 7.

This is best done as the materials are being used by the process. Twenty to thirty data points will be required for each component, so this will be quite a lengthy undertaking.

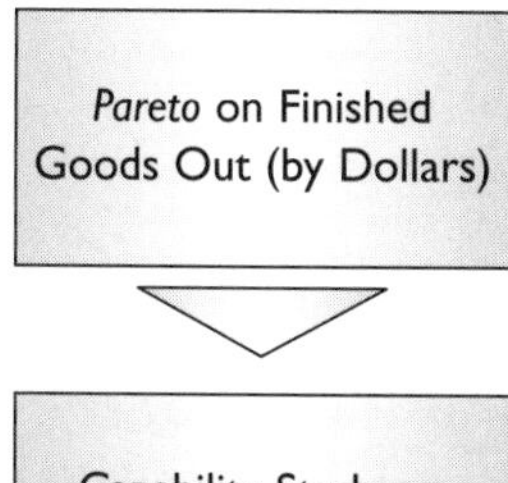

Next, take a similar approach for the Finished Goods. Construct a *Pareto* of the finished (and shipped) goods by cost.

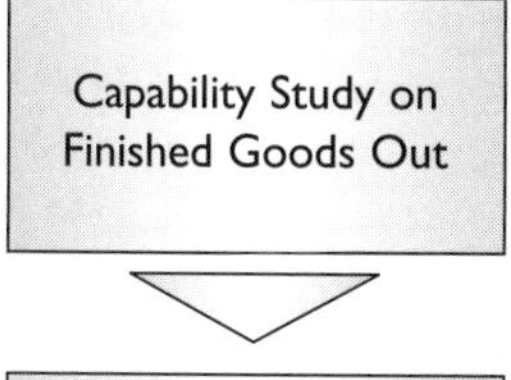

Starting at the top of the *Pareto* of Finished Goods Out, examine stated versus actual contents. Log any significant deviations. A quick hit here would be to mend the filling equipment or system. Overfilling can account for as much as a 5% material loss.

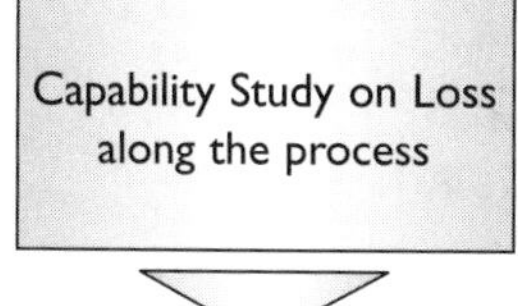

The most significant losses are usually at the front and back end of the process, but measurements taken at key points along the process can identify potential high-loss areas. Measure mass transfer step by step along the process.

After the losses are identified through the process and any quick hits eliminated, the roadmap should be to use the equation $Y=f(X_1, X_2,\ldots, X_n)$, where the Y is the process loss, measured in units of mass. to find the key Xs that minimize the Y.

Go to Section C in this chapter.

U: High Forecast Variation

Overview

Forecasting, from a process perspective, is a guess as to what the process should be doing at a certain point in time in the future. Interestingly enough, this Problem Category seems to be one of the first problems requested early in any deployment of Lean Sigma. This is usually due to a misunderstanding of the root cause of the problem. Often (mistakenly), the belief is that if a perfect forecast can be generated, planning the operation of the process will be straightforward. In fact, the reality is that it is a more responsive process that creates better forecasting rather than the other way around.

Hence, the initial focus should be on reducing Process Cycle Time and more importantly Process Lead Time to reduce the need to plan so far ahead. For more details see "Time—Process Lead Time" and "Time—Global Process Cycle Time" in Chapter 7, "Tools." In parallel with our efforts to make the process more responsive, we should look at the interaction with downstream Customers to smooth variability in demand. Finally, after all opportunity has been captured, we should look at forecasting itself.

EXAMPLES

- **Industrial.** Demand forecasting and operations planning

MEASURING PERFORMANCE

The most common way to measure performance of a forecast is to use variance to forecast. This is measured by taking the aggregate deviation from a cumulative forecast. For example, starting at January, forecast the production volumes as a cumulative number of entities (number of units, Kg of product, etc.) or dollars ($). At any point, the forecasting system is measured by how well it predicted the cumulative actual production. This can be frustrating for all concerned because if the forecast is off in January, it more than likely will be off all year due to the accumulation issue.

A better forecasting metric is to use variation from plan in agreed time buckets.[16] For example, forecast volume weekly and then measure the actual production's deviation from the weekly forecast. If the actual volume is under forecast, log a negative offset (five units under would be written as –5, –5Kg, or –$5). Similarly, if actual is above forecast, log a positive offset. For the column of offsets, take the mean and standard deviation for a rolling period of time (typically this is 6–12 months in the manufacturing industry). The time period needs to be short enough to be considered "current," but long enough that there are enough data points to calculate mean and standard deviation. Thus, to forecast daily, use a rolling month (i.e., calculate the offsets for the past 30 days); if forecasting weekly, use a rolling 6 months (i.e., calculate the offsets for the past 26 weeks).

TOOL APPROACH

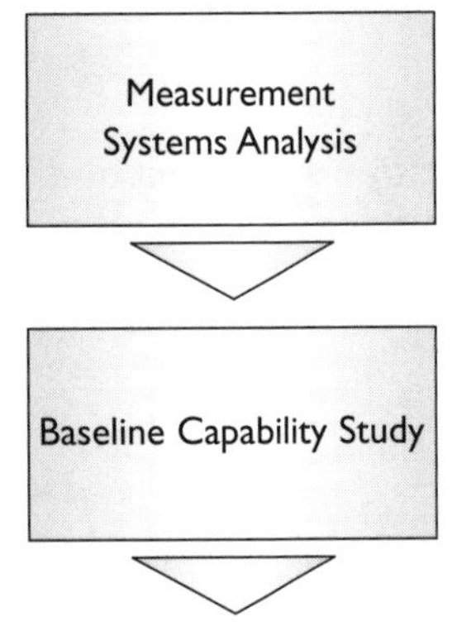

First, the forecast metric itself needs to be agreed upon by the Champion, Process Owner, and Belt. Next, look at validity of the metric; a sound operational definition and consistent measure versus a detailed investigation of Gage R&R will suffice. For more details see "MSA—Validity" in Chapter 7.

Take a baseline measure of variance as defined in the preceding MSA step. This will always be somewhat of a moving target, so pick a point in time and stick to it. For more details see "Capability—Continuous" in Chapter 7.

[16] Another possible approach here is to calculate a simple correlation of the actual volume versus forecasted volume. The closer the Pearson correlation coefficient is to 1, the better the forecast.

The vast majority of improvements to forecasting doesn't come from improving the forecasting processes directly, but by eliminating noise from the operations process itself (i.e. the process that is being forecasted). Eliminating noise in the operations process will dramatically improve our ability to forecast accurately. The question we should ask ourselves is "Why are we asking the forecasting question in the first place?" Data will be needed to answer the question, so apply the following to get the necessary evidence:

Baseline for Delivery Performance, Takt, Process Cycle Time, Days On Hand, and Number of Product Types

Take a baseline measure of

- Delivery Performance, measured as a percentage for On Time In Full (OTIF)
- Level of demand, measured as Takt
- Throughput (capacity), measured as Process Cycle Time
- Inventory levels, measured as Days On Hand
- Number of entity types (products)

Demand Segmentation

Determine the level of demand variation using *Demand Segmentation*. Products with Coefficient of Variation (COV) less than 1 are considered to have smooth demand.

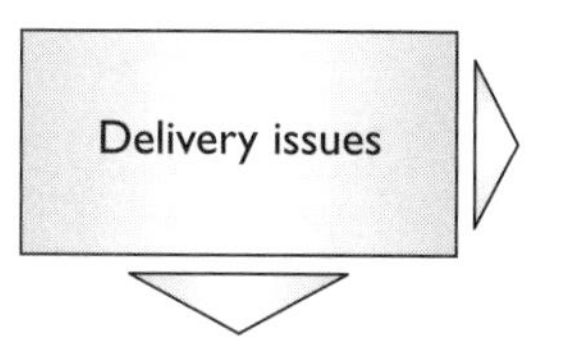

If the delivery performance is poor (low OTIF), the problem should be addressed in the operations process first before looking to the forecasting process. Proceed to Section A in this chapter.

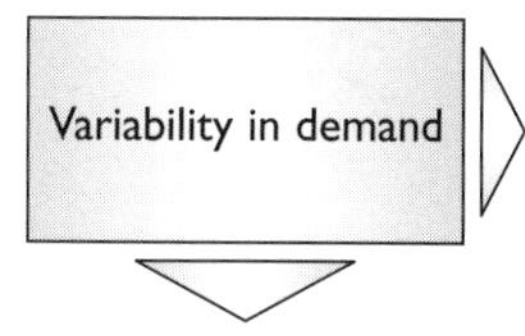

If the *Demand Segmentation* shows large variability in demand, proceed to Section I in this chapter before looking to the forecasting process.

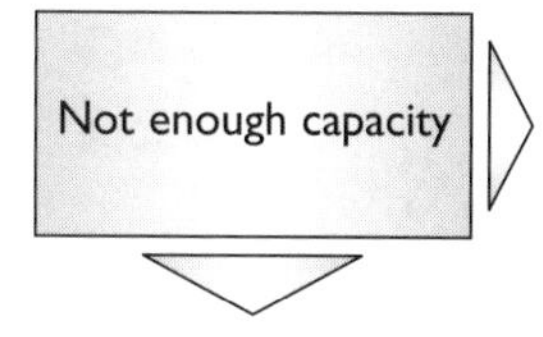

If the Process Cycle Time is above Takt, there is a shortfall in capacity. The problem should be addressed in the operations process first before looking to the forecasting process. Proceed to Section B in this chapter.

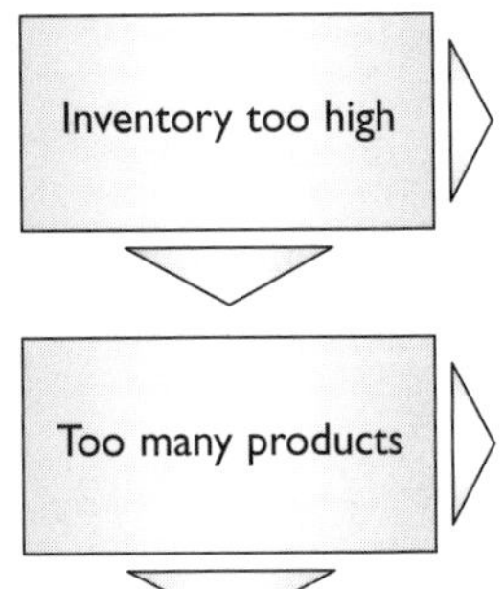

If the inventory Days On Hand is high in the operations process, proceed to Section S in this chapter before looking to the forecasting process.

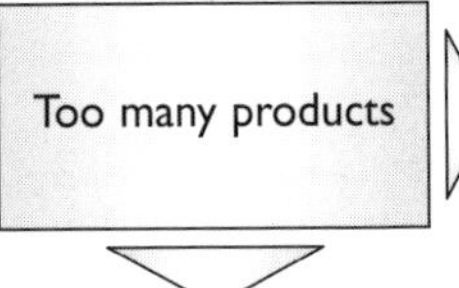

If there are too many entity types (products) in the operations process, proceed to Section J in this chapter before looking to the forecasting process.

After the preceding issues have been addressed (the noise in the process reduced), there might still be genuine reasons to look at forecasting:

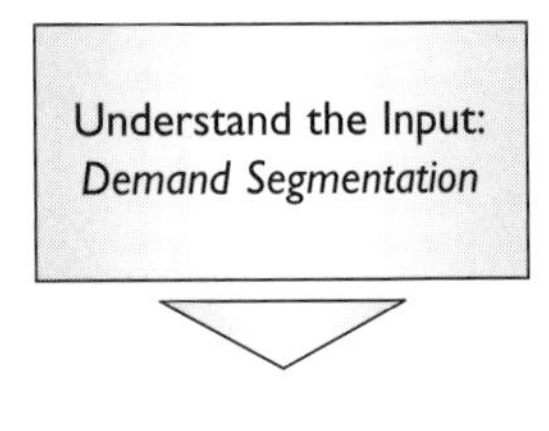

If no changes were made since the *Demand Segmentation* was done earlier, just use the output from that study rather than repeating again here. Forecasting is about predicting demand, but demand is different across entity types. For some types, demand is smooth and therefore easy to predict; others are highly variable and therefore difficult to predict. *Demand Segmentation* separates them for us.

Create the forecast around the smooth-demand entity types first and then treat the variable types separately. It might be suitable to set up the smooth-demand items (sometimes known as *runners* and *repeaters*) on their own operations line(s) and thus run effectively as a separate (high-volume, smooth-demand) business. For the highly variable entity types (sometimes known as *strangers*), consider again

- The **validity** of having them in the portfolio
- The **value** they bring (these should be high-margin or strategic items)

Again, we look to operations and perhaps process these entity types separately on their own stranger line, the core competency of which is the ability to do every entity in a custom way and have rapid changeovers between entity types.

From this point forward, forecasting essentially becomes a matter of mathematical modeling, simply represented by the familiar equation $Y=f(X_1, X_2,..., X_n)$. The tools used form a family known as Time Series Analysis.

In simple terms:

- Developing a mathematical model to describe the behavior of a time series is called *smoothing*.

- Smoothing reduces the effect of purely random fluctuations to reveal any systematic pattern in the available data.
- Using a model to predict future behavior of a time series is called *forecasting*.
- A forecasting model usually includes information from past values of the characteristic of interest (e.g., demand), and perhaps also from leading indicators, but it can also include additional information injected by the forecaster (e.g., knowledge of pertinent forthcoming events, "gut feel," etc.).

To model effectively, the time series data is broken down into its key elements:

- **Current Level.** The mean value at the current time
- **Trend.** The rate of systematic increase (or decrease) in the mean value
- **Seasonal Pattern.** A recurring periodic pattern
- **Random Component.** The portion of behavior that remains unaccounted for after the current level, trend, and seasonal pattern have been identified

These elements are modeled individually using tools such as regression, ARIMA (autoregressive integrated moving average), and S-curves, and the models are laid back on top of each other to create a prediction of future demand.
Although regression is covered in Chapter 7, the combined application of these tools to create a forecast is well beyond the scope of this book.[17]

V: NOT ENOUGH SALES

OVERVIEW

Every business would like more sales. This naturally falls under the realm of Six Sigma for Sales/Marketing,[18] which will not be covered here in detail. There is, however, a simple approach using the standard Lean Sigma roadmaps that can help too.

EXAMPLES

Any business that sells to external Customers (i.e., virtually every business).

---

[17] For further reference, see *Forecasting: Methods and Applications (3rd Edition)* by S. Makridakis, S.C. Wheelwright, and R.J. Hyndman (Wiley, ISBN: 0471532339).

[18] Six Sigma for Marketing is a new field at the time of writing. For more details, see www.sbtionline.com.

Measuring Performance

Standard measures of sales here are applicable (e.g., numbers of units, volume of product, etc.).

Tool Approach

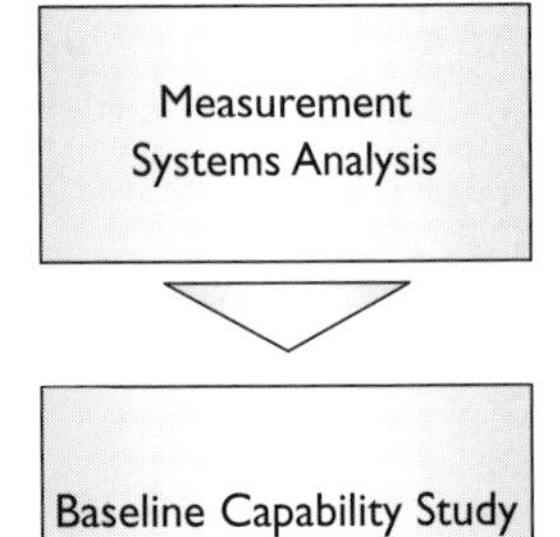

There should be a quick look at validity of the sales metric; a sound operational definition and consistent measure versus a detailed investigation of Gage R&R will suffice. For more details see "MSA—Validity" in Chapter 7, "Tools." Mostly it's just a case of getting the data from the right source.

Take a baseline measure of historical sales as defined in the preceding MSA step. This data should be straightforward to obtain. It will probably be useful to get the data stratified by entity type, as well as the total sales volume. Use a sales period of 3–12 months.

The most common mistake in this type of project is to look to the wrong process for solution. The solution lies not in the internal marketing/sales processes, but instead in the customer decision process.

However, before examining any external processes, it is probably best to consider other Problem Categories first to identify where more readily accessible possibilities for sales growth might exist. This will require a data collection as follows:

Baseline for Takt, Process Cycle Time, Lost Sales, Order Cancellations, and Sales Backlog

Take a baseline measure of

- Level of demand measured as Takt
- Throughput (capacity) measured as Process Cycle Time
- Lost sales measured as a percentage versus total (stratified by major causes)
- Order cancellations measured as a percentage versus total
- Sales backlog measured in days

Sales are restricted due to capacity constraints

If Process Cycle Time is greater than Takt, acquiring extra sales is irrelevant because operations processes cannot meet existing demand. Focus should be on increasing capacity.

Go to Section B in this chapter.

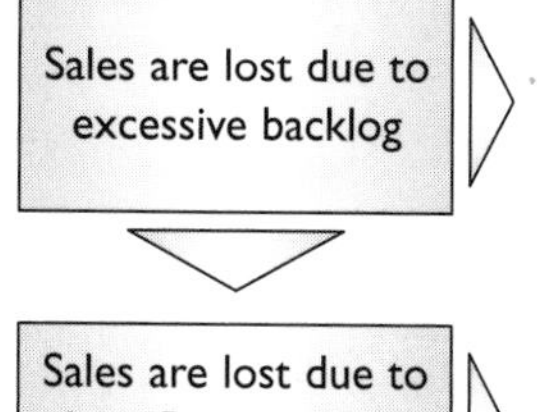

If sales are lost due to the lead times quoted caused by lengthy backlogs in orders, focus should be on reducing the backlog.

Go to Section W in this chapter.

Sales are lost due to long Process Lead Time

If the Process Lead Time is too long, causing lost sales (without a backlog of orders being present), the process is just not responsive enough and the focus should be on reducing the Process Lead Time.

Go to Section G in this chapter.

If the preceding Problem Categories aren't pertinent and the need is simply to sell more, the focus should be turned to the Customers' decision-making process.

The scope should be defined as being from "the time the need arises" to "when the order is placed" (i.e., the decision is finalized).

In this case, use the $Y=f(X_1, X_2,..., X_n)$ approach to determine the key Xs that drive the Customers' selection. The Y would be the dollars spent on our products.

Go to Section C in this chapter.

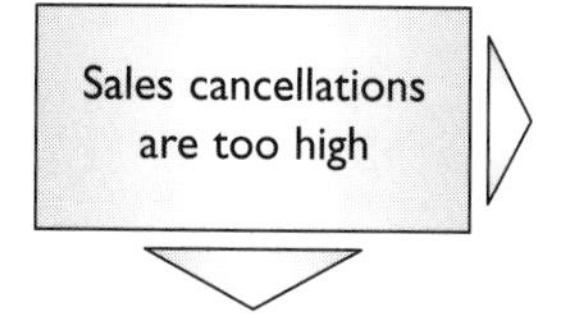

If the Customer is placing orders, but is canceling them later, this is a subtly different problem, but would follow the same roadmap. The scope should be defined as being from "the time the need arises" to "the time the purchase is solidified" (perhaps the product being used, or payment made or similar).

In this case, use the $Y=f(X_1, X_2,..., X_n)$ approach to determine the key Xs that drive the Customers' cancellation. The Y would be cancellations.

Go to Section C in this chapter.

W: Backlog of Orders Is Too High

Overview

Businesses often lose sales because their delivery times are long due to a backlog of sales. Effectively, this is other Problem Categories expressed in a different form, primarily with respect to capacity and Process Lead Time. For more details see "Time—Process Lead Time" in Chapter 7, "Tools."

Examples

Any business process that delivers products or services to external Customers.

Measuring Performance

Although the primary driver for the project is the backlog in sales, this should be augmented with measures of capacity and theirs relationship with demand. Measure backlog (in units of time), capacity (in entities per time period), along with Takt and Process Lead Time. See also "Time—Takt Time" in Chapter 7.

Tool Approach

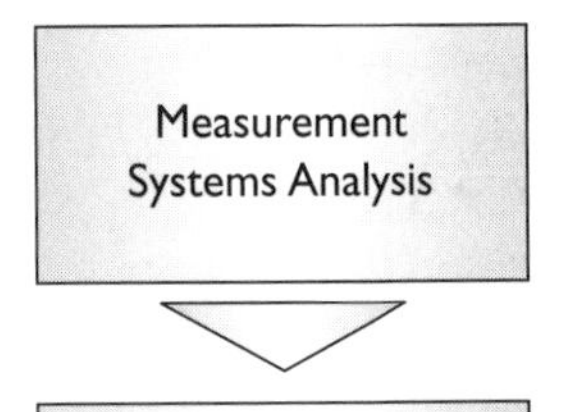

Examine the validity of the order backlog, Takt and Process Cycle Time metrics; a sound operational definition and consistent measure versus a detailed investigation of Gage R&R will suffice. For more details see "MSA—Validity" in Chapter 7.

Baseline Order Backlog, Takt, Process Cycle Time

Take a baseline measure for

- Sales backlog, measured in days and dollars
- Demand level, measured as Takt Time
- Throughput (capacity), measured as Process Cycle Time

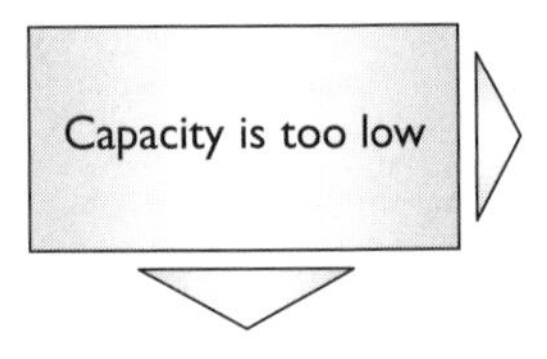

If the Process Cycle Time is greater than Takt, the process will not be able to keep up with demand and focus should be on increasing capacity.

Go to Section B in this chapter.

Process Lead Time is too long

If the Process Lead Time is too long, the process is just not responsive enough. This would cause order backlog, so the focus should be on reducing the Process Lead Time.

Go to Section G in this chapter.

X: Payments Made to Suppliers Not Optimized

Overview

All businesses pay suppliers for products and services in one form or another. At any given time, there is an amount outstanding to suppliers typically known as accounts payable. The goal in this type of project is one of fine balance: keeping hold of the money for as long as possible without going beyond the terms agreed with the supplier.

Examples

Accounts payable for any business that buys products or services from suppliers.

Measuring Performance

Accounts payable is measured in dollars outstanding to suppliers. This can be increased by extending terms with suppliers, which is just a function of the purchasing group. Here the project needs a subtly different approach and examines the payment process. The best metric is to look at performance in terms of time.

For example, if the payment terms are 30 days, move the origin to that point. If payment is made late at say, 32 days, that would be recorded as +2 days. Likewise, if payment is made early, at say 27 days, that would be written as –3 days. This is replicated across all suppliers and agreed terms, resulting in a payment distribution as in Figure 3.2.

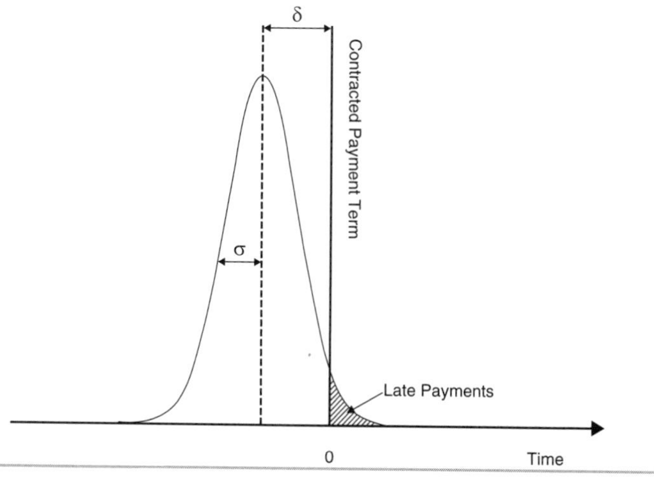

Figure 3.2 Measuring accounts payable: distribution of payments made.

The goal is to never go above zero, but to pay as late as possible (i.e., to record as small a negative number as possible). This is done by minimizing δ, σ and the area of graph overlapping the terms line in Figure 3.2, which in turn maximizes the accounts payable dollars and minimizes failure to meet payment terms.

The secondary metric will be to look at payment accuracy. (Did we pay correctly for what we got?) This is measured as a percentage of payments made correctly (number correct × 100/total).

TOOL APPROACH

Measurement Systems Analysis on Payment Time

Focus should just be on measuring validity—a sound operational definition and consistent measure of Payment Time versus a detailed investigation of Gage R&R. For more details see "MSA—Validity" in Chapter 7, "Tools."

Baseline Capability Study of Payment Time

The data is typically captured over a period of one week to one month (depending on process drumbeat) to get a reasonable estimate of Capability. For more details see "Capability—Continuous" in Chapter 7. Historical data will more than likely be sufficient for the purpose.

Measurement Systems Analysis on Accuracy

Conduct a Measure Systems Analysis on the ability to judge accuracy of payment. This will be an Attribute type MSA. For additional details, see "MSA—Attribute," in Chapter 7.

Baseline Capability Study of Accuracy

Take a sample of 150–200 payments and check for accuracy to give an approximate baseline. This will be an Attribute Capability study. For more details see "Capability—Attribute" in Chapter 7.

Value Stream Map

Construct a Value Stream Map of the process to identify all of the steps in the process from suppliers' product used to payment sent. The Primary Entity can be considered to be the need for payment.

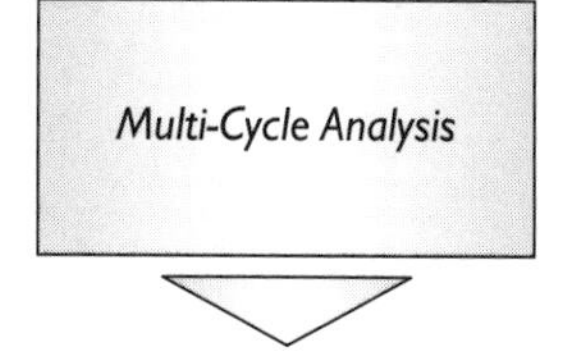

Apply a *Multi-Cycle Analysis* to the *Value Stream Map* to determine where the time is being spent in the process, along with an indication of variation in times. Usually it is the variability in step times that causes the problems.

The solution is usually found in the reduction of NVA activity and subsequently variation in the process, followed by installing a very clear, reliable trigger for payment at the appropriate place in the process.

At this point, it becomes a matter of controlling the process, so the next step is to move to the Control tools in Chapter 5.

Y: Accounts Receivable Are Too High

Overview

Businesses are clearly in operation to sell products and services for money. Unfortunately, payments for those products and services aren't received immediately, due to

- The inability to invoice immediately
- The Customer's understandable unwillingness to pay immediately
- The inability to process payment immediately

The manifestation of this is that at any time a large sum of money is often outstanding, known as accounts receivable. Clearly the business would like to keep this at a minimum.

Examples

All businesses that sell products and services for remuneration.

Measuring Performance

Accounts Receivable (A/R) is measured in Days Sales Outstanding (DSO), which is basically just the average payment duration. It is also possible to measure A/R in *dollar-days*—the amount of dollars outstanding for the amount of days (which represents the true cost to the business of the monies outstanding) because focus should be on both elements. DSO is in fact a better, more objective measure for the project because it is independent of the dollar value of sales made (we could reduce dollar-days by just selling less!).

TOOL APPROACH

Measurement Systems Analysis on DSO

Focus should just be on measuring validity; a sound operational definition and consistent measure of DSO versus a detailed investigation of Gage R&R should suffice. For more details see "MSA—Validity" in Chapter 7, "Tools."

Baseline Capability Study of DSO

The data is typically captured over a period of one week to one month (depending on process drumbeat) to get a reasonable estimate of Capability. Historical data will more than likely be sufficient for the purpose.

After the baseline has been determined for DSO, the project will focus around two driving metrics:

- **Timeliness**, made up of (as a minimum)
  - Invoicing lead time
  - Payment lead time
  - Payment-processing lead time
- **Accuracy**, made up of (as a minimum)
  - Invoicing accuracy
  - Payment accuracy
  - Payment-processing accuracy

Measurement Systems Analysis on Accuracy / Yield

In order to measure Payment Accuracy, first examine the validity of the measure. A sound operational definition and consistent measure of Payment Accuracy, versus a detailed investigation of Gage R&R, should suffice.

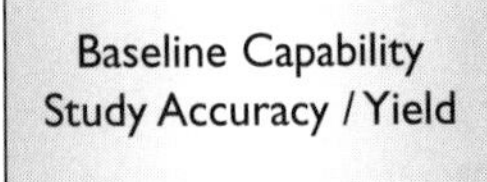

Take a baseline Capability for Accuracy or Yield throughout the process from service delivered to payment received. This is Attribute data and so at least 150 to 200 data points will be needed to determine even an approximation here. For more details see "Capability—Attribute" in Chapter 7.

Detailed Value Stream Map

Construct a high-level *Value Stream Map* of the payment process, showing the high-level steps for the Primary Entity as it progresses through the process, from the time that the service is delivered to the time that the payment is received. This will also serve to determine good operational definitions of timing points for the subsequent *Multi-Cycle Analysis.*

Multi-Cycle Analysis

Apply a *Multi-Cycle Analysis* to the *Value Stream Map* to determine where the time is being spent, along with an indication of variation in times. Usually it is the variability in step times that causes the problems.

It is important to note that, in the middlesection of the process when there is a wait for Suppliers to pay, they are legitimately allowed to withhold payment for an agreed length of time (known as "Terms"). This time should be dealt with separately from the rest of the timing calculations and hence needs to be broken out from the rest of the data.

From both the Payment Timing and Accuracy data collected, the path can be selected from the following:

Payment Timing is too long (excluding terms)

If the overall payment time is too long from the *Value Stream Map* and the *Multi-Cycle Analysis*, the focus should be on reduction of the Payment Lead Time.

Go to Section G in this chapter.

Payment Accuracy is too low

If payments are being made incorrectly along the process, treat a Payment Inaccuracy as a defect and follow the defect reduction roadmap outlined in Section C of this chapter.

Bad debts or Collection problems

For those cases where the problem lies externally, where Customers are failing to pay even within terms, Lean Sigma probably isn't the best roadmap. Some options include perhaps constructing a *Pareto* of the debt. It might be possible even to use $Y=f(X_1, X_2, \ldots, X_n)$ and follow the roadmap in Section C of this chapter with the Y being debt. This might have been done somewhere in the vast multitude of projects out there, but I've never come across it.

Payment Terms are too long

For those cases where the payment terms seem to be too long, Lean Sigma probably isn't the best roadmap. This is a matter for your sales and legal groups to resolve with existing Customers and to set acceptable standards for all new Customers. This is outside the scope of this book.

It is however useful to note that terms are often set based on the perceived value brought by the Supplier. It might be possible to streamline the delivery of entities to the downstream Customer and providethem additional value and thus improve the negotiating position.

4 Individual Step Process Problems

1: A Single Process Step Does Not Meet Takt

Overview[1]

Processes are required to cycle at a rate fast enough to generate entities at a pace to meet customer (or market) demand. Takt Time represents the pace of customer (or market) demand. If the Cycle Time (the actual rate of processing entities) of the process is longer than the Takt Time, then the process isn't cycling quickly enough and inevitably falls behind. For more details see "Time—Takt Time" and "Time—Global Process Cycle Time" in Chapter 7, "Tools."

This category infers that some work has been done to relate the Takt Time to the Cycle Time, usually a *Load Chart*, and also to identify that this single process step is not meeting Takt. If this is not the case, then return to Chapter 3, "Global Process Problems," to select the Problem Category for the process as a whole.

Measuring Performance

The measure used should have been put in place at the Global Problem Category level, but if not, then use the Process Cycle Time as the metric with the goal of being below the Takt Time.

[1] Before reading this chapter it is important to read and understand Chapter 1, "Introduction," particularly the section, "How to Use This Book."

TOOL APPROACH

If this hasn't already been done at a previous step, then:

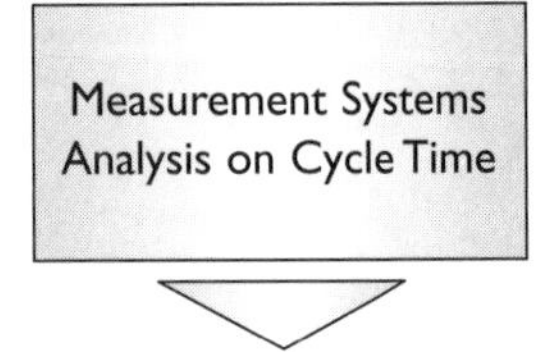

The focus is on measuring validity (a sound operational definition and consistent measure) of Cycle Time versus a detailed investigation of Gage R&R. See "Time—Global Process Cycle Time" and "MSA—Continuous" in Chapter 7, for more detail.

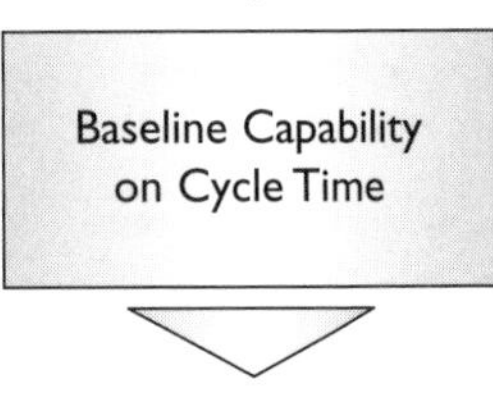

If we were interested in just the average Cycle Time, then a data collection of 10 to 15 points would suffice. However, it is often the case that variability in the Cycle Time is more of a problem, and in that case, it is best to capture 25 to 30 data points at minimum for the baseline. For more details see "Capability—Continuous" in Chapter 7.

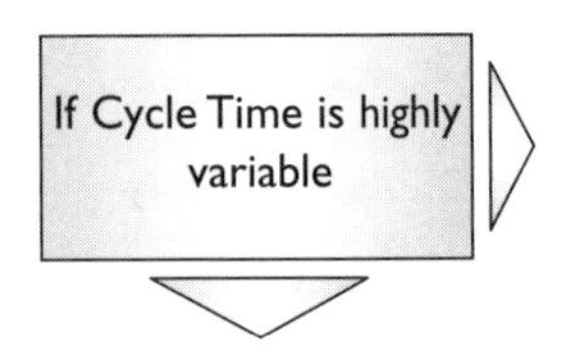

If the data in the Capability Study shows too high a level of variability then it is useful to visit the roadmap in Section 3 of this chapter; however, there are some simple improvements that can be made first. Continue in this roadmap below and return to Section 3 later if needed.

This roadmap is to quickly identify using *Overall Equipment Effectiveness (OEE)*, whether the issue is with quality related problems, the uptime of the process, or the speed of the process when it is running.

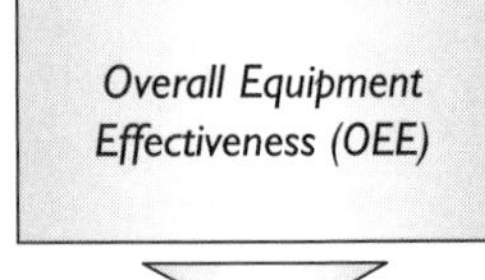

The process would hit minimum Cycle Time if it never went down, did only VA activity, as fast as it has ever run, *and* generated only perfect quality entities.

From the OEE, we gain an understanding of whether the issue is

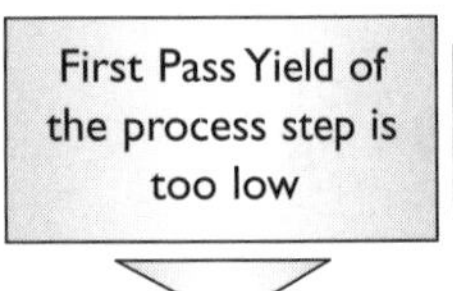

This is a quality-related problem. Cycle Time is effectively increased because the process step spends time generating defective entities.

Continue to Section C in Chapter 3 and follow the roadmap focusing on this single process step.

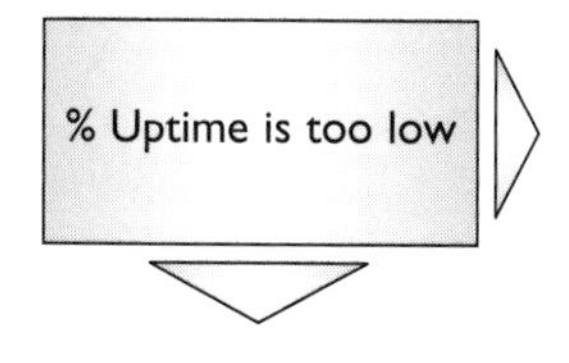

The Cycle Time is extended because the process is spending part of its available time to do VA entity generation doing something else.

Continue to Section D in Chapter 3 and follow the roadmap focusing on this single process step.

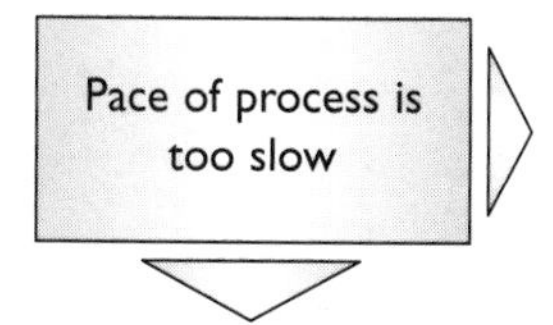

The Cycle Time is extended because the process could do the VA piece of the work more rapidly.

Continue to Section 2 in this chapter.

It is highly likely that the problem is resolved at this time. To confirm this:

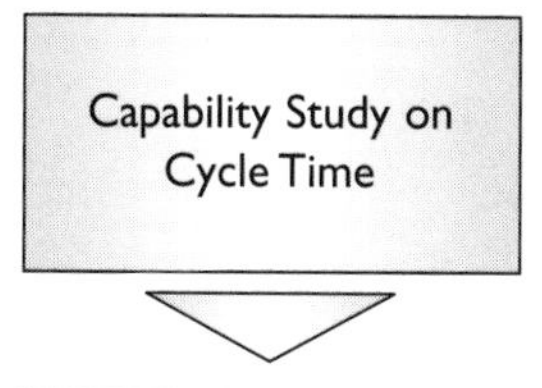

Capture 25 to 30 data points for Cycle Time to understand the new mean, but also the variability in Cycle Time.

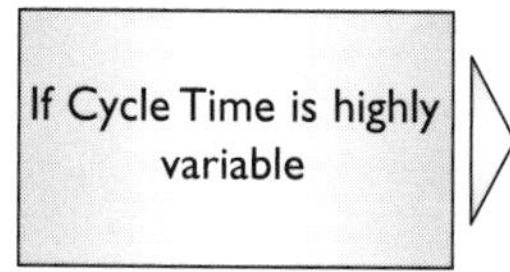

If the data in the Capability Study still shows too high a level of variability then go to Section 3 of this chapter.

At this point, the Cycle Time of the step should be consistently below Takt, in which case proceed to the Control tools in Chapter 5, "Control—Tools Used at the End of All Projects."

If the Cycle Time is still above Takt, look at the process as a whole (versus this single step) and ask

- Why are we doing this?
- Is there any other way to do this?
- Does it make more sense to offload work from this step to other adjacent steps?
- Is it better to add more resources to accelerate the step?
- Is it better to split the step into two or more steps?
- Is there a simple changing of technology or approach to the whole process that would help?

At this point, if the path to improving this process is still not clear, then it might be time to consider designing an entirely new process. A useful methodology at this time would be Six Sigma Process Design.[2]

2: THE PACE FOR A SINGLE PROCESS STEP IS TOO SLOW

OVERVIEW

Sometimes, even though the step is running, it isn't cycling fast enough to process entities to meet downstream demand. Pace is the speed of the process step when it is up and running doing VA work. This is not the average process rate that is traditionally captured. For a detailed explanation of Pace refer to "Overall Equipment Effectiveness (OEE)" in Chapter 7, "Tools."

Plan only to make improvements to the Pace after other categories of NVA work have been improved.

MEASURING PERFORMANCE

Measures of Pace of a process step are usually process-specific but are generally written in a form of number of entities per unit time, for example:

- Kg product/hr
- Patients/hr
- Widgets/hr
- Invoices (or orders)/hr

Similarly, they can be written as an average time to generate or process an entity, such as average processing time.

The data needs to be captured *only* when the process is running and should not take into account the average slowing of pace due to process stoppages, downtimes, and so on. Those elements should be accounted for in the %Uptime metric in OEE. If the primary concern is Uptime, then proceed to Section D in Chapter 3 instead.

---

[2] Six Sigma Process Design (SSPD) is a more recent newcomer to the Lean Sigma world. At the time of writing this, there are no known references for this. See www.sbtionline.com for more details.

Tool Approach

If a Baseline Capability hasn't already been done at a previous step, then:

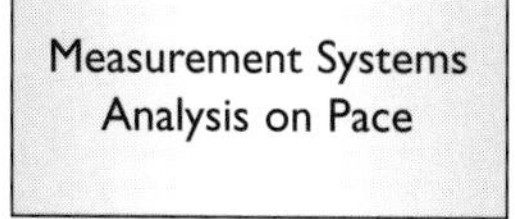

Pace is challenging to determine; so a quick MSA is always worthwhile. The focus is on measuring validity, a sound operational definition, and a consistent measure of Pace while the process step is doing VA work, versus a detailed investigation of Gage R&R. See "Overall Equipment Effectiveness (OEE)" and "MSA–Validity" in Chapter 7.

Baseline Capability of typical Pace versus the instant best

Pace capability is measured by taking the average Pace (over a period of time) as a percentage of the highest instantaneous Pace the process has achieved. See "Overall Equipment Effectiveness (OEE)" in Chapter 7. The data should be captured over a period of typically one week to one month (depending on process drumbeat) to get a reasonable estimate of average pace. There is usually less variation in the VA component of work, and, hence, 15 to 20 data points give a solid average. If there is known variability in the Pace, then capture 30 or more data points.

Even though we often consider a process step of this type to be the smallest, indivisible unit of a process, there are usually sub-steps within each step. To accelerate the Pace, it is useful to subdivide the single step into its sub-steps to see if there are any NVA sub-steps involved at a lower level.

Detailed low-level *Value Stream Map* (VSM) on the process step

This involves an excruciating level of detail for some, but in the majority of cases, it identifies opportunities where the process step waits for something (usually information, materials, testing, and so on).

From the *VSM*, identify actions to remove any NVA activity from the process step. Also, it is important to take advantage of any paralleling of sub-steps that might be available; if this step truly is the bottleneck in the process, then it is worth adding additional resource to it. A useful tool to help here is *Critical Path Analysis*:

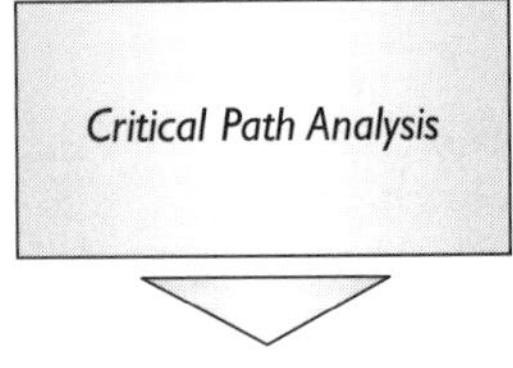

Critical Path Analysis allows the Team to see the longest chain of activity in the process step that drives the Cycle Time. For more details see "Time—Global Process Cycle Time" in Chapter 7. Offloading tasks from the Critical Path reduces the Cycle Time accordingly.

At this stage it useful to take a step back and look at the big picture and ask the questions:

- Why are we doing this?
- Is there any other way to do this?
- Does is make more sense to offload work from this step to other adjacent steps?
- Is it better to add more resources to accelerate the step?
- Is it better to split the step into two or more steps?
- Is there a simple changing of technology or approach to the whole process that would help?

If, after answering these questions, there is still a desire to continue along this path to improve the Pace, then the roadmap to a solution from here on relies on the equation $Y=f(X_1, X_2,..., X_n)$, where the Ys are the Cycle Time and the Primary Performance characteristic(s) of the process. For instance, if in a chemical production process we are interested in the assay of the product as a Primary Performance characteristic, then the Ys are assay and Cycle Time.

Go to Section C in Chapter 3 and focus on this single process step to determine which Xs can be manipulated to gain the best level of performance for the Ys (including Cycle Time).

3: Too Much Variation in the Cycle Time of a Single Step

Overview

Processes are required to cycle at a fast enough rate to generate entities at a pace to meet customer (or market) demand. Takt Time represents the pace of customer (or market) demand. For more details see "Time—Takt Time" in Chapter 7, "Tools." Even if, on average, the Cycle Time (the actual rate of processing entities) of the process is shorter than the Takt Time and the process on average meets demand, the Cycle Time could be variable enough that in the short term it does not meet Takt and thus affects delivery performance. See also "Time—Global Process Cycle Time" in Chapter 7.

This category infers that some work has been done to relate the Takt Time to the Cycle Time. If this is not the case, then return to Chapter 3 to select the Problem Category for the process as a whole.

Measuring Performance

The measures used should have been put in place at the Global Problem Category level, but if not, then use the Process Cycle Time as the metric, calculating both its mean and standard deviation (the aim being to reduce both).

Tool Approach

If this hasn't already been done at a previous step, then:

Measurement Systems Analysis on Cycle Time

Focus should just be on measuring validity (a sound operational definition and consistent measure) of Cycle Time versus a detailed investigation of Gage R&R. See "Time—Global Process Cycle Time" and "MSA—Validity" in Chapter 7 for more detail.

Baseline Capability Study on Cycle Time

For a baseline that takes into account the variability in the Cycle Time, it is best to capture at least 30 to 50 data points. Take the mean and standard deviation of the Cycle Time for the baseline.

Even though we often consider a process step of this type to be the smallest, indivisible unit of a process, there are usually sub-steps within each step. It is often within these sub-steps that there are NVA activities causing the variability in Cycle Time.

Detailed low level *Value Stream Map (VSM)* on the process step

This involves an excruciating level of detail for some, but in the majority of cases identifies opportunities where the process step waits for something (usually information, materials, testing, and so on).

From the VSM, identify actions to remove any NVA activity from the process step. It is less likely to encounter variability in the VA portion of a step than in the NVA elements, so reduction of NVA activity usually significantly reduces variation in Cycle Time.

Major sources of variation were removed

If from the VSM, significant sources of NVA activity were removed and the Team feels the issue might have been resolved, then repeat the Capability Study. If the variability is now at a manageable level, then proceed to Chapter 5.

If not, then continue in this roadmap.

The roadmap to solution from here on relies on the equation $Y=f(X_1, X_2,..., X_n)$, where the Ys are the Cycle Time and the Primary Performance characteristic(s) of the process. For instance, if in a chemical production process we are interested in the assay of the product as a Primary Performance characteristic, then the Ys are assay and Cycle Time.

Go to Section C in Chapter 3 and focus on this single process step to determine which Xs can be manipulated to gain the best level of performance for the Ys (including Cycle Time) and minimize the variability.

5 Control—Tools Used at the End of All Projects

Overview[1]

All projects, regardless to which Problem Category they relate, progress through the same series of tools at the back end of the project (the late Improve phase and through Control to Validation and Signoff). The purpose of this part of the roadmap is to ensure:

- The process is clearly defined and laid out.
- The process control mechanisms are in place.
- The process is done consistently.
- The process roles are defined along with associated competences.
- Process documentation is complete and up to date.
- Process staff are maintained at the right competency level.

All of these ensure the process performs at the new, elevated rate for all time. A project signoff is done to demonstrate performance and its significant difference from the baseline.

[1] Before reading this chapter it is important to read and understand Chapter 1, "Introduction," particularly the section, "How to Use This Book."

TOOL APPROACH

The primary deliverable of any project is the Control Plan. The Control Plan represents all the elements that must come together to control the process so that it performs at the desired level for all time and comprises

- **Critical Parameters.** The key performance characteristics (Ys) by which we measure the process, along with the identified key Xs that drive the majority of the variability in the Ys.
- **Measurement Systems Analyses.** Demonstration of the ability to measure the Critical Parameters.
- **Capability Studies.** For Critical Parameters that demonstrate performance to requirements.
- **Reaction Plans.** For the Critical Parameters if they fail to meet performance requirements.
- **Control Plan Summary.** A single form that documents the summary of the above elements.
- **Customer information.** The levels, profiles, and segmentation of demand that the process experiences.
- **Process Maps.** The definition of the process and its flow.
- **Standard Operating Procedures (SOPs).** Documentation reflecting how the process should run
- **Failure Mode & Effects Analyses.** A history of the risk in the process and how it has been reduced.
- **Continuous Improvement Plan.** Accountability and approach for improving the process (or mending it if it breaks).
- **Training materials and methodology.** The tools to bring the operator competences to the levels required to run the process.
- **Maintenance.** Ongoing maintenance requirements to ensure the process sustains its performance improvement.

The Control Plan ensures that all the elements listed in the overview at the beginning of this chapter are considered.

The following roadmap lists a linear sequence of events that construct the Control Plan and lead up to the signoff of the project. In reality, steps overlap at times and minor refinements can be made to earlier steps after the later work is finished. The sequence of completion is fairly consistent though.

The new process must be clear enough and defined to the point where it can be, or has already been, implemented. The best tool for this purpose is a *Swimlane Map* because it clearly defines sequence of activities by role.

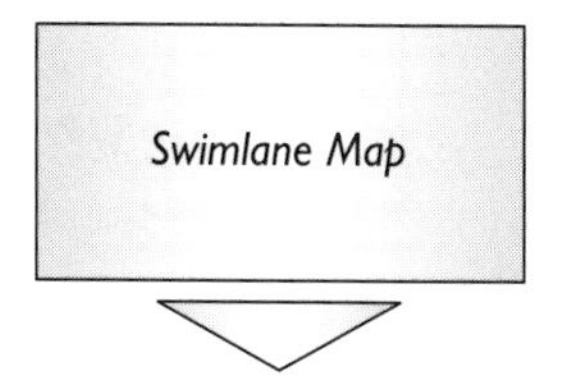

Use the *Swimlane Map* to demonstrate visually how the new process runs and how the roles relate to the process steps.

After this is complete, the Team needs to build in the controls on the critical Xs identified through earlier parts of the roadmap.

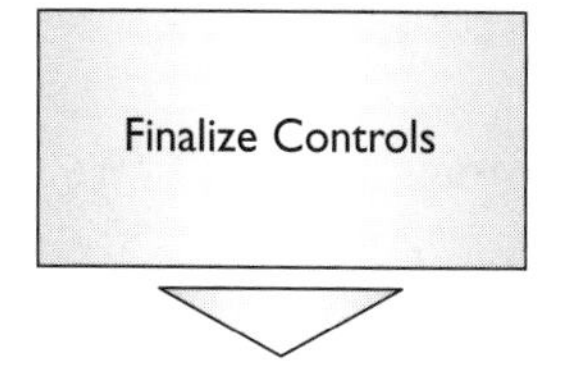

There are a number of options available to maintain control on an X. In descending order of strength, they are

- Design out the X completely. For example, if the operator is the critical X, can we automate the step so that there is no operator involvement?
- Mistake-proof the X so that it cannot be changed inappropriately. This is sometimes known as *Poka Yoke*. For example, for key knobs on machinery, screw the knob down so that it cannot be changed or consider removing it entirely.
- Use *Statistical Process Control* at the workstation to ensure that if the X goes out of control, then it is identified and the appropriate responses are made.
- Post Parameter Cards at the workstation to ensure operators know at what settings to run the Xs.
- Train operators how to maintain the Xs at the prescribed level.

Some combination of all of the above is usually done.

At this point, it is worth reexamining the triggers that ensure each activity begins correctly. These might be some of the key Xs, but this is important to do for the whole process. Triggers are known as *Kanbans* in the Lean Sigma world and ensure that *Pull* is occurring through the process; for example, when you are free to work on something, you have something to work on, unless there is no demand from the Customer. See "Pull Systems and Kanban" in Chapter 7, "Tools," for more detail.

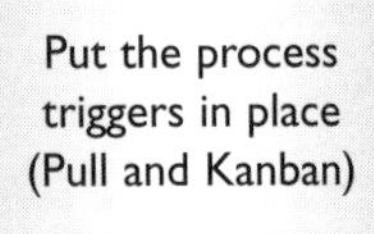

For key points in the process, it is necessary to have signals to upstream steps that subsequent steps are ready to work; for example, in an Operating Room a trigger might go out to feeder processes that the current Patient procedure completes in 10 minutes. This allows the next Patient to be prepped and so on.

This is a Pull System and the trigger is known as a Kanban. Without these, time is lost in the process when entities stall due to lack of readiness.

After the process is defined, it is possible to examine its layout. There are many tools available to aid in this endeavor, such as Cellular Design and Workstation Design. They are not covered here due to the sheer volume of options available, but at this point, the process needs to be laid out correctly based on the business and process requirements.

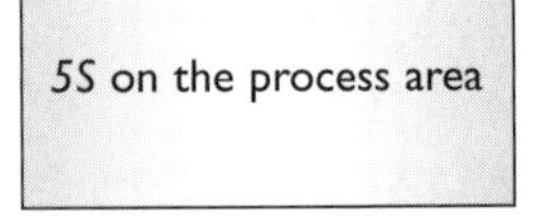

Based on the process layout chosen, the materials, supplies, and so on, need to be organized appropriately. *5S* helps immensely here.

After you define the process, determine the triggers, and lay out and organize the process, the focus becomes ensuring that the process runs like this for all time. The beginning of this is known as Standard Work

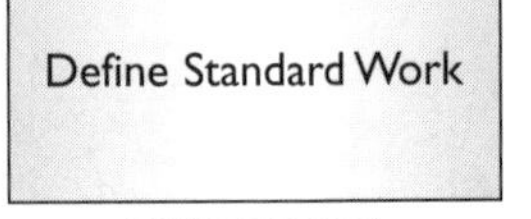

Standard Work is the formal documentation of a group of tools that helps define the day-to-day running of the process. It is made up of

- Takt Time (the pace of market demand) For more details see "Time—Takt Time" in Chapter 7.
- Process Cycle Time (the pace that the process runs) For more details see "Time—Global Process Cycle Time" in Chapter 7.
- Total Work Content (how much labor time is involved in each step)
- *Load Chart* (graphical representation of Takt versus Cycle Time)
- Layout of the process
- Order of the process

- Work sequence (and timings) for the operators
- Work sequence (and timings) for any equipment/machines
- Standard inventory in the process

After the Standard Work is defined, the process can formally begin to perform in its newly prescribed way. To do this the Team needs to set the key Xs to their correct level.

Implement identified Process Settings

For the critical Xs, the levels determined earlier in the roadmap must be set in place. The formalized controls will also be implemented on these Xs at this time. For non-critical Xs, Team focus is on choosing the most appropriate (usually economical) settings.

Install appropriate Metrics and Tracking (includes MSAs and Capability Studies)

The key tracking metrics for the process will be the Ys and critical Xs. For each, if not already done so, a Measurement Systems Analysis will be required. For more details see "MSA—Validity," "MSA—Attribute," and "MSA—Continuous" in Chapter 7. For each MSA, a regular revisit schedule is also required. After the Measurement Systems are deemed fit (the Team must mend them if not), then initial Capability Studies are made. Ideally on-line tracking is instigated or worst-case a schedule is set up to revisit Capability on a regular basis.

Formalize Process Roles

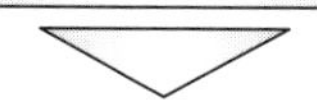

For each role defined in the process, a role checklist or documented approach is created—in effect a "how to do the role." Note that sometimes there is confusion between people and roles. A role is the "chair to sit in"; a person is the "one that sits in the chair." Roles remain; people can come and go.

Create and Finalize Documentation

Construct documentation for the process as a whole. The best documentation involves visuals, so rely heavily on *Swimlane Maps* and so on. Remember to include detail of key Xs, metrics, tracking, and reaction plans (these are sometimes overlooked).

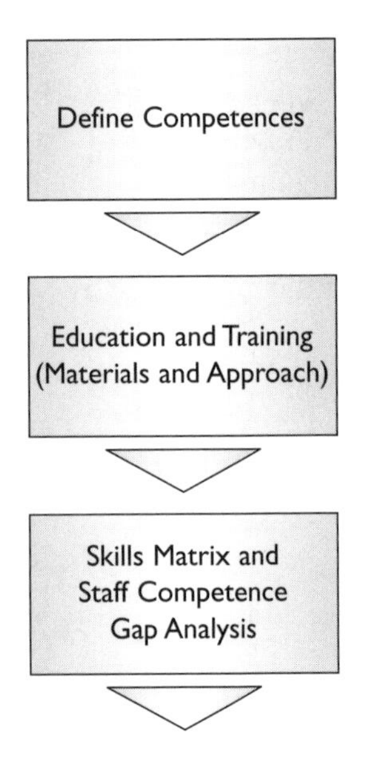

For each role in the process, the Team identifies the competences required to perform the role; think of this like a license. These competences must be written in objective language and their attainment should be readily measurable.[2]

For each competence there needs to be provision of education and training to attain this competence. The actual training is not conducted at this point.

Finally, the existing staff is layered back onto the process by examining their competences. The matrix of staff versus competence is known as a Skills Matrix. The tendency here is to overdo it and break out every tiny skill. The best approach is to keep it grouped to a maximum of 30 to 40 skills across the process or the Skills Matrix becomes too unwieldy.

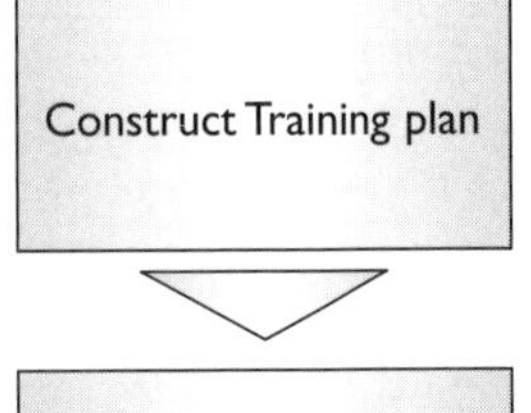

The Skills Matrix is in effect a large Gap Analysis on staff competency. From this, the Process Owner (not the Team) must construct the training plan to bring the staff up to the required levels. This might involve recruitment.

Construct
Implementation plan

Often the solution created is for a single line or entity type and there needs to be a rollout to other lines or entity types. A simple Milestone Plan helps communicate the rollout and improve the likelihood of a timely implementation.

Most processes involve stakeholders who need to be informed of the impending changes. Constructing a Communication Plan ensures the appropriate parties are well informed.

[2] The most common system is known as a 5-Star rating and uses five levels of attainment of competence, usually from "No Knowledge" through "Awareness," "Has Been Involved," and "Led" up to "Can Teach Others."

Execute implementation (Training, Rollout, Communication)

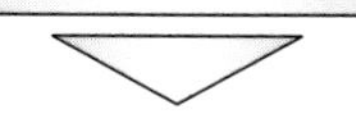

The actual rollout might or might not be considered part of the Lean Sigma project but *is* required by the business. (Signoff for the Lean Sigma Project could be based on the pilot process success.) It might be necessary to form separate implementation teams to take the changes to the rest of the organization.

Construct Final report

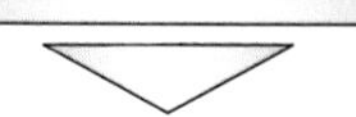

At first glance, the creation of a Final Report might seem like NVA work, but it is important to maintain organizational learning with respect to the project. In future projects, Belts will be able to refer back to the work done here and hopefully avoid unnecessary pitfalls.

Final Review, Validation, and Signoff

Validation is data-based and should be justified by a t-Test, usually a 2-sample t-Test, or similar to show that the changes made to the performance metrics are statistically significant. For example, that the gains are real.

Reviewers should include the Champion, Process Owner, and a technical resource (Master Black Belt). The signoff represents the project coming full-circle and the final handoff to the Process Owner is complete.

All that is left is the important celebration of the results. Remember to include all those involved in the success of the project.

Then take a break; the next project will be along shortly...

6 Discovery—Tools Applied to Identify Projects

Overview[1]

For some processes, it is not obvious where the project should focus. There might be many potential projects across the process and the Team is charged with identifying them prior to selecting one and carrying it to conclusion. This identification approach is known as *Discovery* and follows a slightly abridged version of the roadmap used in Define and Measure in a regular project.

The output of the Discovery Phase is a clearer (albeit high level) understanding of the process, along with a series of identified projects or opportunities for the process.

Tool Approach

The following sequence of tools is applied during *Discovery*:

To initiate the project, the Belt, Champion, and Process Owner meet to construct the preliminary Charter. They determine the initial scope and the Team members who should be used. The initial scope typically is the whole process under scrutiny. At this stage, the likely metrics and potential benefits are probably unknown.

[1] Before reading this chapter it is important to read and understand Chapter 1, "Introduction," particularly the section, "How to Use This Book."

After the initial *Project Charter* is complete, the Project Team is mobilized, and shortly thereafter, the first Team meeting can take place.

SIPOC Map
(+ High-Level Value Stream Map)

To help define the whole process, a *SIPOC* is useful, especially in understanding the scope and purpose of the process. The *SIPOC* is an important tool in Discovery to address, "What really is the question to be answered?"

The central "Process" column of the *SIPOC* can be extracted and rotated 90° to become a High-Level *Value Stream Map (VSM)*. This map is a useful linear representation of the process that can help in slicing up the process into manageable pieces to which to apply projects.

After the *SIPOC* is complete for the whole process, the next step is to determine the Customer Requirements (again for the whole process). This is done in a similar fashion to the regular Define approach by using a number of tools to identify the major customer needs and metrics to determine performance versus those needs.

Brainstorming / *Murphy's Analysis*

To help frame what questions to ask the Customers regarding the process, it is useful to spend time as a Team brainstorming the issues around it. This can be done in the form of traditional *Brainstorming*, but it is preferable to use the tool known as *Murphy's Analysis*. Note that the output of these tools is absolutely *not* the Voice of the Customer (VOC), but rather, these are the Team's internal insights to help guide the structure of the VOC questioning in the subsequent steps.

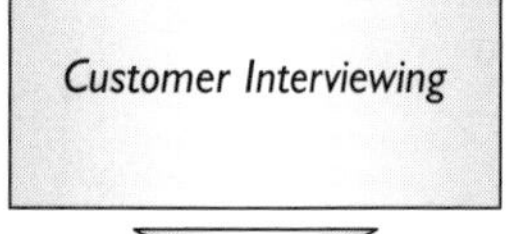

Based on the some of the learning from the internal brainstorming, it should be reasonably straightforward to put together an Interview Discussion Guide and Customer Matrix and then interview Customers. Richness of Customer data is based on diversity of Customers, not on quantity, and thus, 12–20 interviews should be enough to frame even the most complex process. Interviewing can take one to two weeks to complete, so plan the Team's time accordingly.

See "Customer Interviewing" in Chapter 7, "Tools," for more detail.

Customer Surveys

Some processes have a vast multitude of Customers from whom the Team would like some input to ensure that nothing key is missed, but really don't want to spend time interviewing in depth. To gain these Customer inputs, use a survey that is mailed, handed directly to them, or, primarily for internal Customers, posted on the wall where they can write directly on it.

The survey usually will take one week to get the data back.

Surveys are notoriously error prone, so do not rely on this as your main source of Customer input—use the interviews instead.

For more detail see "Customer Surveys" in Chapter 7.

Affinity Diagramming

The VOC is distilled down into useful information by applying an *Affinity Diagram*.

Customer Requirements Trees / 5 Whys

After the *Affinity Diagram* is complete, a simple extraction yields the *Customer Requirements Tree*. This Tree represents a simple hierarchical structure of needs from the highest abstract level down to individual facts and measures. To successfully create the Tree, it is often best to track the noted metrics back to the set of more essential metrics behind the scenes. A useful tool here is the *5 Whys*.

Detailed *Value Stream Map*

Construct a rigorous VSM including all the detailed steps for the Primary Entity as it progresses through the process. At this point, don't map the sub-steps of any VA step. The VSM forms the structure around which the process is sliced to form discrete pieces of process upon which projects are based.

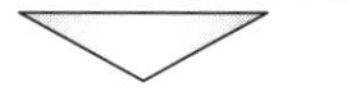

Key Process Output Variables (KPOVs)

Examination of the VSM should help the Team identify several key milestones along the process that are useful points at which to take performance readings. The performance metrics (such as accuracy ratings) are the Ys along the process that are important to the Customer. Determine the Ys taking the output of the *Customer Requirements Tree* and firming up Operational Definitions of the key metrics. For more details see "KPOVs and Data" in Chapter 7. For each metric, a baseline measure is made.

These and other points are the timing points, sometimes known as Time Stamps, used in the *Multi-Cycle Analysis* conducted subsequently.

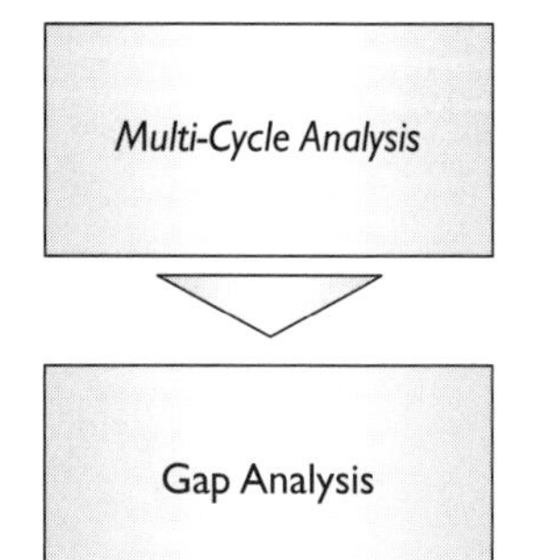

Apply a *Multi-Cycle Analysis* to the VSM to flesh out where the time is spent, along with an indication of variation in times. Usually, it is the variability in step times that causes the problems.

The Team examines the data collected from the previous tools to populate the VSM with data on the timings and the Ys. At this point, the Team, Belt, Champion, and Process Owner meet to identify projects within the VSM and prioritize them.

Projects should target sections of the process for which:

- The Ys are significantly far from entitlement[2] (perfection for the Y).
- Undue time is spent.

For each Project identified, a draft *Project Charter* should be created and then the Projects should be prioritized. Prioritization can be done with a *Cause & Effects Matrix*[3] or just by selecting the areas of greatest pain.

At this point the Team is retasked to tackle the highest priority project and the roadmap follows the regular DMAIC roadmap as outlined in Chapter 2, "Define—Tools Roadmap Applied to the Beginning of All Projects." Obviously a large amount of the early Define and some of the Measure work have been covered, so the Team should make light work of these two Phases.

[2] See "Project Charter" in Chapter 7.

[3] Use of the Cause and Effects Matrix for project selection is described in detail Stephen A. Zinkgraf's book, *Six Sigma—The First 90 Days*, (Prentice Hall PTR, ISBN: 0131687409).

7 Tools

01: 5 Whys

Overview

5 Whys is commonly listed more as a Lean tool than Six Sigma because it isn't rigorously data-based, but it can be particularly useful, especially to Belts that get trapped in the minutia and don't step back to see the bigger picture (see "Other Options"). The simple idea is to keep asking "Why" (usually five times) to ensure that the root cause(s) to the effects are fully understood. The reasoning is that the result of each time the Why is asked gives a different answer, in essence peeling back the onion as follows:

- First Why—Symptom
- Second Why—Excuse
- Third Why—Blame
- Fourth Why—Cause
- Fifth Why—Root Cause

A simple fictitious example demonstrates use of the tool well. A problem in London's Trafalgar Square is that Nelson's Column requires frequent, expensive repairs.[1]

[1] The original source of this example is unknown. The example appears in various guises related to many national landmarks in Europe and the U.S.

First Why—Why does the column need frequent, expensive repairs?

Answer—Frequent washing is damaging the stone.

Obvious solutions in this case include investing in less abrasive cleaning mechanisms and perhaps different detergents. However, this still doesn't reduce the frequency of washing.

Second Why—Why does it need to be washed so much?

Answer—There is a build up of pigeon droppings.

Pigeons are a popular sight in Trafalgar Square and obvious solutions in this case might include investing in a pigeon-scaring device or placing devices to stop pigeons landing on the column. Unfortunately pigeons are arguably part of the tourist attraction itself, so this could be difficult to implement.

Third Why—Why are the pigeons gathering on top of the column?

Answer—The pigeons eat the spiders on the column.

Obvious solutions might include spraying the column regularly with pesticide to kill the spiders. The use of pesticides is frowned upon at the best of times in a highly populated area, but on such a visible landmark there could be major opposition. Also, the impact of the pesticide on the stone itself would have to be examined.

Fourth Why—Why are there spiders on the column?

Answer: The spiders eat the insects on the column.

The solution to the third Why still seems to work. The pesticide would kill both the spiders and the insects.

Fifth Why—Why are the insects there?

Answer—They are attracted to the brightly lit surface in the evening.

Proposed solution—Delay turning on the lights for 30 minutes and they are attracted elsewhere.

A project example of using 5 Whys in the middle of a sequence of root cause analysis is illustrated in Figure 7.01.1. The problem related to delays in the New Product Introduction (NPI) process. After statistical analysis of the data, the issue is identified to be associated with missing parts on the circuit board. At this point, the Team used 5 Whys (with data to back up each Why) to discover a root cause, which is a design error. Clearly this isn't an end in itself, because a solution to "Design Error" isn't immediately obvious and hence subsequent data collection and analysis was used to take the problem to solution.

Other Options

In fact a much stronger use of this tool is to ask "Why do I care?" enough times to relate the issue back to a business level problem. This is incredibly useful in the Define Phase of a project. Commonly when identifying and scoping projects, Belts and Champions latch onto a solution and set that to be the goal of the project, when in reality there might be a better approach to the whole project.

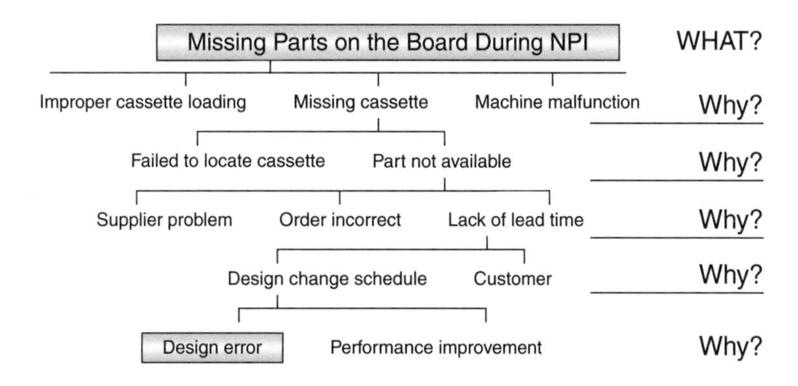

Figure 7.01.1 Example of 5 Whys in a project focused on delays in New Product Introduction (NPI).[2]

To use the same example mentioned in Chapter 2, "Define—Tools Roadmap Applied to the Beginning of All Projects," a hospital client's project was focused on the Prepare visit made by expectant mothers a few weeks prior to the big day. The Team had identified a key Y to be "percent of new mothers given access to the Prepare visit," which is difficult to relate to any major performance metric for the hospital as a whole.

By asking "Why do I care?" a number of times, the thought flow is as follows:

- First Why—More mothers need to go through the Prepare visit.
- Second Why—Education given during the Prepare visit is more readily absorbed than during the birthing visit, and doctors' valuable time is wasted during the birthing visit reworking the education.
- Third Why—New mothers need to understand and retain key learning before they return home.
- Fourth Why—New mothers are the ones who are required to give correct care to themselves and their newborns after they leave the hospital.
- Fifth Why—Sometimes new mothers or babies, after returning home, have to be readmitted to hospital for what are essentially avoidable reasons.

The Prepare visit is in fact a solution to a problem, and the true underlying business issue and associated Y is better defined as the "rate of avoidable returns post partum is too high." This has a clinical impact as well as a business implication to the hospital and clearly brings an entirely new perspective to the problem and hence the project.

[2] Adapted from SBTI's Lean training material.

02: 5S

Overview

5S is a methodology for creating and maintaining an organized, clean, and safe high-performance work environment. It is an unusual tool in the Lean Sigma arsenal because it can be used in the Measure, Analyze, Improve, and Control Phases.

The title 5S refers to five activities within the tool as follows (in brief):

- **Sort (or Separate).** Keep only what is needed in the area to run the process. Remove everything else.
- **Store.** Arrange needed items and identify them for ease of access and use. Organize the area.
- **Shine (or Sweep).** Clean the area (and equipment) regularly to maintain performance as new.
- **Standardize.** Eliminate the causes of dirt and make standards obvious. Standardize the process in the area.
- **Sustain.** Set discipline to maintain the level of performance.

Logistics

Although 5S should be used on an ongoing basis, the first 5S work in an area is typically done as a 5S Event. An event should be targeted on a single process area and the Team should take all five of the Ss to completion during the event. It is not a good idea to try to apply 5S to a whole plant a single S at a time. After a focused area is complete, a subsequent series of events can be applied to roll out to other areas, but again, for each area apply all 5Ss during each event.

The event duration is anywhere from one day for a small process area in a transactional process to perhaps a three to four day event in a larger (probably more dirty) manufacturing process. Typical event duration is two days.

Planning is key to the success of the event and the words "Hey, let's do a 5S event today!" are a surefire indicator for an impending disaster.

The key elements to have in place prior to an event are

- **Support.** A 5S event is not a small affair and significant change will likely be made. This requires solid leadership support to ensure success. The event should have a clear Owner/Champion who kicks off the event and serves as the Customer for the event.

- **Scope.** When conducting an event, the biggest failure is usually due to scope creep, so set a clear, focused, bounded process area to tackle during the event period and don't be tempted to move beyond this area until all 5Ss are in place.
- **People.** The next key element is having the right people involved. 5S is an active team sport and requires people who actually do the work in the target area. Other key functions that also need to be involved are Maintenance, Facilities/Engineering, and probably IS/IT.
- **Communication.**
  - All parties involved in and affected by the event need to know of its existence and implications to them. The other functions listed in the preceding People bullet also need to be informed that work needs to be done "there and then" rather than be put on a To-Do List for future change. 5S follows the Kaizen mentality of doing it *today*, instead of planning to do it over a period of weeks—hence the need for strong Leadership support.
  - The rule in all Process Improvement is to communicate 10 times more than you would expect.
- **Equipment, Supplies, and Ancillary Items.**
  - Most of the event is done in the process area itself, but another quiet area is required to train the Team briefly on the 5S process and to serve as a work area at times through the event.
  - Facilitation equipment and supplies are necessary, such as flipcharts, pens, and perhaps an LCD projector and screen to show the 5S training materials
  - Photographing the area before and after the event is crucial to show the impact of the event. At least one, preferably two or more digital cameras are needed.
  - Work is done on laying out the area. An enlarged blueprint or CAD drawing of the process plan is needed.
  - Items to be removed from the area need to be "Red Tagged" if they cannot be removed immediately (see Step 2 in the Roadmap). A supply of Red Tags is required. An example of a Red Tag is shown in Figure 7.02.1.

Roadmap

The roadmap follows the Ss themselves, so it is quite straightforward to coordinate.

Red Tag

5S Target Area ______________________ Tag Number __________

Area Owner ______________________ Date __________

Item Tagged ______________________________________

____

Red Tag Drop Area ______________________________________

Signed By: ______________________________________

Figure 7.02.1 Example Red Tag.

(zero S)—Baseline

The Team and Champion walk the process and the area to determine the baseline performance. The purpose is primarily to

- Identify all opportunities for 5S (all items that violate the 5Ss).
- Take pictures of the opportunities. For the photos taken, identify picture location and angle on the blueprint, so that the before and after pictures are taken from the same location.
- Perform the baseline 5S audit. The audit is based on the 5S Audit Matrix shown in Figure 7.02.2. Each S has five levels of maturity that can be achieved in the area. The majority of event locations score at level 1 across the board, with perhaps a rare level 2 in the Standardize column. The score from the 5S Audit Matrix is visually displayed in the area as a radar chart, as shown in Figure 7.02.3.
- After the Audit is complete, the Team takes 10 to 15 minutes to prioritize the list of opportunities in the area and revisit the event scope to ensure the maximum impact is made from the event.

1S—Sort

Sort is used to separate all the necessary and unnecessary items. The Team clears the workplace and removes all unneeded items, such as racks, tools, equipment, and excess materials. The reason for this is that

LEVEL 5 *Continuous Improvement*	Cleanliness problem areas are identified and mess prevention actions are in place.	Needed items can be retrieved in 30 seconds with minimum steps.	Potential problems are identified and countermeasures documented.	Proven methods for area arrangement and practices are shared and used.	Root causes are eliminated and improvement actions include prevention.
LEVEL 4 *Focus On Reliability*	Cleaning schedules and responsibilities are documented and followed.	Minimal needed items arranged in manner based on retrieval frequency.	Work area cleaning, inspection, and supply restocking done daily.	Proven methods for area arrangement and practices are used in the area.	Sources, frequency of problems are noted w/ root cause & corrective action.
LEVEL 3 *Make It Visual*	Initial cleaning is done and mess sources are known and corrected.	Needed items are outlined, dedicated locations are labeled in planned quantities.	Visual controls and indicators are set and marked for work area.	Agreements on labeling, quantities, and controls are documented.	Work group is routinely checking area to maintain 5-S agreements.
LEVEL 2 *Focus On Basics*	Necessary and unnecessary items are identified; those not needed are gone.	Needed items are safely stored and organized according to usage frequency.	Key area items are marked to check and required level of performance noted.	Work group has documented area arrangement and controls.	Initial 5-S level is established and and is posted in the area.
LEVEL 1 *Just Beginning*	Needed and not needed items are mixed throughout the area.	Items are randomly placed throughout the workplace.	Key area items checked are not identified and are unmarked.	Work area methods are not always followed and are not documented.	Work area checks are randomly done and there is no 5-S measurement.
	Sorting	***Store***	***Shine***	***Standardize***	***Sustaining***

Figure 7.02.2 5S audit scoring matrix.[3]

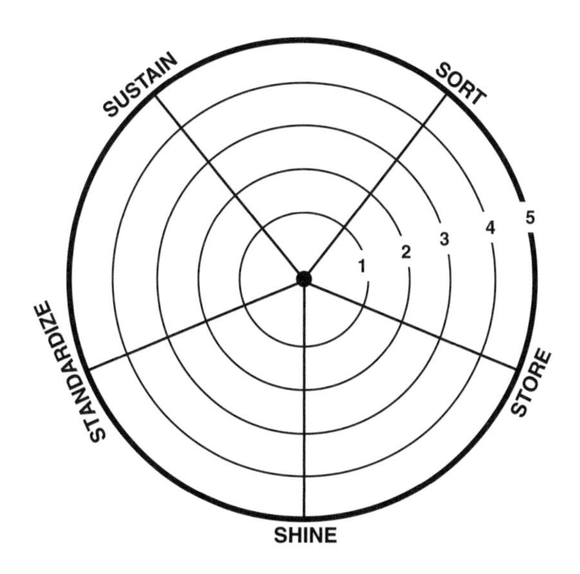

Figure 7.02.3 5S Audit Display Chart.

[3] Source: SBTI's Lean Methodology Training material.

- It removes waste that would otherwise slow the process down.
- It creates a safer work area; there are less trip hazards and dust problems.
- It creates much needed space in the area that allows storage of needed items but also allows the footprint of the process to be shrunk.
- It makes the process easier to visualize.

The approach of Sort is

Step 1: Set up a quarantine area for items that are removed from the process. In the short term, the quarantine area should be close to the process itself to allow items to be easily transferred into it and also, if proved to be required, then moved back again. After completion of the event, a more remote quarantine area should be used to ensure that removed items don't mysteriously return to the area.

Step 2: The Team walks through the process area and discusses removal of items with all persons involved. For each item ask these three questions:

- Is it needed?
- In what quantity is it needed?
- Where should it be located?

For some reason, we all tend to look at items as personal possessions, when in fact they are company possessions. We are only the caretakers of these items. With this in mind, it is best to use an outsider to take the lead in the Sorting activity to take advantage of "fresh eyes."

For items that need to be removed, ensure that the correct decontamination, environmental, and safety procedures are followed as they are moved to the quarantine area. Items that cannot be removed immediately should be tagged with a Red Tag for removal later (refer to Figure 7.02.1). Tag anything not clearly needed and used. If in doubt, ask questions and then tag it!

Step 3: Evaluate the Red-Tagged and quarantined items. If something is moved back to the line, it has to be justified and approved by the Process Owner, as well as having a labeled home before moving back.

Step 4: Document results of the quarantine and Red-Tagging. A list must be made of items and circulated before moving to the longer-term quarantine area (Step 6). Taking photos of the quarantine area serves as a great visual representation of the activity during the event.

Step 5: Items in the quarantine area are examined with the process operators and put into the following categories:

- Unusable items to be scrapped.
- Good, usable items that don't have an obvious recipient and thus, require additional discussion or researching.
- Good, usable items that have an obvious recipient, take them there.
- Items still in question, the Team doesn't know if the items are needed or not. These are the only items that get transferred to the longer-term quarantine area. The purpose of the longer-term quarantine area is to keep other people from taking these items until a decision is made.

A timeframe should be set to either sell, trash, or give away items from the quarantine area.

2S: Store

Store is used to arrange all necessary items, to have a designated place for everything, and to ensure everything is in its place. The reason for Store is to

- Clearly show what is required in the area (and to make obvious what is not).
- Make it simpler to find items or documents.
- Save operator time, not having to search for items.

During the Storing activity there are some simple tricks to remember:

- If items are used together, store them together.
- Everything that can and might be moved has a labeled home.
- The more visual the better, use labels, tape, floor markings, signs, and shadow outlines to indicate item placement.
- Sharable items should be kept at a central location to eliminate excess and have equal access.
- Set limits for all items, excess of items should be made obvious.

During Storing it is also worthwhile to check the following points:

- Are the positions of main corridors, aisles, and storage places clearly marked?
- Are tools divided into specialized use and "regular items"?
- Are all pallets always stacked to the proper heights?
- Is anything stored around fire extinguishers?
- Does the floor have any depressions, protrusions, cracks, or obstacles?

3S: Shine

Shine is used to keep areas clean on a continuing basis, while continuously raising the standards. The reason behind this is that dust and dirt cause product contamination and potential safety hazards. A clean workplace is indicative of a quality product and process, helps to identify abnormal conditions, and improves morale.

The Store activity of "everything has a place, everything is in its place" makes the cleaning activity much easier.

The approach to Shine is

Step 1: Walk the area and identify and keep a list of soiling points and machine defects. Mark the soiling points and defects with orange tape and write on the tape the corresponding number from the list.

Step 2: For the soiling points:

- Repair them.
- Eliminate them so they no longer affect the area.
- If they are hidden, make them easy to see and access.
- Contain them as close to the origin as possible.
- Protect the process area so the soiling points do not affect it.

Step 3: Make cleaning the area easy, so that there is no way an operator would not abide by the rules.

Step 4: Determine how often cleaning needs to happen to maintain the standard. If cleaning is required only once per week, operators can only be held accountable to the standard once per week.

Step 5: Identify individual responsibilities for cleaning and eliminate "no man's land." During the Shine phase it is important to paint and refurbish to get the remaining items into the same condition as when they were new. Some common opportunities during Shine include

- Looking carefully at the floor, aisles, and around the machines to identify oil, dirt, dust, metal shavings, and so on.
- Identifying if any part of the machines, equipment, or desk space are dirty with oil or metal fragments, ink, or parts.
- Identifying if any air, electrical supply lines or gas and water pipes are oily, dirty, or in need of repair and if all computer lines are dressed off the floor.
- Identifying if any oil outlets are clogged with dirt and all filters are clean.
- Ensuring that all light bulbs, reflectors, or shades are not dirty or burnt out.

4S: Standardize

Standardize is used to maintain the workplace at a level that uncovers and makes problems obvious. If 5S is done as a standalone event, separate from a project, then Standardize is just applied to the 5S elements; however, if 5S is done as part of Improve and Control for a project, then Standardize becomes more than that. If the new process is already defined from the Improve Phase, Standardize can envelope the Sort, Store, and Shine, and it can also include the Standard Work for the new process, which means more integrated control of the process.

From a 5S perspective, Standardize is there to develop sorting, storage, and shining activities into the everyday actions. The idea is that the workplace needs to be kept neat enough for visual identifiers to be effective in uncovering hidden problems. Without standardized cleanup, improvements made in the first 3Ss go back to the way they were. Done right, everyone who uses the area knows what the standard is.

To achieve this, some simple approaches pay dividends:

- Use pictures of the Standard to better communicate expectations.
- Create Tool/Equipment/Material lists, and for each, set the expected standards for upkeep.
- Create a shift-by-shift or daily Acton List to maintain Standards (when soiling points cannot be eliminated).
- Have clear accountability for Standards and split the facility into zones for maintaining standards if necessary (avoid no-man's-land).
- Share information using standard terms so that everyone has the same information. Standardize everything and make the Standards visible.

5S: Sustain

Sustain is used to maintain the discipline and to continuously improve by effective regular assessments and actions. Sustain is certainly the most difficult S and needs to practiced and repeated until it becomes a way of life. A wonderful term I learned from a valued client is *Muscle Memory.* Sustain needs to be consciously practiced until it becomes unconsciously part of the daily routine.

There are some simple steps to begin the road to Sustain, as follows:

Step 1: Revisit the 5S Audit Matrix and rescore the area.

Step 2: From Step 1, identify all changes made during the event and prioritize which changes are the toughest to maintain.

Step 3: Based on the prioritized list, create a more specific 5S audit checklist that's no more than one page long and identify who performs the checking and how often (typically this is the Process Owner, once per week).

Step 4: Perform the checking as per Step 3. When something is found to be out of place, it needs to be fixed immediately by the person responsible for the area.

Step 5: The 5S scores are tracked over time and are posted in the area. If multiple shifts work in the area, it is often useful to have 5S built into the shift handoff procedures. Posting 5S scores by shift creates a sense of competition and drives even greater performance.

Step 6: As 5S spreads throughout the site, an additional centralized scoring location creates an even greater sense of accomplishment for Teams and improves performance yet again.

Step 7: People tend to behave the way that they are measured and more specifically the way they are rewarded or recognized. Accountability needs to be built into organizational scorecards and employee appraisals for gains to be long lasting.

As mentioned previously, Sustain is by far and away the toughest S, so try to develop habits that are hard to forget. Encourage leadership to lead by example—follow the rules that you set! A few months ago I was visiting one of my clients in the UK and while waiting in the conference room for the General Manager to arrive, I was taking in the scenery from the conference room window. I watched a gentleman arrive into the car park in an expensive car, get out of the car in an expensive suit, and promptly step across a flowerbed in expensive shoes to pick up some litter. A few minutes later the same gentleman arrived into the conference room and was, you guessed it, the General Manager I was waiting for. There was no way for him to know I was watching his actions earlier, but his actions spoke volumes for his leadership style. Not surprising to me, it was one of the best plants I've toured.

03: AFFINITY

OVERVIEW

Affinity is used in the Lean Sigma roadmap primarily in the context of Voice of The Customer (VOC), to take the fragments of Customer Voice collected from Interviews and Surveys and collate them into a structured usable form. Affinity has a much stronger and more detailed counterpart, known as KJ,[4] which is used extensively in Design For Six Sigma (DFSS) to develop new products and services. The level of detail that KJ

[4] For more detail on KJ see *Commercializing Great Products Using DFSS* by Randy Perry.

provides is obviously better, but is not deemed to give additional worthwhile value in Process Improvement projects for the significant extra effort expended.

Affinity is a simple tool and common to most improvement approaches, but has a few subtleties with which Belts sometimes struggle.

LOGISTICS

Affinity is absolutely a Team activity and should not be attempted by the Belt in isolation. Also, it is sometimes tempting for Belts to try to make a start before a Team meeting (a straw man) to speed things up. This is not a good idea and actually tends to slow things down, as well as potentially undermining the Team's buy-in to the end product.

A session to create a Customer Requirements Affinity Diagram takes about one to four hours total for a typical project. The Team needs an area of wall space to work on, a large sheet of paper (two flipchart pages side by side works well), sticky notes, and pens.

ROADMAP

The roadmap to create the Affinity Diagram of Customer Requirements is as follows:

Step 1: Transfer the Customer Needs (sometimes called "Voices") from the transcripts of *Customer Interviews* and *Customer Surveys* to sticky notes. This is by far the most important step in the process and can potentially cause the biggest problems. It can also take a significant amount of time, usually more than half of the total tool time. It should be possible for a single person to transfer the needs from a transcript to sticky notes provided that person was a Scribe during the interview process and heard exactly what was said during the interview. There are times where input is required from others present at that particular interview to clarify points.

Thus, the transcripts are distributed back to the individual Team members who wrote them. By slowly scanning through the transcript, Customer Needs are highlighted (it is a good idea to do this physically with a highlighter pen) and the Needs are then each transferred to a sticky note. A Need or Voice is an individual thought, idea, or statement that is to be considered and can stand on its own merits. It is a statement of need and should avoid implying a solution. During the Interview process, if a solution was mentioned, the Interviewer should have probed to discover the underlying Need that drove the Customer to conclude that the particular solution was necessary. The Need should be written as said, from the Customer's perspective, and it should not translate or summarize needs.

Write only one Need per sticky note and write them legibly in block letters. There should be no ambiguity in the statement on the note, so write in complete sentences; don't just write a single bullet or a couple of words. It doesn't matter if the Need isn't written absolutely perfectly; it is the concept behind the written Need statement that matters.

If a Need is still not understood by a Team member, seek Team input and if it is still ambiguous then seek clarification from the Interviewee.

After all the Needs from the transcripts are transferred to sticky notes, there should be a large pile in front of each Team member.

Step 2: Prepare the work area using either large chart paper or two flipchart pages side by side. Losing work can be very frustrating so write the project name in large black marker in the upper left corner of both sheets.

Step 3: Place all the sticky notes on the paper, in no particular order.

Step 4: Each Team member should take time to read all of the notes before moving any. This takes time, so everyone should do this in parallel—it's usually a tight squeeze around the work area.

Step 5: Have *all* the Team involved in moving around the notes into logical groups. This is done in silence to begin with and is a skill in itself. The trick is to think from the Customers' perspective, from what was heard in the interviews. Try not to group by internal factors, for example, "those are all our quality-related problems;" rather, try to do it from the external standpoint, from how the Customer perceives the process.

Try to avoid creating huge groups of notes; the meaning behind them might be lost. Groups might be just two or three notes to begin with but make sense to be placed together. There will probably be replication of notes. Remove only *exact* replication. Do not be tempted to keep a loose summary note and assume it would represent a few others. Keep the detail; do not lose information content.

Keep moving sticky notes until the Team comes to silent consensus.

Step 6: Discuss and title each group. The title is based on why that group of notes addresses the same Needs. It is effectively a theme for the small group it represents. The titles are best written in a different colored pen and preferably on a different colored note to distinguish them.

Step 7: It might be obvious at this point that some of the groups themselves should be grouped. Complete this as a Team and title the major groups. The titles are best written in yet another pen color and preferably on a different colored note to distinguish them again.

Step 8: After the Diagram is complete, the Team should sign and date the sheet. The purpose of signing is primarily to ensure that each Team member agrees with the structure and wasn't just "going along with it."

A completed Affinity Diagram is shown in Figure 7.03.1. The example in this case is from a project aimed at reducing Length of Stay in an Emergency Department. The Diagram in this case exhibits only a single level of grouping.

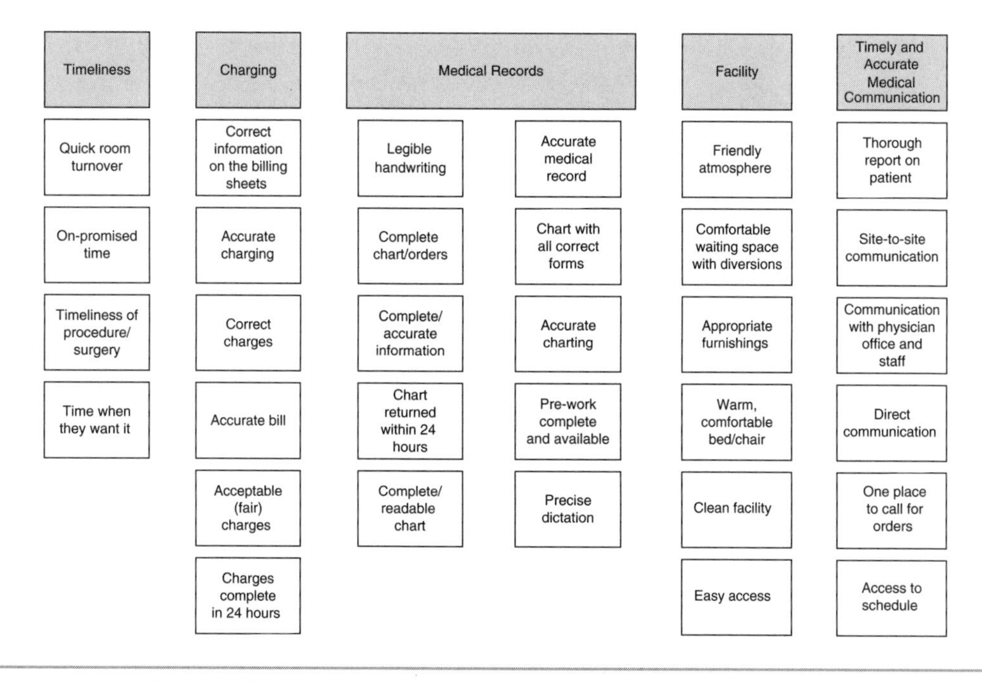

Figure 7.03.1 Example Affinity Diagram showing grouping structure.[5]

Interpreting the Output

Affinity Diagrams are one in a series of tools to capture the VOC. The output of any *Customer Interviews* along with output from *Customer Surveys* are affinitized and then translated into a *Customer Requirements Tree* to identify the Big Ys or Key Process Output Variables (KPOVs) for the process.

04: ANOVA

Overview

One-Way Analysis of Variance (ANOVA) is used to compare the means of two or more samples against each other to determine whether it is likely that the samples could come from populations with the same mean. This is similar to a 2-Sample t-Test except that three or more samples can be examined with ANOVA.

[5] Source: Columbus Regional Hospital, Emergency Department Throughput Team.

ANOVA can also be used to examine multiple Xs at the same time (see "Other Options" in this section), but here the focus is primarily on the One-Way ANOVA, which examines just one X. For example, a Team might need to determine if three operators:

- A single X—Operator
- With 3 levels—3 Operators

Take the same amount of time to perform a task. A data sample would be taken of, for example, 15 points (times in this case) for each operator. ANOVA is used to make the judgment if all the operators' average (mean) task times (as work continues thenceforth) are the same. The level of confidence in the answer depends on how far apart the means of the samples are, how much variability there is in the sample data, and how many data points there are.

This is shown graphically in Figure 7.04.1. The upper curves represent the distributions of all three operators' times (known as the populations). The exact nature of each of these distributions is unknown to the Team, because they represent all data points for all time. What the Team can see, however, are the samples taken, one from each population, shown as the lower curves. ANOVA examines the sample data with the aim of making an inference on the location of the population means (μ) relative to each other. It does this by breaking down the variation (using variances) in all the sample data into separate pieces, hence the name Analysis Of Variance. ANOVA compares the size of the variation *between* the samples versus the variation *within* the samples.

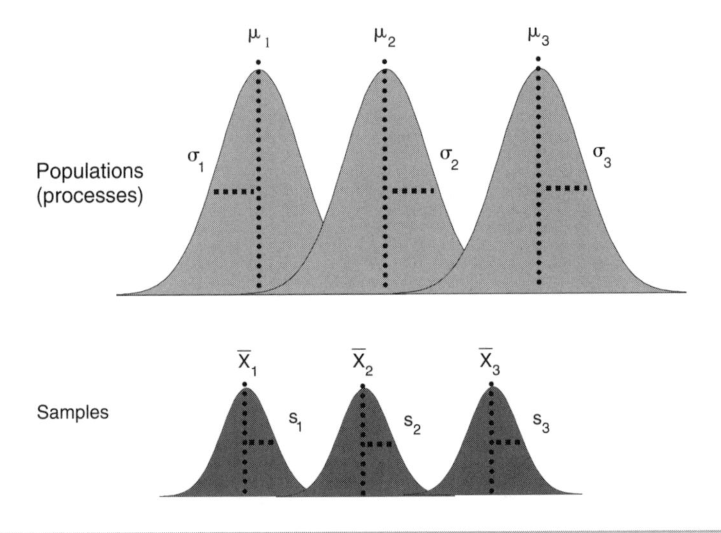

Figure 7.04.1 Graphical representation of ANOVA.

If the variation between the samples is large relative to the variation within the samples, then it means the samples are spread widely (between) compared with the background noise (within), and this would imply that the likelihood of the means of the parent distributions being aligned is low. If the between variation is not large compared with the within variation, then it is likely that the means of the parent distribution are about the same, or more specifically that the test cannot distinguish between them. The result of the test would be a degree of confidence (a p-value) that the samples come from populations with the same mean. In practical terms, the p-value gives an indication of the probability that the mean operator times are the same going forward. If the p-value is low, then at least one of the mean operator times is distinguishable from the others; if the p-value is high, they all are not distinguishable.

Roadmap

The roadmap of the test analysis itself is shown graphically in Figure 7.04.2.

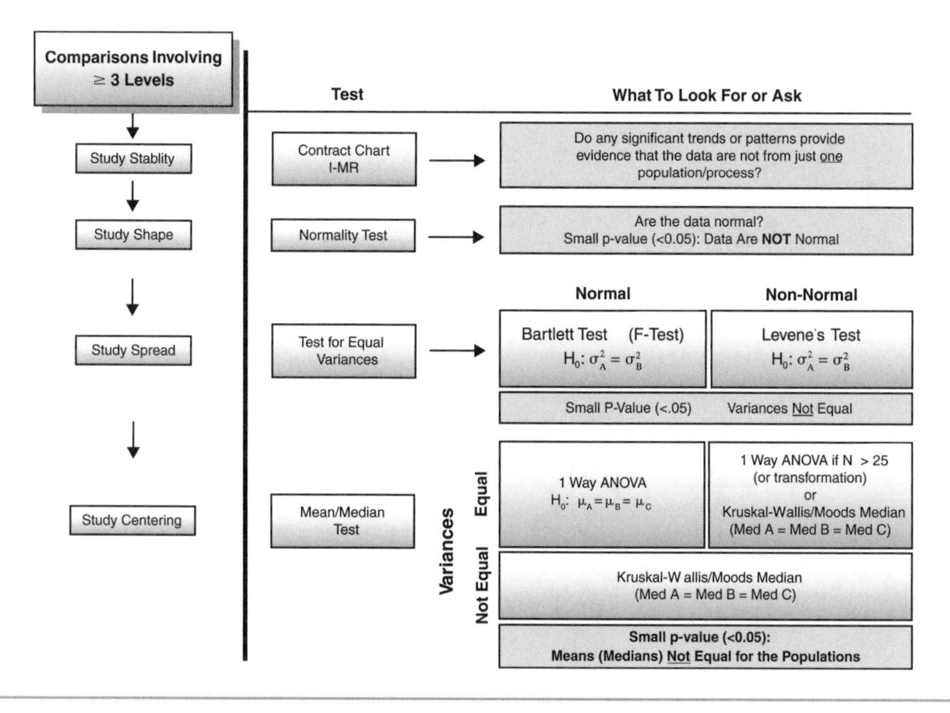

Figure 7.04.2 One-Way ANOVA Roadmap.[6]

[6] Roadmap adapted from SBTI's Process Improvement Methodology training material.

Step 1: Identify the metric and levels to be examined (for example, three operators). Analysis of this kind should be done in the Analyze Phase at the earliest, so the metric should be well defined and understood at this point (see "KPOVs and Data" in this chapter).

Step 2: Determine the sample size. This can be as simple as taking the suggested 15 to 20 data points per level or using a sample size calculator in a statistical package. These rely on an equation relating the sample size to

- The standard deviations (the spread of the data) of each population. This would have to be approximated from historical data.
- The required power of the test (the likelihood of the test identifying a difference between the means if there truly was one). This is usually set at 0.8 or 80%. The power is actually $(1 - \beta)$, where β is the likelihood of giving a false negative and so it might need to be entered in the software as a β of 0.2 or 20%.
- The size of the difference δ between the means that is desired to be detected, that is the distance between the means that would lead the Team to say that the two values are different.
- The alpha level for the test (the likelihood of the test giving a false positive) usually set at 0.05 or 5% and represents the cutoff for the p-value (remember if p is low, H_0 must go).
- The number of levels examined (number of Operators, and so on).

Step 3: Collect a sample data set, one from each level of the X following the rules of good experimentation. If the sample size calculator determined a sample size of ten data points, then ten points need to be collected for each and every level. For example, if the X is Operator and there are three levels (three operators), then $3 \times 10 = 30$ data points are collected in total.

Step 4: Examine stability of all sample data sets using a *Control Chart* for each, typically an Individuals and Moving Range Chart (I-MR). A *Control Chart* identifies whether the processes are stable, having

- Constant mean (from the Individuals Chart)
- Predictable variability (from the Range Chart)

This is important; if the processes are moving around, it is impossible to sensibly decide if they are the same or not.

Step 5: Examine normality of the sample data sets using a *Normality Test* for each. This is important because the statistical tests in Step 6 and 7 rely on it, but in simple terms, if the sample curves in Figure 7.04.1 were strange shapes, it would be difficult to determine if the middles were aligned. In fact, if data becomes skewed, then the mean is probably not the best measure of center (a t-Test is a mean-based test), and a median-based test is probably better. The longer tail on the right of the example curve in Figure 7.04.3 drags the mean to the right; however, the median tends to remain constant. Medians-based tests could in theory be used for everything as a more robust test, but they are less powerful than their means-based counterparts, and hence the desire to go with the mean.

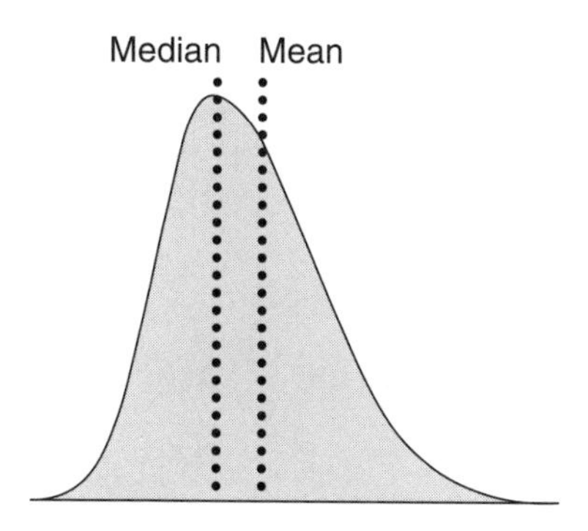

Figure 7.04.3 Measures of Center.

Step 6: Perform a *Test of Equal Variance* on the sample data sets. ANOVA requires the variances of the samples to be approximately the same, and without this, a medians-based approach has to be used instead.

The *Test of Equal Variance* uses the sample data sets and has these hypotheses:

- H_0: Population (process) $\sigma_1^2 = \sigma_2^2 = \sigma_3^2 \ldots$ (all variances equal)
- H_a: At least one of the Population (process) variances is different

Step 7: Perform the ANOVA if all of the sample data sets were determined to be normal in Step 5 and the variances were equal in Step 6. The hypotheses in this case are

- H_0: Population (process) $\mu_1^2 = \mu_2^2 = \mu_3^2 \ldots$ (means equal)
- H_a: At least one of the Population (process) means is different

If the data in either or both of the samples were non-normal then as per Figure 7.04.2:

- Continue unabated with the ANOVA if the sample size is large enough (>25)
- Transform the data first and then perform the analysis, again using the ANOVA[7]
- Perform the median-based equivalent test, a Kruskal-Wallis or Moods Median Test

If the variances of the samples were not equal then as per Figure 7.04.2, perform the median-based equivalent test, a Kruskal-Wallis or Moods Median Test.

The last option often worries Belts, but the medians tests look identical in form to the means test and both return the key item, a p-value (the p-values for a means test and a medians test on the same data are unlikely to be the same though).

INTERPRETING THE OUTPUT

One-Way ANOVA[8] segments the total variation in the data into two pieces:

- Variation *within* levels of the X (basically the background noise in the process)
- Variation *between* levels of the X (the signal strength due to the X)

It then calculates a ratio of the signal (variation due to the X, the "between") relative to the noise (any other variation not due to the X, the "within"). If the signal-to-noise ratio gets large enough then this would be considered to be unlikely to have occurred purely by random chance and the X is thus considered statistically significant. This is achieved by looking up the signal-to-noise ratio in a reference distribution (F-Test), which returns a p-value. The p-value represents the likelihood that an effect this large could have occurred purely by random chance even if the populations were aligned.

Based on the p-values, statements can be generally formed as follows:

- Based on the data, I can say that at least one of the means is different and there is a (p-value) chance that I am wrong
- Or based on the data, I can say that there is an important effect due to this X and there is a (p-value) chance the result is just due to chance

Example output from an ANOVA is shown in Figure 7.04.4.

7 Transformation of data is considered beyond the scope of this book.

8 The technical details of ANOVA are covered in most statistics textbooks; *Statistics for Management and Economics* by Keller and Warrack makes it understandable to non-statisticians.

One-way ANOVA: Time versus Operator

Source	DF	SS	MS	F	P
Operator	2	80.386	40.193	44.76	0.000
Error	87	78.116	0.898		
Total	89	158.502			

S = 0.9476 R − Sq = 50.72% R − Sq (adj) = 49.58%

```
                                 Individual 95% CIs For Mean Based
                                 on Pooled St Dev
Level    N    Mean    St Dev  ----+---------+---------+---------+-----
Bob     30  24.848    0.869   (----*---)
Jane    30  25.446    0.988         (---*---)
Walt    30  27.084    0.981                          (----*---)
                              ----+---------+---------+---------+-----
                              24.80     25.60     26.40     27.20
Pooled St Dev = 0.948
```

Figure 7.04.4 ANOVA results for a comparison of samples of Bob's vs Jane's vs Walt's performance (output from Minitab v14).

From the first table in the results:

- The average variation due to Operator was 40.193 units (SS ÷ DF in the Table[9]).
- The average variation due to Error (everything else not including Operator) was 0.898 units.
- The signal-to-noise ratio is therefore 40.193 ÷ 0.898 = 44.76.
- The likelihood of seeing a signal-to-noise ratio this large (if the populations were perfectly aligned) is 0.000% (p-value), which is well below 0.05, and thus, a conclusion that at least one of the trio is performing significantly differently from the others.
- The X "Operator" explains 50.72% of the variation in the data (the R-Sq value).
- R-Sq (Adj) is close to R-Sq; so there are no redundant terms in the model (if this value drops much lower than R-Sq, which commonly occurs in a multi-way ANOVA, then it is likely that an X is having no effect—here the X clearly has a marked effect).
- 49.28% of the variation in the data is coming from something other than Operator, and thus, presents a possible opportunity (100% − R-Sq).

[9] From a practical standpoint, a discussion of Degrees of Freedom (DF) and Sequential Sums of Squares (SS) is usually just confusing to Belts. If you are so inclined, then refer to any standard Statistics text. The favorite *Statistics for Management & Economics* by Keller and Warrack will prove useful—however, I caution Belts that if they are so wrapped up in the statistics, they are probably missing the bigger practical picture.

From the bottom table in the results:

- A sample of 30 data points was taken for each operator.
- Bob's sample mean is 24.848, Jane's is 25.446, and Walt's is 27.084.
- Bob's sample standard deviation is 0.869, Jane's is 0.988, and Walt's is 0.981.
- The text graph shows the 95% confidence intervals for the locations of the population means for each of the trio.

The p-value of 0.000% in the upper table indicates that at least one of the trio is performing differently from the other two. There is no overlap in 95% confidence intervals in the bottom table between Walt's performance and the other two; therefore, it is clearly Walt who has a different mean.

OTHER OPTIONS

The preceding description gave the roadmap for a single X (One-Way ANOVA), however ANOVA can also be applied to multiple Xs and breaks down the variation accordingly. An example of this is shown in Figure 7.04.5.

Analysis of Variance for Seal Defects, using Adjusted SS for tests

Source	DF	Seq SS	Adj SS	Adj MS	F	P
Shift	1	36.000	24.600	24.600	10.40	0.032
Operator	1	12.250	0.344	0.344	0.15	0.722
Cement Lot	3	242.750	151.973	50.658	21.42	0.006
Flame Temp	3	4.645	6.211	2.070	0.88	0.525
Line Speed	3	2.645	2.645	0.882	0.37	0.778
Error	4	9.460	9.460	2.365		
Total	15	307.750				

S = 1.53786   R − Sq = 96.93%   R − Sq (adj) = 88.47%

Figure 7.04.5 Example multi-way ANOVA for defects in Seals (output from Minitab v14).

From the ANOVA table:

- The p-value for Shift and Cement Lot can be seen to be significant (p-values less than 0.05), but all other Xs are not statistically significant (p-values well above 0.05).
- 96.93% of the variation is being explained by the Xs (R-Sq), but some Xs are meaningless (the R-Sq (Adj) value is much lower than the R-Sq value).

These two bullets would drive us to re-run the ANOVA, but only include the two significant Xs, see Figure 7.04.6.

Analysis of Variance for Seal Defects, using Adjusted SS for Tests

Source	DF	Seq SS	Adj SS	Adj MS	F	P
Shift	1	36.000	36.000	36.000	13.66	0.004
Cement Lot	3	242.750	242.750	80.917	30.69	0.000
Error	11	29.000	29.000	2.636		
Total	15	307.750				

S = 1.62369 R − Sq = 90.58% R − Sq (adj) = 87.15%

Figure 7.04.6 Example multi-way ANOVA re-run including only significant Xs (output from Minitab v14).

From Figure 7.04.6, the amount of variation explained (R-Sq) has dropped slightly which is to be expected because a number of terms (Xs) have just been removed. The R-Sq (adj) value is much closer to the R-Sq value indicating that all the terms (Xs) included actually do something. The p-values have also decreased representing an increase in significance for the remaining Xs.

No additional re-runs would be required at this point and focus would turn to the practical implication of the Shift and Cement Lot having an impact on the defect rate.

05: Box Plot

Overview

The Box Plot is a graphical representation of data, as shown in Figure 7.05.1.

The graph is composed of elements describing the distribution of the data.

- The y-axis is the output being measured; for instance if the data represents purity of a product, then the y-axis will be purity.
- The middle horizontal line in the box is the median of the data, that is the fiftieth percentile of the data (a common mistake here is to assume it is the mean, not the median).
- The bottom of the box lies on the twenty-fifth percentile of the data.
- The top of the box lies on the seventy-fifth percentile of the data.

- The box in its entirety thus represents the middle 50% of the data; meaning, the 25% of the data that lies either side of the median from the twenty-fifth percentile to the seventy-fifth percentile.
- The bottom whisker runs down from the twenty-fifth percentile to usually the tenth (or the fifth depending on software settings). Anything below this is considered a *possible* outlier. It cannot be thrown out or discounted without investigation.
- The top whisker runs up from the seventy-fifth percentile to usually the ninetieth (or ninety-fifth depending on software settings). Anything above this is considered an outlier.
- Asterisks represent the outliers above the end of the top whisker or below the end of the bottom whisker.

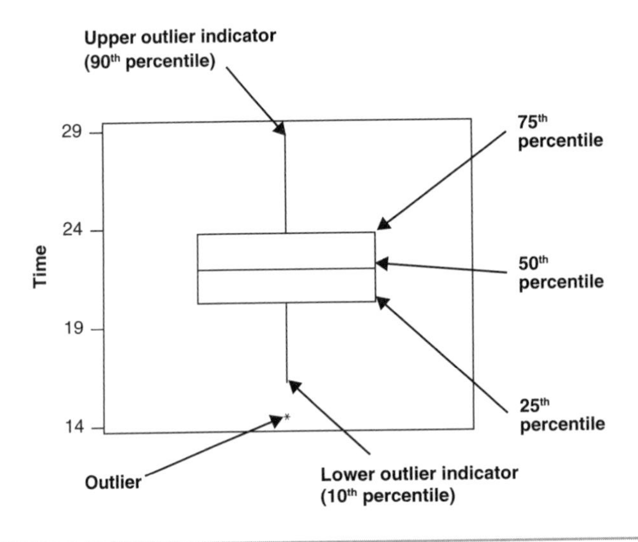

Figure 7.05.1 Composition of a Box Plot.

Box Plots are dull when applied to a column of data in one lump. They become far more useful when the data is cut by an X. For example in Figure 7.05.2, the data is cut by the X Operator, and there are three boxes representing the Operators Bob, Jane, and Walt.

INTERPRETING THE OUTPUT

Box Plots should not be used in isolation. Figure 7.05.2 seems to show a difference in the time each Operator takes to process an entity (the boxes aren't all aligned). The key word here is *seems*. A Box Plot can help identify if there *might* be differences between levels of an X, such as the Operator, but those differences might not be statistically significant. Any graphical tool such as this should be followed by the appropriate statistical tool to understand the likelihood of seeing the difference purely by random chance.

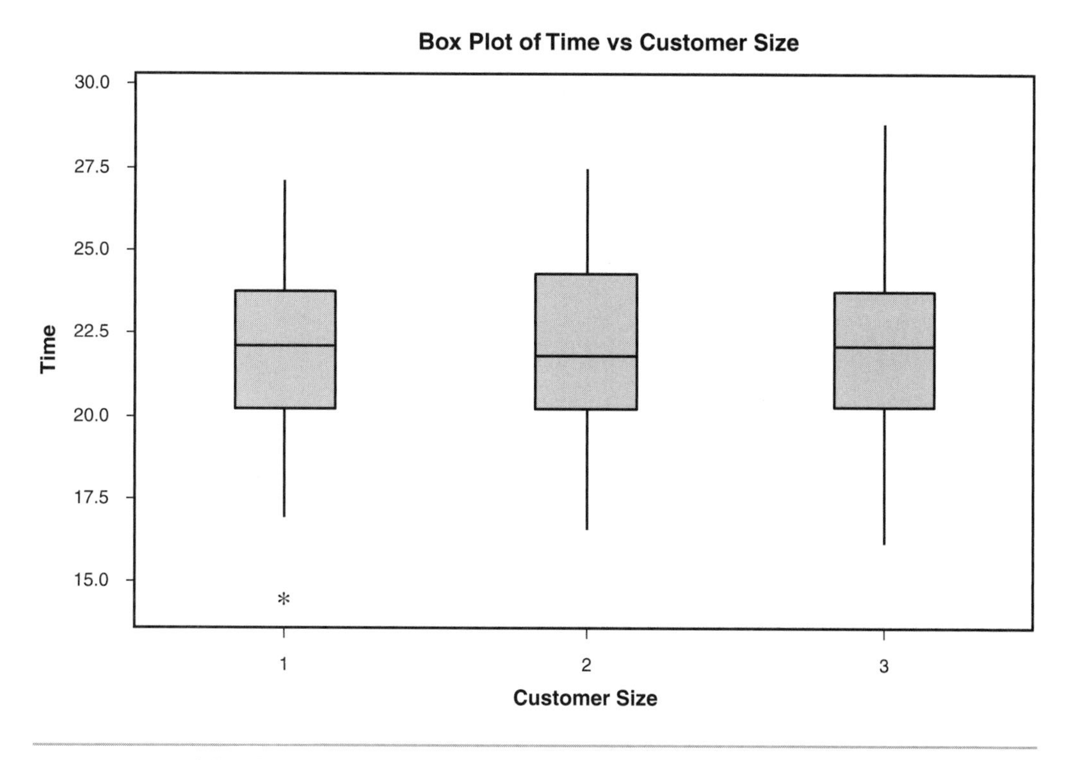

Figure 7.05.2 A Box Plot applied to data cut by Operator.

In fact Box Plots are best used as part of a *Multi-Vari Study* where data is collected and analyzed systematically using multiple graphical and statistical tools. Factors, such as the shape of the data (normal or non-normal) and the number of data points, along with a host of others, play a role in understanding the data.

06: CAPABILITY—ATTRIBUTE

OVERVIEW

Early in any project (in Define for goal setting or more formally in Measure) it is crucial to understand the current level of performance of the process prior to making any changes. Many Champions and Belts mistakenly believe this is only to show how much is saved by the project or how big an improvement is made to justify continuation of the Lean Sigma program. These things are important but are only a small piece of the picture. The primary use of the measure is to ensure the gains are sustained after the improvements are in place. If the change is unmeasured it often is undone later (with all the best intentions) because it is not fully understood. However, if a measured and verified performance change is made, then there is less likelihood for future damaging "tweaks."

There are many performance metrics available (throughput, *Overall Equipment Effectiveness*, quality, and so on) of which Capability is just one. Capability comes in two forms depending on the type of data to which the tool is applied, but both look at performance versus specification(s). Here the focus is on the application to Attribute type data.

As always, whenever tools are applied to Attribute data, the tools are weaker, and here the same problem applies. Attribute Capability is expressed as Defects per Unit (DPU), calculated as:

$$\text{DPU} = \frac{\text{Total Number of Defects}}{\text{Total Number of Units}}$$

It is necessary, therefore, to define the Unit and the Defect. A Unit is the physical output from the process (described as an entity in previous sections). It is something that is inspected, evaluated, or judged to determine "suitability for use." It is something delivered to Customers or users. Examples might include

- Invoice
- Shipment
- Customer Call
- Order

A Defect is anything that does not meet a critical Customer requirement or established standard. Examples might include

- Typographical error on an invoice
- Incomplete shipment
- Customer call dropped
- Missing order information

There can be multiple defect types for each entity and multiple defects of each type per entity.

DPU thus considers the total of all the individual defects (or errors) on each unit, or it can be applied to individual defect types, thus identifying which defect type is creating the largest loss. It is generally used to identify opportunities by prioritizing based on the commonest defect type (typically using a *Pareto Chart*).

DPU does not however take process complexity into account; to do this, the measure used is Defects per Million Opportunities (DPMO), defined as

$$\text{DPMO} = \frac{\text{Total Number of Defects}}{(\text{No. of Units})^{*}(\text{No. of Opportunities})} \times 1{,}000{,}000$$

The definition of Unit and Defect is as before. To understand Opportunities, it is useful to mention the difference between Defects and Defectives. A Defective Unit is any Unit containing a Defect. There can be multiple Defects in one Defective Unit and the Defectives Units are a result of Defects. It is impossible to reduce Defective Units without reducing the number of Defects. Opportunities therefore are the number of potential chances within a Unit to be Defective. It can often be difficult to identify all opportunities and it generally depends on the Customer. Examples might include

- **Purchase Orders.** Opportunities = Number of critical fields × 2 because the fields can either be empty or incorrect.
- **Customer Call.** Opportunities = Number of defect reason codes (for example, a missed call, a dropped call, an unresolved call, a call sent to a manager, a call that requires a call back, and a call that lasts longer than 15 minutes).

Opportunities are notoriously open to abuse. A process could be "artificially improved" if there are more Opportunities considered at the end of the project than at the beginning, so it is very important for the Team to be consistent in how they are defined.

ROADMAP

The roadmap to calculating the Capability for Attribute type data is as follows:

Step 1: For the metrics in question, define the goals and specifications. Ensure the validity of the metric (see "MSA—Validity" in this chapter).

Step 2: Define the Unit, Defect, and Opportunity as per the descriptions in "Overview" in this section.

Step 3: Collect process data. At least 100 data points are required if the proportion of defects is greater than 5%. If the defect rate is less than 5%, then the sample size needs to be increased accordingly or it might be better to switch to a reliability-type metric, such as days between defects.

Step 4: Calculate DPU and DPMO as per the equations in "Overview" in this section. From the DPMO calculate the sigma rating from a lookup table (see "Interpreting the Output" in this section).

INTERPRETING THE OUTPUT

Capability is required to sign off on the Measure Phase of the project and is revisited in Control to validate performance.

DPMO is the capability measure primarily used to calculate process Sigma Rating. The infamous 3.4 Defects per Million representing "Six Sigma performance" is actually 3.4 DPMO. Some novice Belts worry about the required Sigma Rating and assume the project is only successful if the process has a Six Sigma defect rating when the project is complete. This is untrue; Six Sigma as an initiative is about breakthrough performance improvement. If a change is significant in terms of the savings it generates or the additional capacity or revenue it creates, then the Sigma Rating is really a secondary concern.

07: CAPABILITY—CONTINUOUS

OVERVIEW

Early in any project (in Define for goal setting or more formally in Measure) it is crucial to understand the current level of performance of the process prior to making any changes. Many Champions and Belts mistakenly believe this is purely to show how much the project saves or how big an improvement is made to justify continuation of the Lean Sigma program. These things are important but are only a small piece of the picture. The primary use of the measure is to ensure the gains are sustained after the improvements are in place. If the change is unmeasured, it often is undone later (with all the best intentions) because it is not fully understood. However, if a measured and verified performance change is made, then there is less likelihood for future damaging "tweaks."

There are many performance metrics available; for example throughput, OEE, quality, and so on, are typically represented as single numbers based on conformance to some goal. A better performance metric is to look at performance versus a specification(s), known as Capability.[10]

The simplest form of Capability for continuous data is known as C_p and is calculated as the ratio of the specification range divided by the process width:

$$C_p = \frac{USL - LSL}{6s}$$

[10] For more details see *How To Construct Fractional Factorial Experiments, Vol. 14* by Larry Barrentine.

where:

- s is the short-term process standard deviation
- USL is the Upper Specification Limit
- LSL is the Lower Specification Limit

The process width denominator is chosen as 6 standard deviations because this is deemed to a reasonable representation of the width of the process.[11]

C_p suffers from one obvious flaw as depicted in Figure 7.07.1; it doesn't take into account the centering of the process. Figure 7.07.1 shows 3 graphs with same C_p but different process centering.

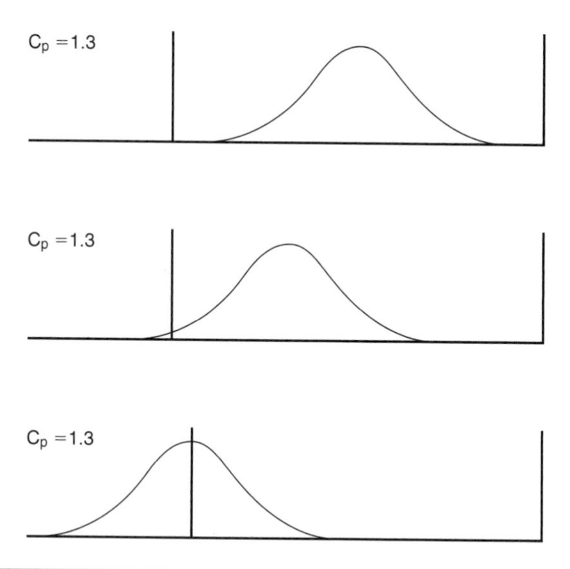

Figure 7.07.1 The effect of process centering on C_p.

A second metric needs to be introduced to counter this; known as C_{pk}, it is defined as

$$C_{pk} = \text{Min}\left\{\frac{\bar{X} - \text{LSL}}{3s}, \frac{\text{USL} - \bar{X}}{3s}\right\}$$

[11] 99.73% of data points lie between ±3 standard deviations in any normally distributed data.

C_{pk} represents the distance of the center of the process to the nearest specification limit in units of process width (in fact it is half the width because only one side of the process curve is considered at a time).

C_{pk} is positive when the mean of the process is inside the specifications; it drops to zero as the mean hits the USL or LSL. In fact, it becomes negative as the process mean moves outside the specification range, as shown in Figure 7.07.2.

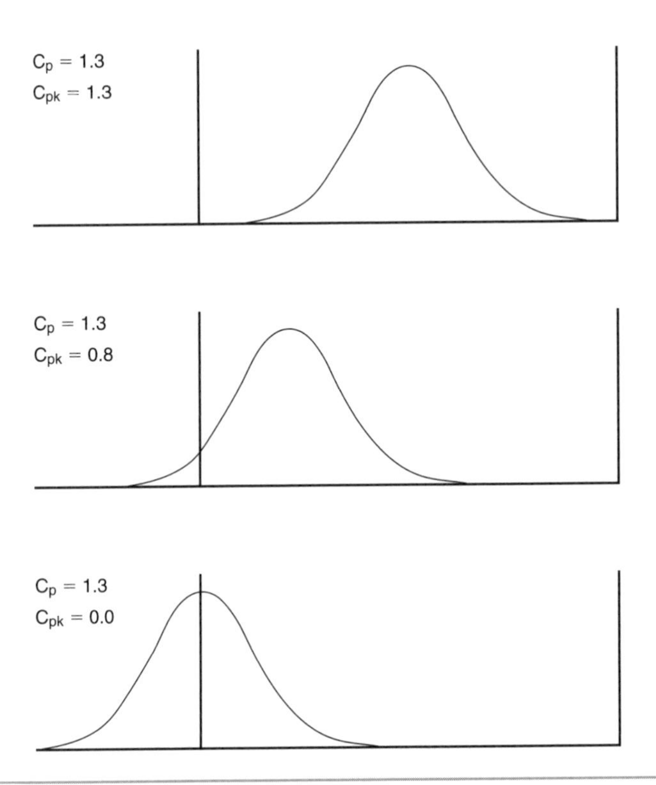

Figure 7.07.2 The effect of process centering on C_p and C_{pk}.

ROADMAP

The roadmap to calculating the Capability for continuous data is as follows:

Step 1: For the metrics in question, define the goals and specifications.

Step 2: Collect process data. At least 30 data points are required.

Step 3: Check the process stability (see "Control Charts" in this chapter). To be considered capable, the process needs to be stable. The definition of stability is having

- Consistent centering (mean)
- Predictable variation

It becomes clear that a process could not consistently meet Customer requirements if its mean were moving around and its variation were changing unpredictably. To check stability, Belts can simply apply the appropriate *Control Chart* to the data prior to calculating its capability.

Step 4: Check for Normality. The Capability calculations of C_p and C_{pk} rely on the data being normal. To check the normality, use a *Normality Test* as described in the section "Normality Test" later in this chapter. For non-normal data see "Other Options" in this section.

Step 5: Calculate C_p and C_{pk} as per the preceding equations. Most statistical software packages do this readily.

INTERPRETING THE OUTPUT

Example output for a Capability Study is shown in Figure 7.07.3. The key metrics to focus on are

- Lower Specification Limit (LSL) as entered by the user
- Target value if entered
- Upper Specification Limit (USL) as entered by the user
- Sample Mean
- Potential Capability C_p and C_{pk} as defined by the equations in "Overview" in this section

Capability (expressed as C_p and C_{pk}) is intended to represent short-term behavior of the process. In reality processes tend to shift and drift over time; the variation stays reasonably consistent, but the mean moves to and fro.[12] Taking this into account, the longer-term variation is actually larger than short-term and so the "capability" is lower in the long term than the short. Long-term "capability" is known as Performance and the equations are identical to those for Capability (short-term), but a longer-term standard deviation (σ instead of s) is used

[12] Empirical process studies show that most processes tend to shift and drift about 1.5 standard deviations. Lean Sigma Belts really only need to know that it happens, rather than the equations to justify why.

$$P_p = \frac{USL - LSL}{6\sigma}$$

$$P_{pk} = \text{Min}\left\{\frac{\bar{X} - LSL}{3\sigma}, \frac{USL - \bar{X}}{3\sigma}\right\}$$

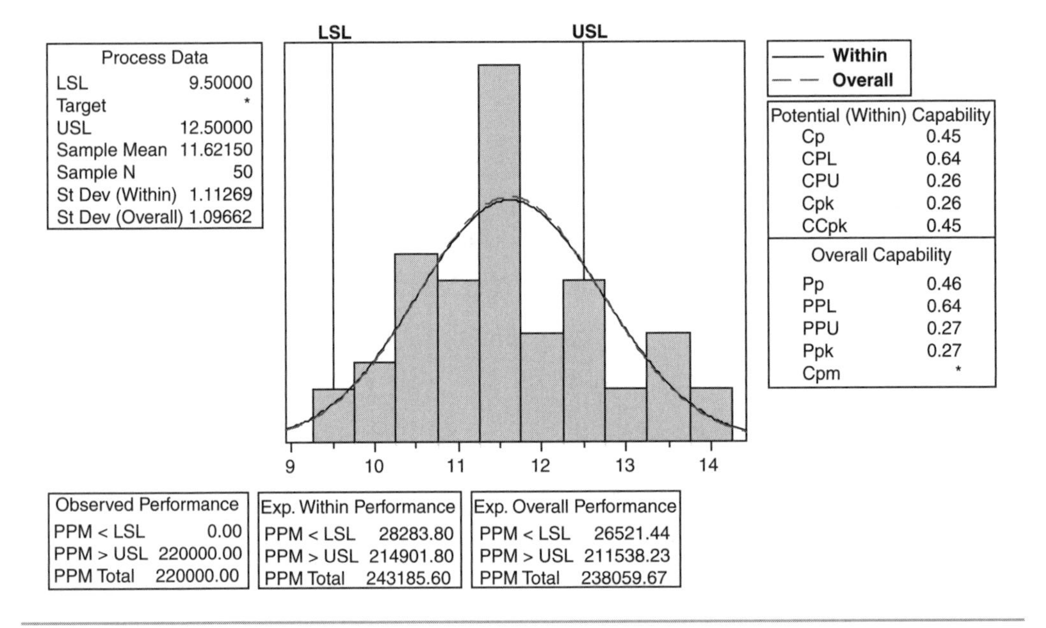

Figure 7.07.3 An example of a Capability Study (output from Minitab v14).

Most software packages try to emulate this short-term versus long-term standard deviation by measuring it in two different ways; for long-term the regular standard deviation of all the data is used and for short-term the value comes from an equation involving the Moving Average across the data. In Figure 7.07.3, the *within* standard deviation represents short term and the *overall* standard deviation represents long term. The within value is used to calculate the C_p and C_{pk}, whereas the overall value is used to calculate the P_p and P_{pk}.

The target value for C_p in Lean Sigma is 2.0 and for P_{pk} it is 1.5. These are not absolute requirements in any way, but if a process exhibits Capability at this level then it can be considered to be performing very well.

At the bottom of Figure 7.07.3 are boxes explaining likely performance of the process in terms of Parts per Million defective (PPM):

- The Observed Performance represents the PPMs of the actual data points below the LSL, above the USL, and the total of both. If no points fall outside of specification during the data collection, then the PPMs here are zero.
- The Within Performance represents the PPMs as calculated from a normal curve with the sample mean and *short-term* standard deviation. The calculated curve hangs over the LSL and USL, and thus, the area under the curve outside of the specification limits gives the PPMs. These are the expected defectives on a short-term basis.
- The Overall Performance represents the PPMs as calculated from a normal curve with the sample mean and *long-term* standard deviation. The calculated curve hangs over the LSL and USL, and thus, the area under the curves outside of the specification limits gives the PPMs. These are the expected defectives on a long-term basis.

Of course, all the values are calculated from the sample of data taken. If a subsequent sample were taken, it is almost certain that different answers will arise; thus, it is inappropriate to quote PPMs to anything more than two or possibly three significant figures. Belts often take great delight in including numbers in reports to many decimal places, which is acceptable provided that they realize that the numbers will be different next time.

Other Options

There are a number of variants to the preceding (standard) approach to calculating Capability that generally depend on the shape of the data and the behavior of the specification limits.

Non-Normal Data

As mentioned in the Roadmap, normality of the data is a key consideration and is often a potential failure point for Belts. If the data are non-normally distributed, for example they are skewed to one side, then the C_p calculated is fallacious and could be misleading to the Belt.

If the data is non-normal and unimodal (just one hump in the distribution), then the situation can be remedied by transforming the data. If the data is multimodal (more than one hump), it is likely that the process is unstable and hopefully there are simple special causes that can be identified and eliminated. Remember that in process improvement, a poor process is best seen as a big opportunity.

The approach of transforming data is not dealt with here, but there are a number of transformations available in most statistics software packages, such as Box-Cox or Johnson. In fact in the more user-friendly statistical software packages there are specific non-normal C_p tool options.[13]

Single-Sided Specifications

Many processes have both an Upper Specification Limit (USL) and a Lower Specification Limit (LSL); some have just one of these. For example, a metric, such as strength, might have an LSL in that there is a minimum strength to be considered acceptable by the Customer, but a USL doesn't make sense. In these cases, the C_p is no longer available as a metric because both specifications are required to calculate it:

$$C_p = \frac{USL - LSL}{6s}$$

Instead the C_{pk} is more appropriate:

$$C_{pk} = Min\left\{\frac{\bar{X} - LSL}{3s}, \frac{USL - \bar{X}}{3s}\right\}$$

For a single-sided specification, only half of the C_{pk} equation makes sense, so new metrics, known as C_p(upper) and C_p(lower), are used

$$C_{pL} = \frac{\bar{X} - LSL}{3s} \qquad C_{pU} = \frac{USL - \bar{X}}{3s}$$

The same rules apply to C_{pU} and C_{pL} as they do to C_{pk}. The target value is 1.5 or above. If the mean of the data lies on the specification limit, then the associated value of C_{pU} or C_{pL} is zero. If the mean is outside of the specification, then the C_{pU} or C_{pL} is negative.

Bounds

For some processes a lower or upper bound exists on the performance metric, for example:

---

13 For more details of applying transformations to Capability, see *How To Construct Fractional Factorial Experiments, Vol. 14* by Larry Barrentine

- Scrap cannot go below zero
- Yield cannot go above 100%
- Radius cannot go below zero

In this case, the bound cannot be considered to be a specification limit, so the same approach is used here as is used for a single-sided specification and the bound is effectively omitted. Any predicted points below a Lower Bound are ignored, as are any above an Upper Bound.

08: Cause & Effect (C&E) Matrix

Overview

The *Cause & Effect (C&E)* matrix is a mechanism to narrow down the number of Xs by using the existing knowledge of the Team. The tool is not used to rank the highest priority Xs, but to remove some of the trivial Xs with low priority. Construction requires the whole Team to take part.

Logistics

This is a team sport and should include representation for all the key areas affecting the process. Choose those individuals who live and breathe the process every day, not just Process Owners who might not be close enough to it to know the details.

The activity typically takes a morning to complete. It is usually best to create the matrix directly in Excel, projected up on a screen using an LCD projector, rather than creating it by hand on a wall chart and then transferring it to a spreadsheet later.

A common concern here is that the Belt wishes to create a strawman version prior to the working session with the Team. This is a mistake because the Team members usually just nod and agree to what the Belt has done and the matrix truly doesn't have a Team input. It is best to create it from scratch in the Team meeting.

Roadmap

If there are more than six process steps, consider using a two-Phase approach as listed in "Other Options: 2-Phase C&E" in this section.

The stages to the construction are as follows:

Step 1: Starting with a blank C&E matrix (see Figure 7.08.1), enter the process responses (measures of performance) critical to the Customer (both internal and external). Each should be accompanied by a rating score of 1 to 10, because not all responses carry the same weight in the eyes of the Customer. Two or more responses might carry equal ratings if appropriate, but not everything can be scored a 10.

0,1,3,9		Rating of Importance to Customer	10	8	7	9	5	10	
			1	2	3	4	5	6	
			Low turnover	Hire in less than 2 weeks	Process cost	Get the best candidates	Top schools	Position is filled	Total
	Process Step	Process Input							
1	Create Requisition	[illegible]pt needs	[illegible]	[illegible]	1	9	3	9	295
2	Create Requisition	Budget	9	1	1	9	3	9	291
3	Create Requisition	Responsibility/ Authority	9	1	1	9	3	9	291
4	Post Requisition	Headhunters	3	9	0	0	0	0	102
5	Post Requisition	Temporary Agencies	3	9	3	3	1	9	245
6	Screen Resumes	Resumes	3	9	9	3	3	9	297
Total			690	552	266	648	160	780	

Figure 7.08.1 Single Phase Cause & Effect Matrix.

Step 2: Enter the list of process steps and their accompanying inputs (Xs) from the *Process Variables Map*. Do not remove any Xs in the transfer; let the C&E matrix perform the task.

Step 3: The Team rates the strength of the relationship between each input and each response using the scale:

- 0: No relationship
- 1: Weak relationship
- 3: Medium relationship
- 9: Strong relationship

A common mistake here is to use a different scoring mechanism, say 0,1,3,5. This causes the matrix to give poor discrimination between Xs, and it is difficult to find a suitable cutoff point in Step 6. Also, don't try to second-guess the scoring to make an X more important, let the matrix do the work.

Step 4: The rating of each input is calculated by summing the multiples of relationship X response rating, for example, for the X called Budget in Figure 7.08.1, $291 = (9 \times 10) + (1 \times 8) + (1 \times 7) + (9 \times 9) + (3 \times 5) + (9 \times 1)$.

Step 5: The whole matrix is then sorted by the Rating column to give the ranked list of Xs.

Step 6: Team judgment is made on where to make a cutoff line, below which the inputs are considered unimportant and, therefore, not carried over to the next step in the roadmap. Obviously the trick here is to find the right cutoff point. If the cutoff is drawn too high, then potentially important Xs are eliminated from further investigation. If it is drawn too low, then many unimportant Xs are carried forward, and it will take time and effort to investigate them. Clearly the former would be much more of a problem than the latter.

One solution is to work slowly down the matrix from the top row until the Xs are deemed by the Team to be noise—draw the cutoff line above the first noise X. Then move slowly up the matrix from the bottom row until the first possibly important X is found and draw the cutoff line just below it. Now reconcile the two lines.

A major mistake that is often made in this step is to assume that the C&E matrix is used to prioritize the top few Xs. It is not. The tool is used to eliminate unimportant Xs from the bottom.

INTERPRETING THE OUTPUT

The C&E matrix provides a reduced list of Xs to carry forward to more powerful data-driven tools. The most important Xs (those that have the strongest effect on the Customer, based on the strength of the relationship between the Xs and the Ys and the corresponding rating of each Y) drift to the top. The least important drift to the bottom.

The individual ranking of Xs is irrelevant; the cutoff line is the primary element here. Xs below the cutoff line are deemed unworthy of further investigation.

OTHER OPTIONS: 2-PHASE C&E

For processes with many steps, it is usually best to reduce the number of steps first (Phase 1) and then use a second matrix to reduce the Xs (Phase 2).

The stages to the construction are as follows:

Step 1: Starting with a blank C&E matrix, enter the process responses (measures of performance) critical to the Customer (both internal and external). Each should be accompanied by a rating score of 1 to 10, because not all responses carry the same weight in the eyes of the Customer. Two or more responses might carry equal ratings if appropriate.

Step 2: Enter only the list of process steps from the *Process Variables Map*, not the Xs (see Figure 7.08.2). Do not remove any steps in the transfer; let the C&E matrix perform the task.

		Rating of Importance to Customer	8	4	10	3	5	9	
			1	2	3	4	5	6	
	Process Step	Process Inputs	Timely shipment	Low cost	Correct shipment	Inventory accurately updated	Order accurately placed	Customer charged correct amount	Total
1	*Place Order*		3	1	1	9	9	3	137
2	*Process Order*		3	1	9	9	9	9	271
3	*Fulfill Order*		9	0	9	0	0	0	162
4	*Pack Order*		9	3	9	0	0	0	174
5	*Ship Order*		9	3	9	9	1	1	215

Figure 7.08.2 Phase I of a 2-Phase C&E Matrix.

Step 3: The Team rates the strength of the relationship between each process step and each response using the scale:

- 0: No relationship
- 1: Weak relationship
- 3: Medium relationship
- 9: Strong relationship

Step 4: The rating of each step is calculated by summing the multiples of relationship X response rating, for example, for the step called Place Order in Figure 7.08.2, $137 = (3 \times 8) + (1 \times 4) + (1 \times 10) + (9 \times 3) + (9 \times 5) + (3 \times 9)$.

Step 5: The whole matrix is then sorted by the Rating column to give the ranked list of process steps.

Step 6: Team judgment is made on where to make a cutoff line, below which the steps are considered unimportant and, therefore, not carried over to the next step in the roadmap.

Step 7: Starting with a second blank C&E matrix, enter the responses and ratings as before.

Step 8: Carry over from the Phase 1 C&E matrix only those process steps above the cutoff line.

Step 9: For this reduced set of process steps, list the associated Xs from the *Process Variables Map* (similar to Figure 7.08.1 but with a reduced set of Xs).

Step 10: As a safety net, check the omitted process steps from Phase 1 for any obviously important Xs and carry those over to the second Phase C&E also. Do not carry many over, just the obvious ones.

Continue as per the 1-Phase C&E from Step 3 of the 1-Phase C&E roadmap onwards to reduce the number of Xs.

09: Chi-Square

Overview

Chi-square is one of the statistical tools in the *Multi-Vari* approach and is both the simplest and least powerful. Chi-square helps determine the statistical significance of a relationship between an Attribute X and an Attribute Y in $Y = f(X_1, X_2, ..., X_n)$.

Y Data \ X Data	Attribute	Continuous
Attribute	**Chi-Square**	**Logistic Regression**
Continuous	**ANOVA Means/ Medians Tests**	**Regression**

The approach used is to assume the variables are independent and set up the hypotheses as follows:

- H_o: Data are Independent (Not Related)
- H_a: Data are Dependent (Related)

The output of the test is a "p-value" that indicates the likelihood of seeing a relationship this strong in a sample purely by random chance; that is, there is no relationship at the population level, it just happened by fluke in selecting the sample from the population. As in most statistical tests, if the p-value is less than 0.05, then the null hypothesis H_o should be rejected. In English, if the p is less than 0.05, then the likelihood of seeing a relationship this strong is less than 5%, and, therefore, there is a good chance that it is real; if p is greater than 0.05 then the conclusion should be that there is no relationship.

As with any statistical test, Chi-square comes with its set of "could be" and "might be" statements.

For example if the Personnel Department wants to see if there is a link between age and whether an applicant is hired, then both Age (old and young) and Got Hired (did or didn't) are attribute type data. A Chi-Square test would be applicable to answer the questions:

- Are age and hiring decisions dependent or independent?
- Does an association exist between age and hiring practice?
- Is one age-group more likely than the other to get hired, or is the chance of getting hired independent of age?

The hypotheses would be

- H_o: Age and Hiring Decisions are independent
- H_a: Age and Hiring Decisions are dependent

As with all statistical tests, a sample of reality is required, similar to that shown in Table 7.09.1 where about 455 data points were taken and the data distributed amongst the four possible outcomes.

Table 7.09.1 Sample of Reality for the Relationship between Age and Hiring

	Hired	Not Hired
Old	30	150
Young	45	230

The data can then be analyzed in a statistical package, such as JMP or Minitab. The software calculates the expected values in each box, and compares the observed with the expected frequencies to produce a signal-to-noise type ratio (how far the observed is from the expected) using

$$\chi^2 = \sum \frac{(O-E)^2}{E}$$

Where O is the observed frequency and E is the expected frequency in a box.

The software then looks up the χ^2 (the sum of all the discrepancies) value in a statistical table to discover the likelihood of seeing a difference that big.[14] As a Belt using the tool practically, all that is important, after the data has been captured using robust data collection methods, is the output of the tool, which should be similar to that shown in Figure 7.09.1.

Chi-Square Test: Hired, Not Hired

Expected counts are printed below observed counts
Chi-Square contributions are printed below expected counts

	Hired	Not Hired	Total
1	30	150	180
	29.67	150.33	
	0.004	0.001	
2	45	230	275
	45.33	229.67	
	0.002	0.000	
Total	75	380	455

Chi-Sq = 0.007, DF = 1, P-Value = 0.932

Figure 7.09.1 Results of the Chi-Square test for the relationship between Age and Hiring Practice (output from Minitab v14).

The first place to look is the p-value, which in this case p = 0.932. In this instance, the p-value is not low (not below 0.05); thus, the conclusion should be that the relationship between Age and Hiring Practice is not significant for the sample of data taken.

Chi-Square can be applied in virtually any transactional processes where attribute data usually abounds. For example:

[14] For more detail on exactly how this is calculated see *Statistics for Management and Economics* by Keller and Warrack.

- HR—Number of sick days by employee or department
- Accounting—Number of incorrect expense reports by employee or department
- Sales—Number of lost sales by account or region or country
- Logistics—Number of deliveries late by distribution center or country
- Call Center—Number of missed Customer calls by associate or shift
- Installation—Number of repeat service calls by field technician
- Purchasing—Number of days delivery-time for orders by supplier
- Inventory—Number of parts by distribution center

ROADMAP

The roadmap to setting up and applying a Chi-Square test is as follows:

Step 1: Understand the question at hand. There should be a clear relationship in question; does X affect Y? The relationship needs to involve data for both X and Y that is attribute or discrete valued. There should be a business reason for asking the question in the first place, that is the question "Why do we care?" needs to have been answered.

Step 2: Set up the hypotheses in the form:

- H_o: Data are Independent (Not Related)
- H_a: Data are Dependent (Related)

Step 3: Determine a data collection method and a sample size. The sample size is based on the expected values in each box in the data collection table. To have a reasonable confidence in the result of the test, there needs to be an expected value in each box greater than 5. Thus, to calculate sample size, identify the lowest potential proportion likely in any of the boxes. Divide 5 by this proportion to give an approximate bare minimum number of data points to collect. For example, if the expected proportion in one box is 2.5% (0.025), then dividing 5 by 0.025 gives 200 data points. This is a little hit and miss, but gives a ballpark approximation. Typically the approach is to double this number to be on the safe side. The Chi-Squared test is data hungry, with sample sizes often above 500.

A simple Tally Sheet is enough to capture most test data, placing check marks in the appropriate box as a data point is collected.

Step 4: Collect the sample of reality. Ideally the data is available historically, or available quickly in large quantities or otherwise the project might need to idle while data is collected.

Step 5: The data is entered in the form of a table into a statistical package and analyzed.

INTERPRETING THE OUTPUT

The first place to look during analysis is to the p-value. If the p-value is higher than 0.05 then the conclusion is that the X and Y are not dependent based on the sample of reality taken (similar to the Age versus Hiring Practice example in "Overview" in this section).

However, if the p-value is low (less than 0.05) then there is reason to believe that the X and Y are dependent in some way and the distribution of data points within the table isn't as expected if everything was based on random chance.

To demonstrate this, consider the data in Table 7.09.2, which represents loan approval or rejection decisions on different days of the week. The bank in question clearly would like loan decisions to be independent of the day the loan was processed.

Table 7.09.2 Loan Decision Data by Day of Week

	Rejected	Approved
Monday	9	27
Tuesday	8	21
Wednesday	11	25
Thursday	7	24
Friday	25	23

The results of the Chi-Square Test analysis are shown in Figure 7.09.2.

Looking immediately to the p-value of 0.028 (less than 0.05), it is clear that the likelihood of seeing a distribution of data in the boxes like this purely by random chance, given that there was no relationship, is slim. Thus, the conclusion is that the null hypothesis should be rejected and the alternate "Data are dependent" should be accepted instead. In English, we conclude there is something fishy going on, because the chances of getting a loan varies by day of the week.

To understand how this is manifested, the next step is to look to the contingency table in the analysis output as shown in Figure 7.09.2. To read this table, the numbers 1 to 5 down the left side represent the days Monday to Friday. The first number in each box is the observed data; that is how the decisions were actually made. The second number in each box is what would be expected if the decision were independent of the day. The final number is the Chi-Square statistic for the difference of the Observed versus the Expected values; the larger the number, the bigger the signal that is present.

Looking at the table it is clear that Friday has the larger deviation from expected, and this is contributing most of the Chi-Square total (5.063 + 2.531 out of 10.888). Looking at the observed versus expected values in the Friday row shows for Rejection an expected of 16 and an observed of 25; it seems on Friday a significant number of people get rejected more than we would expect.

Expected counts are printed below observed counts

Chi-Square contributions are printed below expected counts

	Rejected	Approved	Total
1	9 12.00 0.750	27 24.00 0.375	36
2	8 9.67 0.287	21 19.33 0.144	29
3	11 12.00 0.083	25 24.00 0.042	36
4	7 10.33 1.075	24 20.67 0.538	31
5	25 16.00 5.063	23 32.00 2.531	48
Total	60	120	180

Chi-Sq = 10.888, DF = 4, P-Value = 0.028

Figure 7.09.2 Chi-Square Test analysis results for loan data (output from Minitab v14).

The next steps would be to

- See if, by omitting Friday's data and analyzing just the Monday to Thursday data, the other days of the week show a problem (in fact they don't).
- Seek to understand the practical implications of Friday being different; that is, what causes the Friday phenomenon (perhaps the staff doesn't want to get bogged down in paperwork on a Friday afternoon and wants to get out early). The Belt can't jump to any conclusions here and should play detective to determine the real reasons.

10: Control Charts

Overview

Most novice Belts confuse *Statistical Process Control (SPC)* with Control Charts and vice versa.[15] SPC is the use of Control Charts to help control a process. Control Charts are used in a number of places in the Lean Sigma roadmap to test whether a process is stable and in control, that is having

- Consistent centering (typically the mean)
- Predictable variation

For example, the tool is used

- To test for stability before performing a Capability Study in the Measure Phase
- To test stability before performing a test of means in the Analyze Phase
- In *Statistical Process Control* in the Control Phase
- To validate stability after project completion

SPC, on the other hand, is solely a Control tool and appears only in the Measure Phase (to check what is currently being controlled) and the Control Phase (to control what should be controlled going forward).

Control Charts[16] typically take the form shown in Figure 7.10.1. Data is plotted over time across the x-axis, with the height on the y-axis representing the level of the characteristic in question. For example, if the Control Chart were placed on the purity of end product, then the y-axis would be Purity.

From the data in the chart, Control Limits are calculated and represent the boundaries of likely behavior within the process. A point landing outside of these boundaries is considered out of the ordinary. The Control Limits are calculated from the process data itself and depend on the data type.

For example, if the sample data from the process is Continuous and furthermore is normally distributed, then it is possible to predict where points should lie on the Control Chart applied to the sample as per Figure 7.10.2. The Control Chart used in this case is

---

[15] Use of Control Charts for Process Control dates back to the 1920s and Dr. Walter Shewhart of Western Electric.

[16] For extensive detail on the structure and theory of Control Charts see *Statistics for Management and Economics* by Keller and Warrack.

known as an Individuals Chart. The figure shows that by plotting the data there would be expected numbers of data points within certain bands given by the standard deviations from the center line (mean) as follows:

- From –1 to +1 standard deviations either side of the mean—68% of data points (60–75% in practice)
- From –2 to +2 standard deviations either side of the mean—95% of data points (90–98% in practice)
- From –3 to +3 standard deviations either side of the mean—99.73% of data points (99–99.9% in practice)

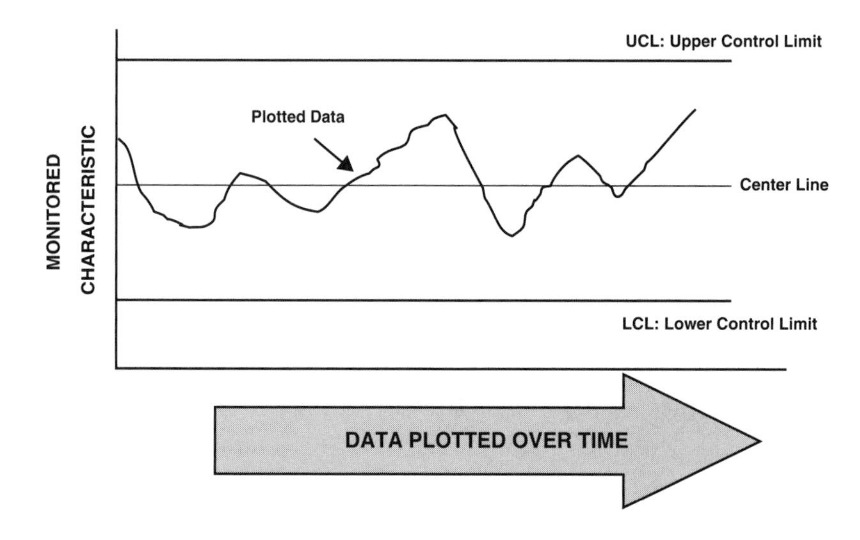

Figure 7.10.1 Structure of a Control Chart.

The lines at ±3 standard deviations are known as Control Limits, and an expected 99.73% of the process variation is expected to fall between these limits. In other words, the likelihood of seeing a point outside of these limits is 0.27% or about 370:1 against. This is considered an unusual event. Obviously in a process that generates hundreds or even thousands of entities, the occasional point falls outside the lines. Two in a row almost guarantees that something highly unusual has occurred or that the process has changed in some way (the foundation of SPC).

There are other unusual occurrences that could be detected using the data distribution in Figure 7.10.2 with odds greater than 250:1 against

- Nine points in a row on the same side of the center line
- Six points in a row all increasing or decreasing
- Fourteen points in a row alternating up and down
- Two out of three points more than two standard deviations from the center line on the same side
- Four out of five points more than one standard deviation from the center line on the same side
- Fifteen points in a row within one standard deviation of the center line on either side
- Eight points in a row more than one standard deviation of the center line on either side

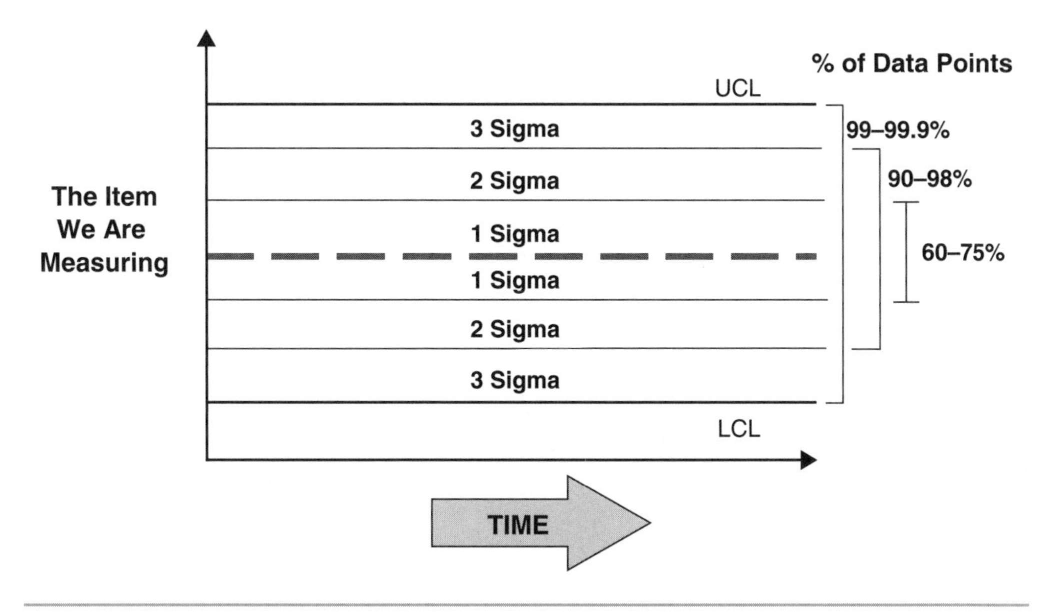

Figure 7.10.2 Distribution of data points with respect to standard deviations in a Normal Distribution.

There are probably more that could be dreamt up, but these are the ones found in most statistical packages. Fortunately, none of these tests have to be run by hand; the statistical software package does this automatically and highlight any points that break the rules.

Stability is determined by having

- Consistent centering (typically the mean)
- Predictable variation

The Individuals Chart primarily tackles the former of these. To cater for the variation element, a Moving Range Chart usually accompanies the Individuals Chart. As shown in Figure 7.10.3, the Moving Range Chart shows the distance between adjacent points from the Individuals Chart. If there is a small change in the Individuals Chart, then the point on the Moving Range Chart is lower; if there is a large change, then the point is proportionately higher. If there is a very large change, then an unusual event has occurred and the point would register out of the Upper Control Limit (UCL). Similar to the list of unusual events for the Individuals Chart, the Moving Range Chart has its equivalent set.

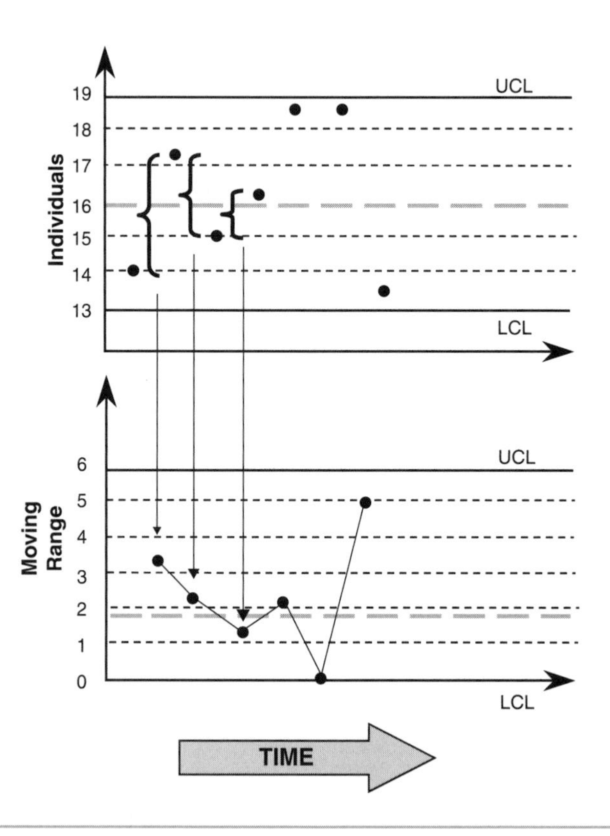

Figure 7.10.3 Relationship between the Individuals Chart and the Moving Range Chart.

Together the Individuals Charts and Moving Range Charts express both characteristics of stability (centering and variation), so they are generally found together as a pairing known as Individuals and Moving Range or I-MR.

The Individuals Chart and Moving Range Chart are only two of many types of Control Charts. The Chart used for a particular application depends on the data type of the sample it represents.

Roadmap

The roadmap to select the appropriate Control Chart is shown in Figure 7.10.4 and is based on the data type of the sample concerned (for more information see "KPOVs and Data" in this chapter):

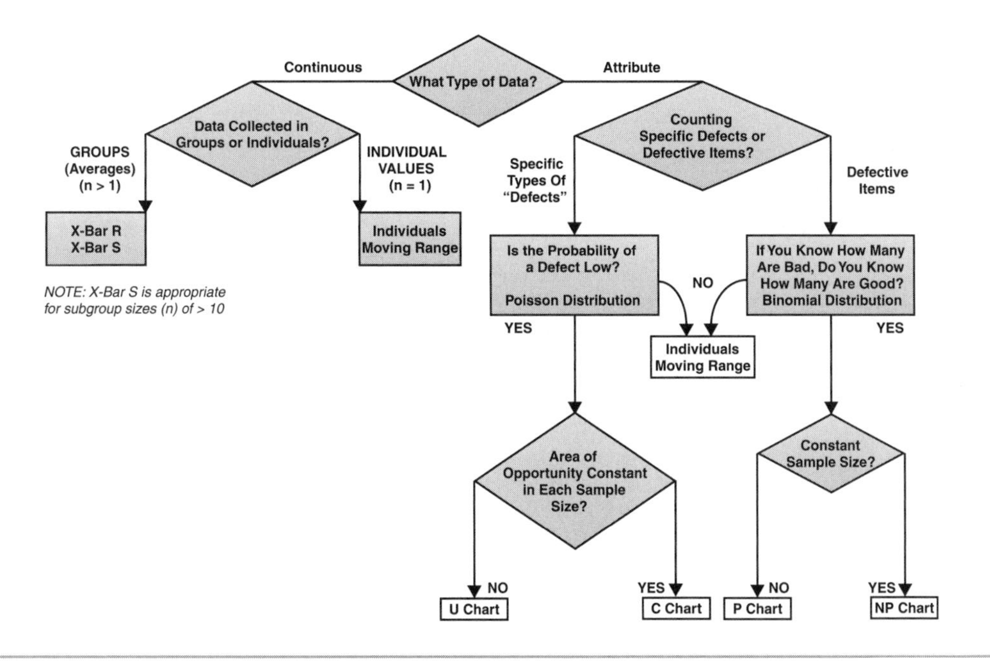

Figure 7.10.4 Control Chart selection roadmap.

- **Continuous.** The data is continuous (measured). Data of this type results from the actual measuring of a characteristic, such as diameter of a hose, electrical resistance, weight of a vehicle, and so on. Continuous data has two classifications of its own:
  - Individuals—Data points are captured and recorded individually.
  - Groups—Data points are captured in groups of, for example, five entities and recorded together. An explanation of why this is useful is given in "Other Options" in this section.

- **Attribute.** The data is discrete (counted). Data of this type results from using go/no-go gages, or from the inspection of visual defects, visual problems, missing parts, or from pass/fail or yes/no decisions. Attribute data has two classifications of its own:
  - Defect—An individual error on an entity (scratch, blemish, a typographical error, and so on).
  - Defective—The whole entity is not acceptable to the Customer. One defective entity might have multiple defects.

INTERPRETING THE OUTPUT

As mentioned in "Overview" in this section, all Controls Charts take on the same form as shown in Figure 7.10.1. An example Control Chart for calls going to a call center is shown in Figure 7.10.5. The data type in this case is Attribute, defective items are being measured (the whole call), and there is a varying sample size (total number of calls in a period). So from the roadmap in Figure 7.10.4, the Control Chart used is a P-Chart.

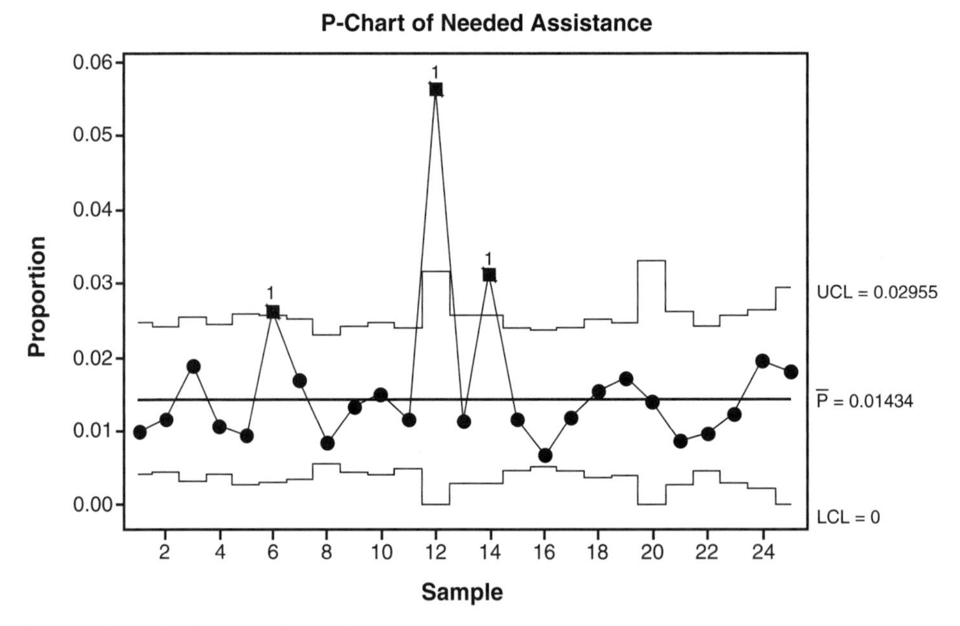

Figure 7.10.5 P-Chart of callers needing assistance from a varying Total Number of Calls (output from Minitab v14).

Although the Control Limits are not horizontal straight lines (due to the varying sample size), the way to read the chart is exactly the same as any other. If data points lie outside the Control Limits (points 12 and 14 for example), then something unusual occurred at that point in time (and similarly for any of the other "unusual" conditions listed in "Overview" in this section).

The key to understanding Control Charts is an appreciation of the different types of variation in a process,[17] specifically:

- **Common Cause.** The inherent variation present in every process, produced by the process itself. This is effectively the background noise in the process and is removed or lessened only by a fundamental change in the process, usually the process physics, chemistry, or technology. A process is Stable, Predictable, and In-Control when only Common Cause variation exists in the process.
- **Special Cause.** The unpredictable variation in a process caused by a unique disturbance or a series of them. Special Cause variation is typically large in comparison to Common Cause variation, but it is not part of the underlying physics of the process and can be removed or lessened by basic process control and monitoring. A process exhibiting Special Cause variation is said to be Out-of-Control and Unstable.

Thus, Control Charts are used to find the signals (special cause variation attributable to assignable causes) in amongst all of the background noise (common cause variation).

The biggest mistake Belts and Champions make is to confuse Control Limits with Specification Limits. Control Limits are calculated based on data from the process itself, they are not determined by anyone but the process itself. Product Specification Limits are *not* found on a Control Chart at all. Understanding how the process matches up against Customer requirements *is* important to know though and this is done with a separate tool called Process Capability.

OTHER OPTIONS

Control Charts are based on an understanding of where data points should lie and thus indicating if, for some reason, points seem to be out of place. This clearly relies on an ability to *know* where data points should lie. This is typically done from known distributions, such as the Normal Distribution for the Individuals Chart, or a Binomial or Poisson Distribution for Attribute type data. The difficulty arises when the distribution of the data doesn't fit a standard.

---

[17] A wonderful reference here, written in plain language rather than Statspeak, is Donald Wheeler's book *Understanding Variation—The Key to Managing Chaos.*

To counter this, there is a useful tool in statistics called the Central Limit Theorem,[18] which relates to taking small groups of data points together rather than examining points individually. These small groups are known as subgroups. The useful property here, which stems from the Central Limit Theorem, is that even if the process itself is creating non-Normal data, the *means* of subgroups from the data is approximate normal data. This might be strange but it is true and extremely useful across many statistical tests.

This is useful to the Control Chart because, rather than taking data and dealing with its strange distribution, perhaps having to transform it (to make it Normal) and then plotting a Control Chart from it, it is actually possible to take small subgroups of, for example, five data points and plot the means of the subgroups instead. The means tend to be more Normally distributed and the Control Chart is actually valid. This type of Control Chart is known as an X-bar Chart (the reason being that the symbol for the mean of a sample is an X with a bar over it). In fact another property of the Central Limit Theorem also takes effect here. The Control Limits (±3 standard deviations) are tighter on an X-bar Chart than would be expected for the raw individuals data,[19] but don't worry about how the Control Limits are calculated, just use the Chart.

As mentioned previously, stability relies on two properties:

- Consistent centering (typically the mean)
- Predictable variation

The X-bar Chart takes care of the first of the two conditions; the second is usually done with a Range Chart. The Range Chart plots the actual range of the data in each subgroup (maximum minus minimum values) instead of a Moving Range as in the IM-R. The two graphs together become the Xbar-R Chart, see Figure 7.10.6.

The first point in the top graph (X-bar Chart) is actually the mean of the points 1 to –5; the second point is the mean of the points 6–10, and so on. The first point in the bottom graph (Range Chart) is the range of the points 1–5; the second is the range of the point 6–10, and so on.

It might seem complicated at first, but it is easier to manage in practice for Statistical Process Control than a complex transformation. The Control Chart is interpreted just the same way as any other.

---

18 For more information, see *Statistics for Management and Economics* by Keller and Warrack.

19 From the Central Limit Theorem, the variation in the means of subgroups is smaller by a factor of $1/\sqrt{n}$ (where n is the subgroup size), so for a subgroup size of 5, the variability in the means is $1/\sqrt{5}$ of the variability in the individual data points.

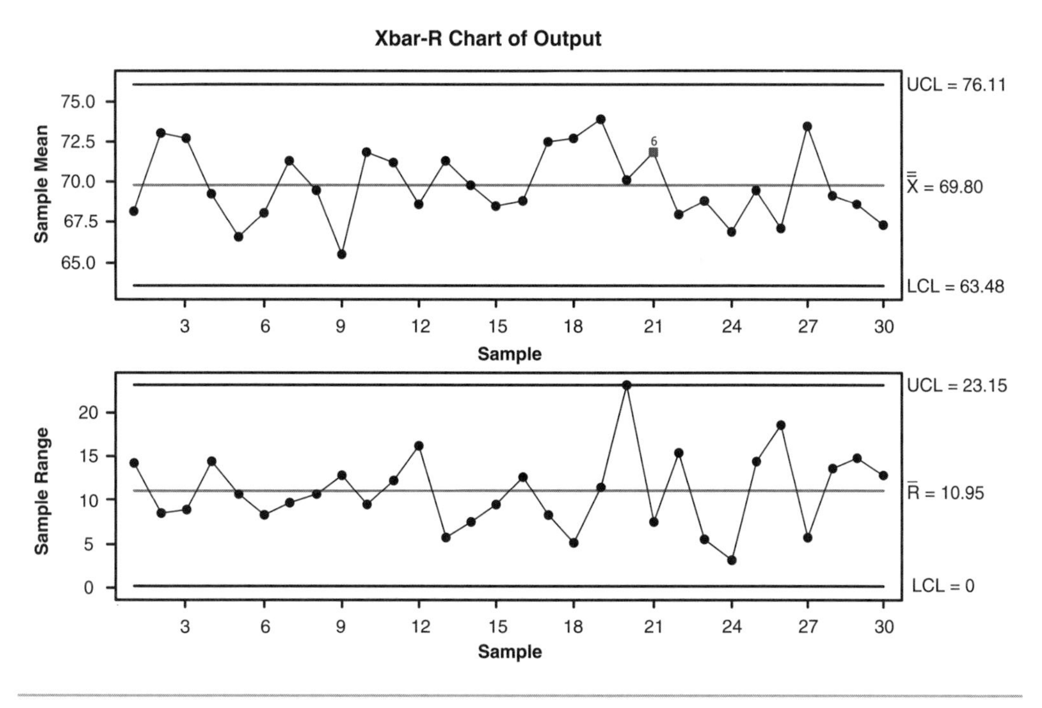

Figure 7.10.6 An example of an Xbar-R Chart (output from Minitab v14).

11: Critical Path Analysis

Overview

Critical Path Analysis in Lean Sigma is a simplified version of the tool originating from Project Management. It is primarily used in *Rapid Changeover*, but is applicable to any complex system of process steps to minimize *Process Lead Time*.

The basis of the tool in Project Management is the network diagram (sometimes known as a PERT Chart), which is a structured graphical representation of the tasks in a project. In Lean Sigma this is completely analogous to the steps in a detailed Process Map representing the activities in a process.

The network diagram is structured on the premise that every task must have at least one successor and one predecessor. It is possible, therefore, that tasks can be conducted in parallel as in Figure 7.11.1.

After all the tasks are connected, the structure is to that shown in Figure 7.11.2. The Critical Path is defined as the longest chain of time through the network (as shown in grey).

The Critical Path is so called because any time lost on it is lost on the project (or process) as a whole. In terms of Process Improvement, any opportunities for reduction of the Process Lead Time should center on the Critical Path, particularly in removing NVA activity.

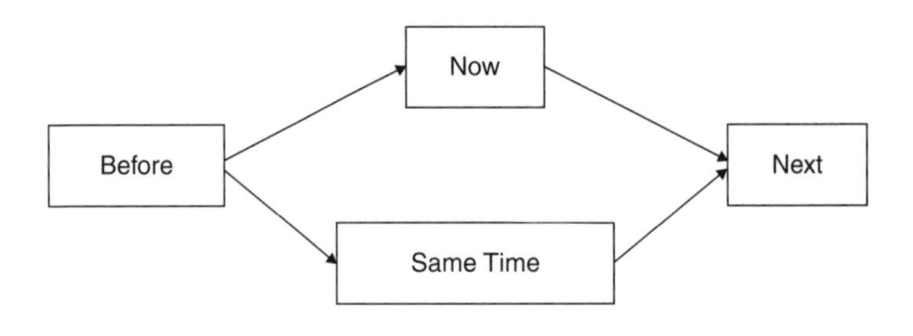

Figure 7.11.1 Simple network diagram rules.

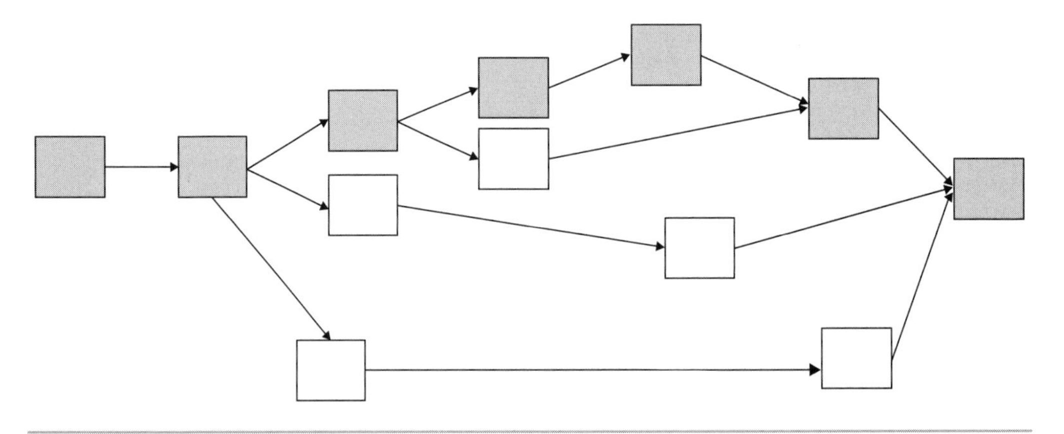

Figure 7.11.2 Network diagram.

Roadmap

The roadmap for applying Critical Path Analysis is as follows:

Step 1: Using a Project Management tool such as Microsoft Project™, and a previously constructed *Swimlane Map* or Detailed Process Map, enter the process steps and durations. This can be done in a network diagram or in a Gantt Chart.

Step 2: Starting at the beginning of the process, identify and enter the process step dependencies, working systematically across the map from left to right. Process step dependencies are typically created by highlighting two dependent tasks and clicking a link-icon to connect them.

Step 3: After the dependencies are created, in the Gantt Chart, format the tasks to include the Critical Path. In Microsoft Project™ this is done using the Gantt Chart Wizard. An example is shown in Figure 7.11.3.

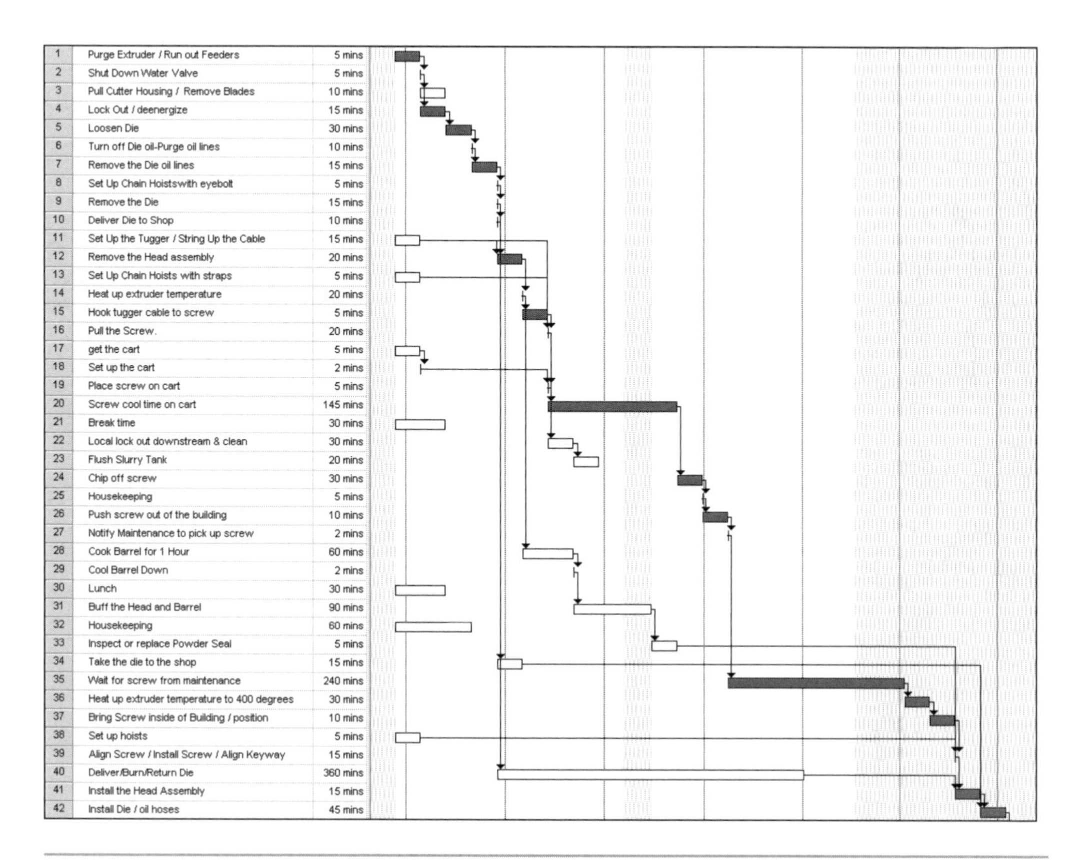

1	Purge Extruder / Run out Feeders	5 mins
2	Shut Down Water Valve	5 mins
3	Pull Cutter Housing / Remove Blades	10 mins
4	Lock Out / deenergize	15 mins
5	Loosen Die	30 mins
6	Turn off Die oil-Purge oil lines	10 mins
7	Remove the Die oil lines	15 mins
8	Set Up Chain Hoistswith eyebolt	5 mins
9	Remove the Die	15 mins
10	Deliver Die to Shop	10 mins
11	Set Up the Tugger / String Up the Cable	15 mins
12	Remove the Head assembly	20 mins
13	Set Up Chain Hoists with straps	5 mins
14	Heat up extruder temperature	20 mins
15	Hook tugger cable to screw	5 mins
16	Pull the Screw.	20 mins
17	get the cart	5 mins
18	Set up the cart	2 mins
19	Place screw on cart	5 mins
20	Screw cool time on cart	145 mins
21	Break time	30 mins
22	Local lock out downstream & clean	30 mins
23	Flush Slurry Tank	20 mins
24	Chip off screw	30 mins
25	Housekeeping	5 mins
26	Push screw out of the building	10 mins
27	Notify Maintenance to pick up screw	2 mins
28	Cook Barrel for 1 Hour	60 mins
29	Cool Barrel Down	2 mins
30	Lunch	30 mins
31	Buff the Head and Barrel	90 mins
32	Housekeeping	60 mins
33	Inspect or replace Powder Seal	5 mins
34	Take the die to the shop	15 mins
35	Wait for screw from maintenance	240 mins
36	Heat up extruder temperature to 400 degrees	30 mins
37	Bring Screw inside of Building / position	10 mins
38	Set up hoists	5 mins
39	Align Screw / Install Screw / Align Keyway	15 mins
40	Deliver/Burn/Return Die	360 mins
41	Install the Head Assembly	15 mins
42	Install Die / oil hoses	45 mins

Figure 7.11.3 Example Critical Path Analysis applied to an extruder changeover (output in Microsoft™ Project 2000).

Interpreting the Output

After the Gantt Chart is complete with the Critical Path highlighted, it gives focus for the Team to make reductions in the Process Lead Time. On the Critical Path the Team should strive to

- Eliminate all NVA activities
- Eliminate or reduce other activities
- Move activities off the Critical Path

Off the Critical Path strive to free up resources by eliminating or reducing activities.

12: CUSTOMER INTERVIEWING

OVERVIEW

Customer Interviews[20] are the primary means to collect qualitative VOC information during a Lean Sigma project. Although *Customer Surveys* can be useful to capture quantitative data for a larger number of individual Customers, Customer Interviews give much greater returns in terms of information content. They allow collection of data from a variety of Customer channels and the data is from the source—the actual Customer!

The key to interviewing is in the planning, ensuring a structured approach to interview a few key, diverse individual Customers to gain as much insight into the process as possible.

LOGISTICS

Customer Interviewing is a Team activity and should not be attempted by the Belt alone. No matter how objective we think we are, each of us has our own set of biases that can influence how we interact with and elicit information from a Customer. Thus, all planning, preparation, the interviewing itself, and the post-interview processing must be done by a balanced Team to ensure non-biased VOC data.

Planning and preparation can take as much as a day; the interviewing itself is comprised of 15–25 individual interviews, each last from 30 to 90 minutes depending on the Customers' involvement with the process. If all the planets align and sub-teams are formed to interview, the whole process could be done during a one-week period. Typically, however, it usually spills over into a second week, sometimes more if key Customers are unavailable.

Processing the results takes from 3 to 6 hours depending on the volume of data captured.

---

[20] The level of detail here should be enough for a Lean Sigma process improvement Team to capture the VOC data required to complete a project. However, it probably does not go to the level of detail required for a product development project, for that it will be useful to refer to *Commercializing Great Products Using DFSS* by Randy Perry.

ROADMAP

The roadmap for planning, preparing, conducting and post-processing interviews is as follows:

Planning and Preparation

Step 1: Define the purpose for the Customer Interviews. This is vital step. If the Team cannot articulate amongst themselves the reason for interviewing, then they will not be able to articulate to the Customer why that person should give up valuable time to speak with the Team. Also, interviews can falsely raise the expectations of the interviewee, so a consistent well-formed Team message is crucial.

The Purpose typically contains the following:

- What the Team needs to learn from the Customer and why—At the highest level to be able to explain the whole point of the Interviewing. This is used to help gain access to the Customer in Step 6.
- What information the Team is trying to gather—Using the Team's brainstormed Customer requirements and input from the *Murphy's Analysis* helps here.
- What actions the Team takes after the data is in hand—This helps shape the interview questions.

Step 2: Develop an Interview (Discussion) Guide to have a consistent approach to interviewing. This is used so that during an interview, no matter which Team member is conducting it, the interview focuses on the Team objectives and investigates, as a minimum, the same few key areas. The Guide is absolutely *not* a questionnaire or a survey. It generally contains from six to ten open-ended lead-in questions on key topics to direct the flow of the interview. Below each lead-in question are follow-up questions (topics) to be used as needed in case the Customer conversation flow slows. There might be multiple variations for different Customer types.

The simplest way to generate the Interview Guide is, as a Team, to

- In silence, individually write as many open-ended questions you can think of to elicit the data you seek.
- Write each question legibly in black on a sticky note, one question per sticky.
- While doing this, keep the Purpose Statement close by to make sure all questions tie in to it.
- After each Team member has run out of ideas, place all the sticky notes on a large sheet of paper on the wall.

- As a Team conduct a "net-touching" exercise to group similar questions:
  - Using one person to lead, put a single question off to one side and everyone reads it.
  - Gather others that are similar.
  - Remove any *exact* replication, do not remove any information content (the tendency is often to want to keep just a summary statement, and lose the detail—this is the wrong way around, keep the detail).
  - Continue, until all questions are grouped.
  - For each group, write a question on a separate sticky note in red that is a lead-in question into the area of all questions in that group.

At this point, there are usually from six to ten major groupings with their associated lead-in question. The red lead-in question is the primary question to guide the Interview in a certain direction; the black lower level questions might become bullets if needed for prompting the Customer.

The best questions evoke images of experiences and needs, an emotional response, rather than an intellectual one. Some suggestions might be

- Would you walk me step-by-step through your process as you use our product/service?
- What are its key features, functions, and measures of success?
- What is important for you about this process?
- What do you like about the current product/service?
- What do you dislike about it?
- What changes would you recommend? Why?

Questions should also focus on

- Perceptions of weaknesses or problems, by eliciting past experiences with the product/service
- Current usage considerations, perhaps even a competitive comparison
- Possible future enhancements or even "delighters" for the product/service

There are some types of questions to stay away from, for example:

- Closed ended, such as, "Would you like…?"
- Leading, such as "Wouldn't you agree that if we could…?"

- The words "are," "do," and "can" are usually part of closed-ended questions, which might get only a "yes" or "no" answer

Questions that work well in an open-ended way include

- What—This tends to focus conversation on events, such as "What problems have you experienced?"
- How—This tends to focus discussion on the process, such as "How do you use...?"
- Why—This is useful if any explanation is required, such as "Why is that?" However, it might elicit a defensive reaction if over-used.
- Could—This is usually perceived as a gentle approach and open, such as "Could you give an example?"

Finally, it is a good idea to keep questions short and to break complex issues into a series of short questions. Always avoid technical jargon, because it is more pervasive than we might think.

Step 3: Create a Sampling Plan, sometimes known as a Customer matrix. A common Belt mistake is to select interviewees as the interviewing proceeds and select them based on availability. It is important to identify the right group of Customers to interview up front. Some are difficult (perhaps lost Customers) and are not the most available, but they do provide valuable insight for the project. The key here is diversity. If the Customers are selected well, then the first few yield the vast majority of information the Team is seeking, as in Figure 7.12.1. For a Lean Sigma project, somewhere between 12 to 25 interviews is enough, if the interviewees are well chosen and diverse.

As a Team, determine whom to interview, remembering to get representation from all appropriate Customer Segments and Types, such as by

- Function
- Location
- Role
- Size

The sample size (number of interviews conducted) depends somewhat on the size of the Customer or Segment population, as per Table 7.12.1.

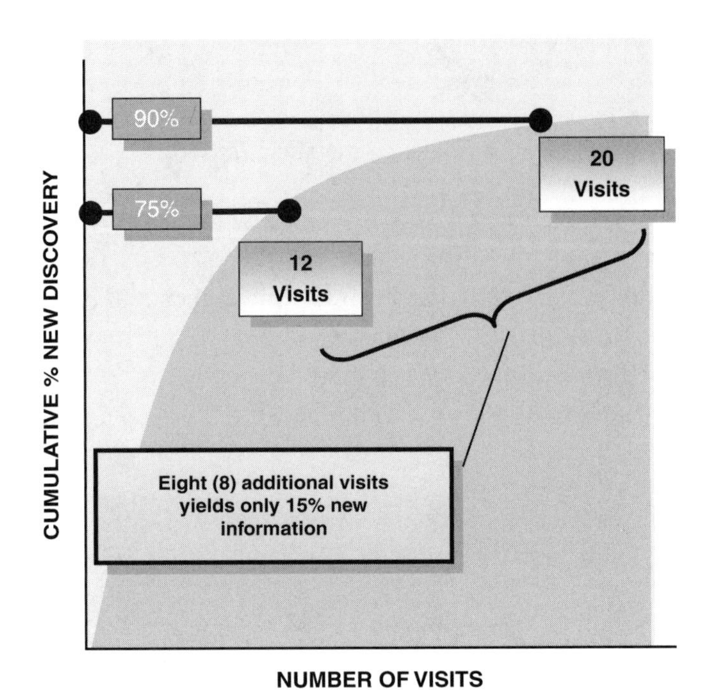

Figure 7.12.1 The relationship between information gained and Customers interviewed.[21]

Table 7.12.1 Sample Size as It Relates to Customer Segment Population

# Customer (by segment)	# to Interview
<5	All
5-10	5
10-20	7
21-50	8
51-100	10
>100	10-15

[21] Source: "Voice of the Customer," Marketing Science Institute Working Paper, Report Number 92-106 by John Hauser and Abbie Griffin.

Step 4: Form interview Teams. It is unwise to interview as an individual; no matter how objective we all think we are, each of us has our own set of biases that can influence how we interact with and elicit information from a Customer. It is best to interview in pairs or threesomes, with each person taking a specific role.

- **Interviewer.** Builds rapport with the Customer, asks the questions, and leads the interview. Manages the discussion and covers all key topics.
- **Scribe.** Takes detailed notes, verbatim when possible, and acts as a backstop for the interviewer.
- **Observer (optional).** Soaks up impressions and listens "between the lines" (watches for body-language signals) and supports Interviewer if the need arises. Supports the Scribe to ensure all questions are documented (the Scribe and the Observer usually take turns scribing).

Step 5: Practice! Interviewing is not a common strong point in Teams, but can be learned. Start out practicing within the Team, role-playing the Customer. Then during the course of Interviewing, progress to easy (perhaps internal) Customers first and finally to the trickier ones.

Step 6: Set up the Interviews. This might require visits to Customer locations. Contact the Customers in advance and schedule the interview. Keep in mind that people might be concerned about why you want to talk to them; the Purpose Statement is valuable here to explain exactly why the interview is requested. Be sure not to conflict with other schedules. If necessary, discuss the interview with the target's supervisor to avoid problems for the Interviewee later. Explain that the interview is short and ask them ahead of time to collect any forms, documents, lists, or procedures that are normally used.

The typical problems that occur here are

- The Customer wants personnel to be interviewed as a group to cut down on time
- A supervisor wants to sit in on an interview

It is strongly advised that interviews are conducted individually. Group interviews tend to only generate middle-of-the-road consensus answers and vital information is lost. If a senior person is present, then all answers tend to subordinate to that person. It is imperative to get interviewees alone. Polite explanation of these concerns, usually allows access to the individuals.

Conducting the Interview

Step 7: Conduct the Interviews. The interview itself generally has three phases:

- The beginning creates rapport:
  - Use open introductions and small talk to put them at ease; perhaps discuss the weather, something from the news, a previous visit, or perhaps family if they are known to you.
  - Review the purpose and explain the note taking to ensure no data is lost.
  - Describe how the information will be used.
  - Emphasize the importance of their contribution.
- The middle follows the Interview Discussion Guide to capture data
  - Develop all major topics.
  - Elicit data from the Customer.
- The closing ensures rapport is maintained and future visits are possible if needed
  - Finish early, don't promise 30 minutes and take 60.
  - Allow time for questions.
  - Thank them.

The whole point of the Customer Interviews is to experience their world, to explore the nature of their work, explore their responses to relevant products and services, and most importantly, to learn something new! This is best done in the context of their work environment, so if at all possible have them walk the Interviewing Team through the steps of their job and how they do it. While watching them work, let them do the talking. After a new understanding is gained, check it with them. Accept surprise, if the VOC was completely understood, then Interviews would be redundant. It is crucial to remember that satisfaction can be defined only by the Customer, so do not challenge what they are saying, accept everything, take in whatever is offered, and suspend judgment—you are there to listen and learn.

There are fortunately a few techniques that can help extract the needed information more readily:

- **Active listening.** Most listening is done passively, in that the listener sits back and takes in whatever is said. This is useful, but after a while the speaker needs encouragement. Reflecting on what you hear signals hearing and understanding. It also helps to clarify points and encourages them to go further and deeper. Examples include
  - I think what I heard you say...
  - It sounds like...
  - Let me make sure I understand what you mean...
  - So, based on your experiences...

- **Probing.** Responses to questions often explain what the Customer wants (a solution), when the Team really is looking for the need (the underlying problem). Also, Customers tend to talk in abstractions (conclusions), rather than the underlying facts that brought them to that conclusion. The Team needs to identify specific, concrete, and actionable facts. Probing helps get to these because it moves beyond the judgmental language ("this is good") to factual language ("this shortens my processing time"). Some examples of probing include
  - When you say...could you give/show me an example of what you mean?
  - Could you explain a little more?
  - Anything else?
  - What specifically?
  - Why would that be a good solution?
  - Could you give me an example of when that occurred?
- **Silence.** A seasoned interviewer knows not to fill silence but let it ride a few seconds until the Interviewee continues to speak. Silence can be uncomfortable but helpful to encourage the Customer to continue. Obviously this must be moderated.

At the end of the interview, the Team needs to leave knowing the Customer's simplest, most specific issues expressed as statements of need versus solutions. The language needs to be that of the Customer, not a translation or summary; therefore, the Scribe has to take as close to verbatim notes as possible. Interviewing pitfalls include

- A superficial or confused response—Acknowledge what you understand, clarify and probe, and stay on track.
- The Customer runs off on a tangent—Let one run, wait for a small pause, then interrupt. It is sometimes easiest to take control by using the Customer's name as you interrupt.
- The Customer contradicts an earlier statement—It is necessary to resolve the contradiction immediately, otherwise it compromises the integrity of interview. It is best to blame oneself for misinterpreting an earlier statement, rather than blaming Customers for contradicting themselves.
- An argumentative Customer—No matter what, don't argue back. Remember this is the Customer voice and perception; therefore, the Customer is right. You cannot win. Allow the person to vent, and perhaps ask for an example to get to the facts.
- Doing harm to the Customer relationship—Don't forget you are there representing your company.
- Sensitive information—Assume everything you say might be quoted.

- Referring to competitors—If there is a reason to make reference to a competitor company or product, speak only positively about them. It is a projection of your credibility to do so.
- Price—If price of the current product comes up, then seek to understand the issues. Understand the cost of using your product and service.
- Don't be defensive—Even if the Customer directly criticizes your work, continue to use positive body language, listen carefully, and seek to understand the issues.

After the interview is complete, be sure to thank the Customer's manager or supervisor for letting them meet with you (they can't be doing their job if they are talking to you).

Step 8: Debrief the interview as soon as possible, preferably immediately afterwards. With this in mind it is best to avoid scheduling interviews back-to-back. As an interviewing Team, discuss general observations and read the notes carefully together, filling in gaps with the Customer's actual words. Discuss and note insights about the Customer, their environment, and needs. Make a copy of any notes as soon as possible.

Assess the interview guide and the interviewer's skills to identify improvements for the next interviews.

Step 9: Follow up on any actions identified during the interview. Honor any commitments made with the Customer. Send a note to the Customer to thank them for participating.

Post Processing

Step 10: As soon as possible, while details are still fresh, get together as a whole project Team and review the interviews. Discuss general observations and spend time talking through each conversation noting any unusual response or new learning.

Step 11: Extract key insights and transfer to large Sticky Notes in complete sentence format and then use an *Affinity Diagram* or KJ[22] to distill them into meaningful, usable information.

Validation

Step 12: After any affinity or KJ work is complete, it is recommended that the Team validates the findings with a few of the people interviewed. This ensures that the information is captured correctly and allows any substantial oversights to be corrected. Also, after any process improvements are made, it helps the Customer buy-in to process changes.

[22] Randy Perry covers KJ in detail in *Commercializing Great Products Using DFSS*. Affinity is covered in "Affinity" in this chapter.

INTERPRETING THE OUTPUT

Customer Interviews are one in a series of tools to capture the VOC. The output of any interviews along with output from *Customer Surveys* are affinitized and then translated into a *Customer Requirements Tree* to identify the Big Ys or KPOVs for the process.

13: CUSTOMER REQUIREMENTS TREE

OVERVIEW

A Customer Requirements Tree is the last in a series of tools used to capture the VOC. The output of any *Customer Interviews,* along with output from *Customer Surveys,* are affinitized and then translated into a Customer Requirements Tree to identify the Big Ys or KPOVs for the process.

The tool is used to graphically show the major Customer Needs and associated metrics for the process, so that the Belt, Champion, and Process Owner can agree on the major metrics that the project will address.

An example Customer Requirements Tree is shown in Figure 7.13.1.

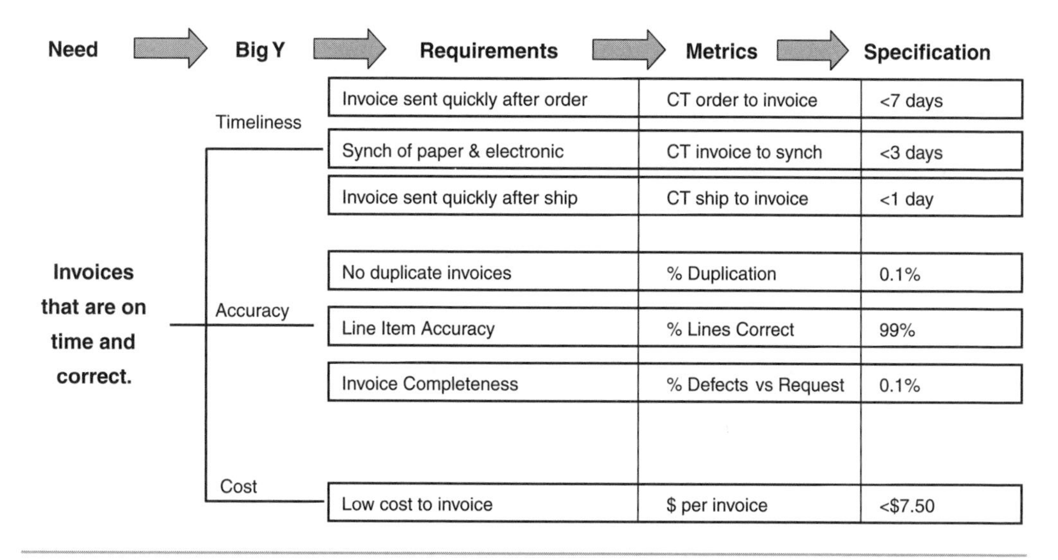

Figure 7.13.1 Example Customer Requirements Tree for an invoicing process.[23]

[23] Adapted from SBTI's Transactional Process Improvement Methodology training material.

The tree helps translates broad Customer requirements into specific critical requirements and then to detailed specifications.

Logistics

It speeds up completion of the Tree if the Belt does some draft construction prior to a working session with the Team. The primary input to the Tree is the Customer Requirements *Affinity Diagram*, so this needs to be readily accessible during construction.

A working session to construct the Tree in the Define Phase typically takes 1–2 hours, but often requires follow up on specific elements, which could theoretically go on well into Measure if no usable metrics exist or little is known about them. Signoff for Define, however, cannot be done until the Tree is complete and the subsequent KPOVs are defined.

Roadmap

The roadmap to constructing the Customer Requirements Tree starts much earlier, with the outputs from interviews and surveys being organized logically in an *Affinity Diagram* as shown graphically in Figure 7.13.2.

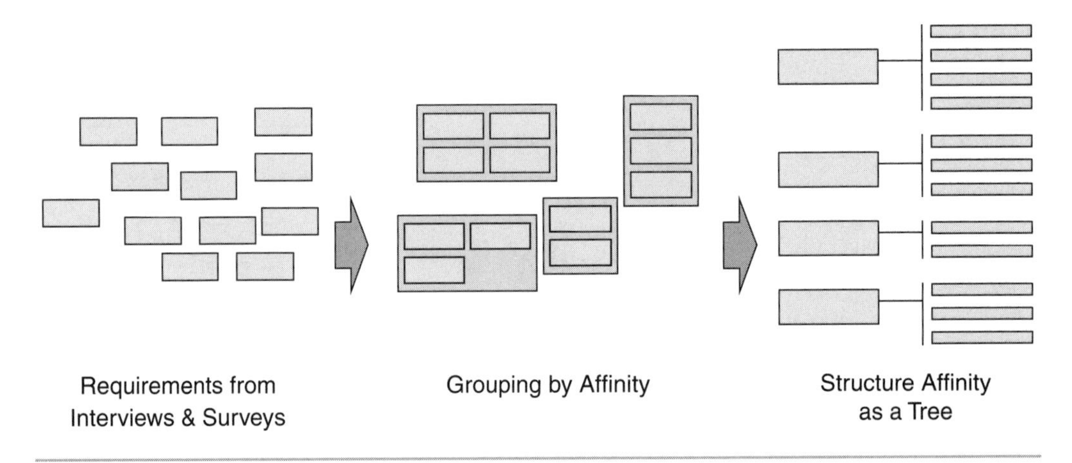

Figure 7.13.2 Construction of a Customer Requirements Tree.

Step 1: Convert the Customer Requirements *Affinity Diagram* into a tree structure:

- The highest level Categories in the *Affinity* become the Drivers (major branches)
- The individual Voices or Needs become the Requirements

An example of a completed version of this is shown in Figure 7.13.3.

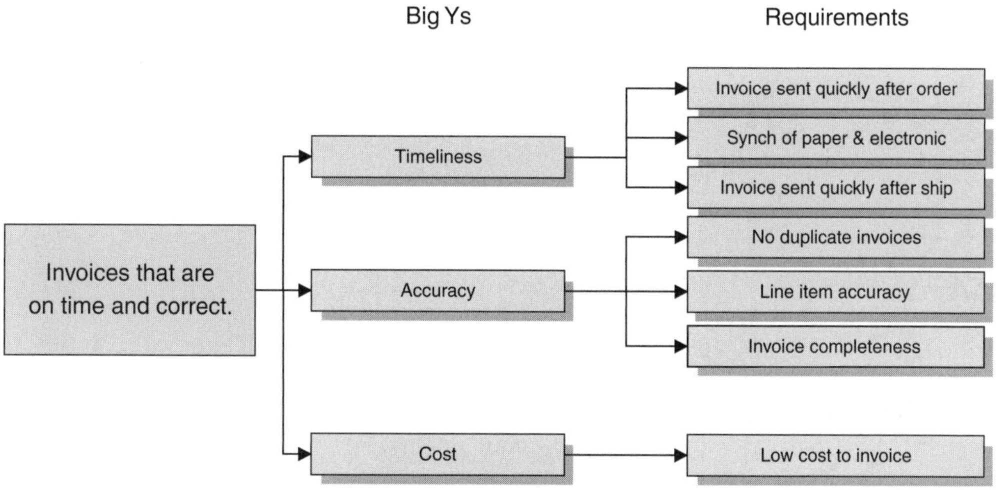

Big Ys
Requirements
Invoices that are on time and correct.
Timeliness
Accuracy
Cost
Invoice sent quickly after order
Synch of paper & electronic
Invoice sent quickly after ship
No duplicate invoices
Line item accuracy
Invoice completeness
Low cost to invoice

Figure 7.13.3 Initial Tree Structure from Affinity Diagram.

If the *Affinity* has more than one layer of grouping (the Team grouped the groups), then the tree has the following structure:

- Drivers (major branches)
- Sub-Drivers (sub-branches)
- Requirements

Step 2: For each requirement, identify the appropriate metric or metrics. Later, if the metric is deemed to be one of the KPOVs or Big Ys for the project, then some Measurement Systems Analysis (MSA) work is required to verify the reliability of the metric and specifically the ability to measure it.

Step 3: For each metric list the specifications if they are known. The specifications can be determined only by the Customer, or from interaction with the Customer. Specifications might be known for a few metrics, but commonly are not known and some additional investigative work is required.

INTERPRETING THE OUTPUT

The Customer Requirements Tree is used to identify the superset of major metrics for the process in question. Only a subset of these metrics can be efficiently tackled in a Lean Sigma process, so the Tree (in context of all the *Customer Interviews* and *Customer Surveys*) is used to guide the Belt and Champion in a discussion about what the selected target metrics should be for the project.

14: CUSTOMER SURVEYS

OVERVIEW

Although *Customer Interviews* are the primary means to collect VOC information during a Lean Sigma project, Customer Surveys can be extremely useful to capture additional data for a larger number of individuals outside of the group of Customers interviewed. Even though Customer Surveys tend to yield less return in terms of information content, they allow collection of data from a variety of Customer channels and the data is from the source—the actual Customer! Surveys also tend to be less time consuming than interviews and can touch more customers than interviews.

The key to surveying is in the preparation of the document itself and the management of the distribution and collection.

LOGISTICS

The creation of a Customer Survey requires extensive Team input and should not be attempted by the Belt in isolation. However, after the surveys start to be returned, the Belt can work individually to collate and process the data and then feed the results back to the Team.

Creation of a robust Survey can take as long as eight hours if done from scratch. In the Lean Sigma roadmap, it is always done alongside *Customer Interviewing*, and the work done to create the Interview Guide reduces time to create a Survey to perhaps four hours at most. It is advisable to use help from someone with experience in creating surveys if this is the first time for the Belt.

ROADMAP

The roadmap to create a Customer Survey is as follows:

Step 1: Define the Survey objective. The objective is key to define the scope, define the population, keep the survey focused, and ensure the survey meets the needs of the project. The objective can be derived from two questions:

- What is the problem to be solved?
- What new information is required to solve it?

Survey objectives often include the words:

- *Describe* or *understand* a population's characteristics
- *Explain* the relationship between two characteristics
- What are the most important variables to *predict* some other variable
- *Compare* variables and levels to find differences that aren't due to chance

For example, for a problem description, "Based on mistakes in receiving transactions, there might be a need for some training for personnel on the docks. It is not obvious who needs it most, what they need or what alternatives make sense." Some prioritized survey objectives might be

1. Describe experience level and prior training of anyone who does at least two transactions a week.
2. Determine the most common needs for training.
3. Find out if people who fit (1) are satisfied with their current skill and knowledge levels.
4. Identify whether people who fit (1) are willing to participate in off-hours training.
5. Determine if increased training increases confidence and satisfaction with the job.

The survey objective can lead to collecting both qualitative data and quantitative data, depending on the problem in hand.

Step 2: Design the Survey. There is no surefire approach to this, but some simple rules apply:

- Keep the audience and time in mind.
- Use simple, common language and take out acronyms, jargon, and abbreviations. Can a co-worker fill it out without asking for instructions?
- Vague questions tend to get vague answers, so be concrete and specific.
- Ask one simple question at a time.

- Watch out for "hot buttons" or prejudicial words that might provoke an unwanted response.
- Use one of the standard scales (described later in this section).
- Don't try to put too much on the page, keep a simple, clean layout.

A survey typically is comprised of

- Demographics, for example:
  - How many years with the company
  - Division
  - Job function
- Close-ended questions, for example:
  - Yes/No (circle one)
  - Anchored scales (described later in this section)
  - Simple answers (circle one)
- Open-ended questions such as
  - What did you like most about the course?
  - What is your top priority for performance improvement?

Surveys invariably involve some kind of subjective assessment, which can be fraught with danger if not handled correctly. The simplest approach is to use what is known as an anchored scale to give better consistency of data over time and different people. To create an anchored scale:

- Specify the measure you are using.
- Assign points to the scale. This is usually an odd numbered scale, because even numbers force a choice.
- Anchor the high and low points with some simple language. If needed, fill in an intermediate point with simple language too.

For most uses, the data can be treated from an anchored scale as continuous data or interval data.

Some examples of anchored scales include

- 1 = Poor; 2 = Fair; 3 = OK; 4 = Good; 5 = Excellent

- 1 = Strongly Disagree; 2 = Disagree; 3 = Neutral; 4 = Agree; 5 = Strongly Agree
- 1 = Low; 2; 3; 4 = High

The language at one end has to match the language at the other end and the scale has to be a continuum. Intervals between scale points have to be equal so that the distance between 1 and 2 is the same as the distance between 2 and 3, and so on.

Some inappropriate scales include

- 1 = Disagree; 5 = Strongly Approve
- Would you say your current age is 1) too young, 2) about right, 3) not old enough, or 4) too old?

Figures 7.14.1a and 7.14.1b show both pages of a survey used by SBTI for course evaluation. The form includes multiple anchored scales questions about the class and Program, along with some open-ended questions on Figure 7.14.1b. No demographic information is gathered to preserve anonymity.

Step 3: Determine the data collection method. There are a number of approaches to collect survey data, the most common being

- One-on-one can be used to tackle complex questions and is beneficial if it is likely that people won't respond by phone or mail. However, it does require trained, experienced interviewers (to avoid bias) and people can be reluctant to honestly respond face to face. It also tends to take as long as an interview, in which case an interview approach might be better.
- Phone surveys produce the fastest results and have the best quality control because calls can be monitored. However, there is more risk of the interviewer influencing responses than a mail survey and sometimes respondents give quick answers due to time pressure. It also requires trained interviewers and a high callback rate to ensure sample integrity.
- Mail surveys require the least amount of trained resources and can be much lower cost. They are the best approach to reach a large number of respondents and have a lower risk of the interviewer influencing responses. Quality control is tough; there might be skipped questions or misunderstandings and a risk of bias from non-response. The response rate is typically low at around 5% to 20% for an external survey and around 30% to 50% for an internal survey.

The approach taken is best determined by the survey objective. To gain a better return rate, talk to Customers individually explaining the purpose before you give them the survey. Response rates are much higher if the purpose and value are understood.

SBTI Course Evaluation

Course Name: ____________________ Date: ______________

Instructors' Names: ____

Please circle the number that best expresses how you rate each of the evaluation criteria listed.
Note: 5 is the highest score; 1 is the lowest.

Instructional Quality	LOW				HIGH
1. Event objectives for the week were clear	1	2	3	4	5
2. Event objectives were met	1	2	3	4	5
3. Material was of practical value for my job	1	2	3	4	5
4. Material was generally presented clearly and accurately					
Ian Wedgwood	1	2	3	4	5
Instructor 2	1	2	3	4	5
5. Instructor was enthusiastic about subject					
Ian Wedgwood	1	2	3	4	5
Instructor 2	1	2	3	4	5
6. Instructor was knowledgeable about the subject					
Ian Wedgwood	1	2	3	4	5
Instructor 2	1	2	3	4	5
7. Instructor allowed me to ask questions and be involved in the discussions					
Ian Wedgwood	1	2	3	4	5
Instructor 2	1	2	3	4	5
8. Instructors gave me adequate time to provide feedback	1	2	3	4	5
9. The event sequence was logical	1	2	3	4	5
10. Amount of time allotted for event was appropriate	1	2	3	4	5
11. Facilities were adequate for the session	1	2	3	4	5
12. The degree of confidence I have that I'll use the skills and knowledge gained	1	2	3	4	5

Figure 7.14.1a SBTI Course Evaluation (c2001) page 1.

Step 4: Create a Sampling Plan. A sample (for more detail see "KPOVs and Data" in this chapter) is the collection of only a portion of the data that is available or could be available from a whole population. From the characteristics of the sample, statistical inferences (predictions, guesses) can be made about the population as a whole. Due to the typically much smaller size of the sample versus the whole population, it is a faster, less costly way to gain insight into a process or large population. Surveys are useful to get input from 10–1000 Customers, but if you are looking at a number greater than 100, consider a sampling approach rather than surveying everyone.

Implementation Quality	LOW				HIGH
1. There is clear leadership support for events	*1*	*2*	*3*	*4*	*5*
2. Your team gets the right support from other operational or functional groups	*1*	*2*	*3*	*4*	*5*
3. The probability of success of implemented actions is:	*1*	*2*	*3*	*4*	*5*
4. What is the updated forecast of the financial value of your project?		*Now*		*$*	
		After actions		*$*	

What two to three things did you like best about this event?

What two to three things did you like least about this event?

Please make any additional comments or recommendations in the remaining space or on the back of this form.

Figure 7.14.1b SBTI Course Evaluation (c2001) page 2.

The Sample Plan is affected by two sample properties:

- Sample size—To determine if it is necessary to identify the size of the total population in question. For example, does "Customers" mean current, former, hoped-for, or all of the these? Sometimes no sampling is required. The population might be small enough to survey every data point, known as a census. The Sample Size also depends on the data analysis to be performed on the resulting data. Table 7.14.1 shows minimum needed sample sizes for some common analyses. In general, if the data to be collected is Continuous (see "KPOVs and Data" in this chapter) then only 30 data points are enough; for Attribute data 100 data points are required. Note that when analyzing subgroups within a sample, the sample size is effectively reduced.

Table 7.14.1 Sample Sizes Needed for Common Analyses

Tool or Statistic	Minimum Sample Size
Average	5–10
Standard Deviation	25–30
Proportion Defective (P)	100 and $nP \geq 5$
Histogram or Pareto	50
Scatter Diagram	25
Control Chart	20

- Sample Quality—A good sample is a miniature version of the population; it is just like it, only smaller. There are a number of ways to make a mistake, so plan to avoid the following:
  - Coverage error—People in the sample aren't really representative of target Customers.
  - Sampling error—Using a sample always means an estimation. This is a fact of life; however, this error can be quantified and minimized.
  - Measurement error—Errors or noise are introduced with the survey tool, from the interviewer or from the respondent.
  - Non-response error—People in the sample who didn't respond are different in an important way from those who did.
  - Selection bias—"But I only wanted to talk to people who looked nice."

Step 5: Conduct or send out the Survey. The survey usually takes one week to get the data back. Stress the objective of the study and the importance of their involvement and include the name and telephone number of a Team member who can be contacted for assistance.

Step 6: Create an Analysis plan, include reporting. Determine how the data will be analyzed, and how it will be reported and to what audience.

Step 7: Follow-up with a postcard, letter, or phone call to thank the respondents for their participation.

INTERPRETING THE OUTPUT

Customer Surveys are one in a series of tools to capture the VOC. The results of the surveys, along with output from *Customer Interviews,* is affinitized and then translated into a *Customer Requirements Tree* to identify the Big Ys or KPOVs for the process.

15: DEMAND PROFILING

OVERVIEW

Demand Profiling is simply the application of a Time Series Plot to the Customer demand on a process. It is considered a tool in its own right because

- Understanding demand is so important in any Lean Sigma project
- Interpretation of the plot is the key to success

Figure 7.15.1 shows a simple Demand Profile plot. Demand data is captured over time at uniform[24] time intervals and plotted with an x-axis of Time and a y-axis of Demand. The plot highlights the likely average demand on the process and also the likely variation in demand to which the process has to respond.

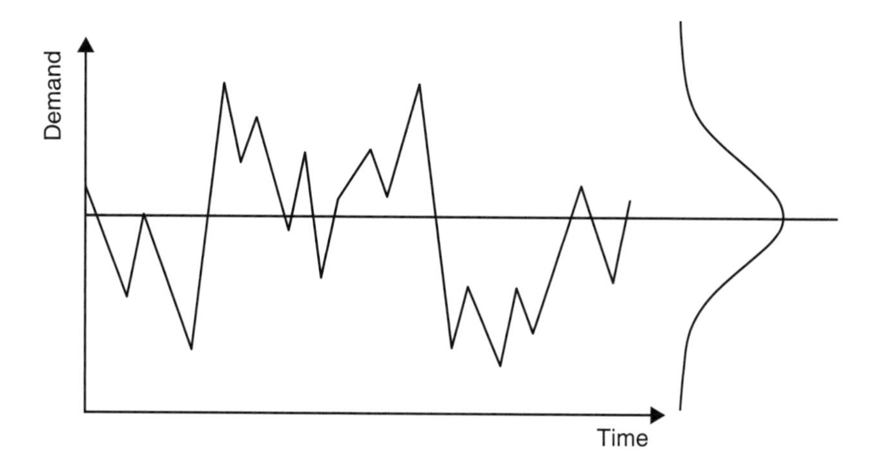

Figure 7.15.1 A simple Demand Profile.

LOGISTICS

As a simple analysis tool this can be applied by the Belt without the rest of the Team, however the data might have to come from multiple sources and often requires Team involvement to collect it. The analysis itself is done entirely in a spreadsheet or statistical package and can be done in a matter of a few minutes after the data is in the correct format.

[24] Uniformity isn't imperative but helps greatly in interpretation.

Initially, data is historical, but after the organization understands the application of the tool, forecast data can also be used.

ROADMAP

Step 1: Identify which entity type or types are to be examined. Demand Profiling represents demand for either one entity type or the sum across a few entity types. If the Team needs to understand the volume and variation across many entity types then it is probably best to look at a *Demand Segmentation* instead of a Demand Profile.

Step 2: Identify the time increments. It is necessary to have at least 25 increments of captured demand to have a useful graph. The increments themselves should be meaningful, and it is useful to take the typical replenishment cycle as the driver. For example, if Customers were replenished weekly, then a week would be a reasonable time increment.

Step 3: Data is collected from the downstream Customer for the entity type for each time increment. The most common mistake is to look to the Process Planning group for when we *decided* to make the entity, not when it was actually demanded by the Customer. Customer demand rates are typically much smoother than we care to admit, and in fact, their usage rates are even smoother. Internally, we tend to batch entity processing into large lots, which we make on an infrequent basis; so it shows a much higher variation in demand than is actually there. For Demand Profiling, we need to look at *demand* patterns, not our own planned process patterns.

Step 4: After the demand data for the entity type across multiple time increments is collected, create the graph similar to the one shown in Figure 7.15.1, taking the time as the x-axis and the volume of demand as the y-axis.

INTERPRETING THE OUTPUT

There are some important points to consider when examining the Demand Profile. The purpose of creating the profile was to understand, from historical data, future volume and variation in demand. Using historical data might be misleading; so the first question to consider is whether future usage patterns are expected to be similar to those in the past. It is always useful to study the market and technology changes and factor these into the analysis.

In effect, the Demand Profile is being used as a rudimentary forecasting tool, but rather than using the data to determine how much to generate on a short-term basis, it is used to create a more responsive and capable process.

Demand Profiles can exhibit one or combinations of a multitude of patterns, the most common of which are shown in Figure 7.15.2. See the "Interpreting the Output" section in "Demand Segmentation" in this chapter. Graph A represents a Zone 1 Demand Profile, whereas Graph D represents a Zone 3 Demand Profile.

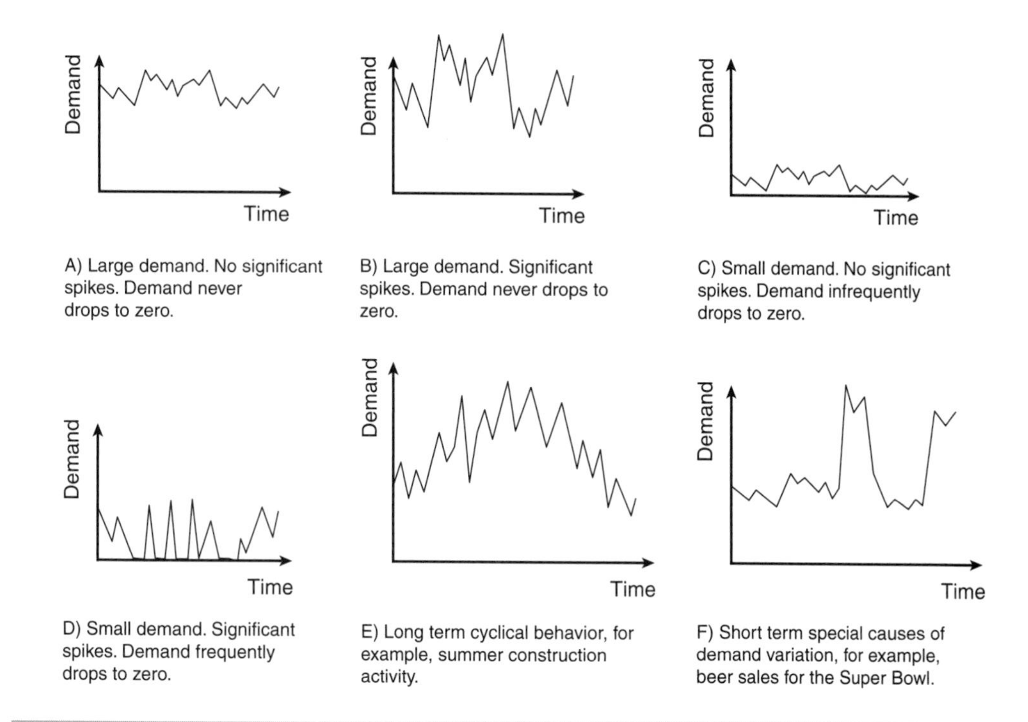
Demand
Time
A) Large demand. No significant spikes. Demand never drops to zero.
Demand
Time
B) Large demand. Significant spikes. Demand never drops to zero.
Demand
Time
C) Small demand. No significant spikes. Demand infrequently drops to zero.
Demand
Time
D) Small demand. Significant spikes. Demand frequently drops to zero.
Demand
Time
E) Long term cyclical behavior, for example, summer construction activity.
Demand
Time
F) Short term special causes of demand variation, for example, beer sales for the Super Bowl.

Figure 7.15.2 Interpreting the Demand Profile.

The value that Demand Profiling brings above and beyond *Demand Segmentation* is that cyclical patterns become visible. Rather than just knowing that there is variation, the Belt obtains an understanding of how the demand is varying. After this is understood, the process can be laid out and resourced accordingly. For example, if there are significant spikes in demand at the end of each day:

- More staff could be used at that point (effectively increasing capacity)
- Inventory could be built ahead of time to serve the demand
- Later staff hours could be used to spread the demand
- Customers could be encouraged not to batch their demand and perhaps spread it to earlier hours, and so on

However, the demand is dealt with from the Demand Profile, without visibility of the variation from the graph, the process is always at its mercy.

OTHER OPTIONS

Although Demand Profiling is simply the use of the graphical tool the Time Series Plot, an obvious extension is the application of statistical analysis to the plot data. Analysis of this form relies on interpolation techniques similar to *Regression*. Although extremely valuable as an enhanced version of Demand Profiling, it is a difficult and complex subject considered outside the skill set of Black Belts and Green Belts and is considered beyond the scope of this book.

In simple terms the interpolation techniques break down the time series data into its component parts, namely:

- Current Level—The mean value at the current time
- Trend—The rate of systematic increase (or decrease) in the mean value
- Seasonal Pattern—A recurring periodic pattern
- Random Component—The portion of behavior that remains unaccounted for after the Current Level, the Trend and the Seasonal Pattern have been identified.[25]

16: DEMAND SEGMENTATION

OVERVIEW

Typically, when examining Customer and Market demand, the approach is to calculate the figures based on averages. In reality this gives only half of the picture, because *variation* in demand can have as big (and in fact usually much greater) an effect on a process. Demand Segmentation[26] examines both the average and the variation in demand in one graph, as in Figure 7.16.1.

Products in the top left corner of the graph exhibit smooth high volume demand. Products in the bottom right-hand corner exhibit variable, low volume demand. Most production planning, forecasting, and process leveling approaches mistakenly consider these in the same way, which misses opportunities to create more predictable, responsive processes and reduce inventory all at the same time.

[25] For further reference see *Forecasting: Methods and Applications (3rd Edition)* by S. Makridakis, S.C. Wheelwright, and R.J. Hyndman.

[26] For more detail see *Manufacturing for Survival: the how to guide for practitioners and managers* by Blair R. Williams.

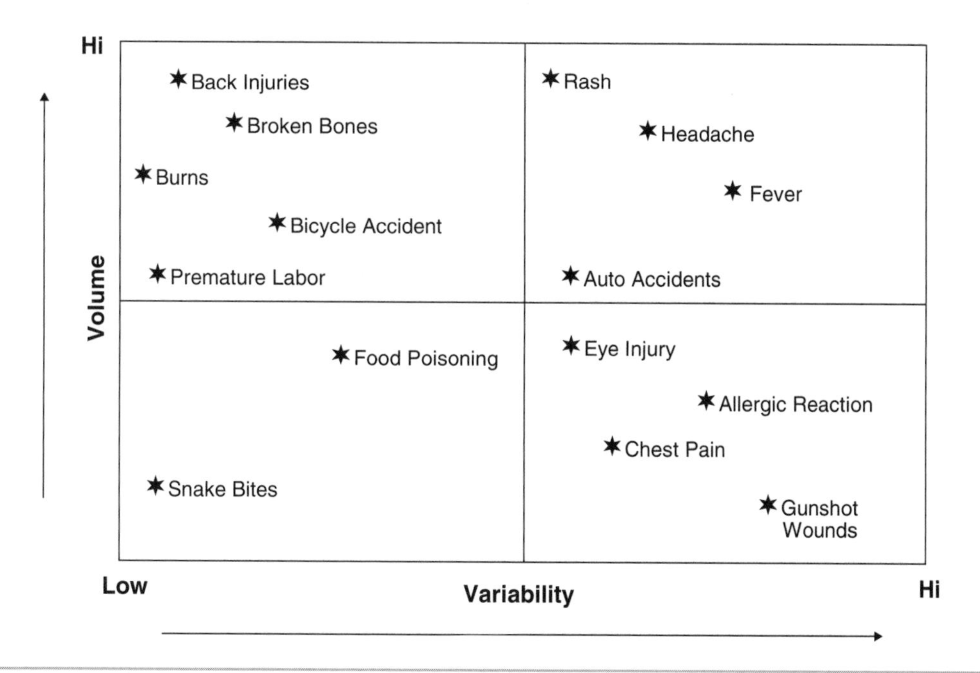

Figure 7.16.1 Volume of Demand versus Variability in Demand.[27]

As I tell Belts in my classes, the statistics tools are great, but Demand Segmentation is probably the most powerful tool in the Lean Sigma arsenal.

Project examples include

- Industrial—Product rationalization, raw material rationalization, warehouse stocking
- Healthcare—On-floor medications inventories
- Service/Transactional—Product rationalization

Logistics

As a simple analysis tool this can be applied by the Belt without the rest of the Team; however, the data might have to come from multiple sources and often requires Team involvement to collect it. The analysis itself is done entirely in a spreadsheet, such as Excel, and can be done in a matter of a few minutes after the data is in the correct format.

Initially, data is historical, but after the organization understands the application of the tool, forecast data can also be used.

[27] Source: SBTI's Lean Sigma for Healthcare training material.

Roadmap

Step 1: Identify which entity types are to be examined. Demand Segmentation works best with at least five entity types to make the graph sensible. If fewer entity types are examined, it is probably best to look at a simple *Demand Profile* for each, which gives more detail.

Data is captured for each entity type separately. Identifying the entity types is not as easy as it first appears. The trick is to ask what problem is to be resolved. For example, in Production Planning there might be 25 different Customer products and you would like to schedule production based on their demand. At first glance, you might jump to the conclusion that there are 25 different entity types. Perhaps the 25 products are only six differently labeled manufactured products for different Customers. If you look at Production Planning, you would choose to use the six entity types rather than the 25, because those are the products you manufacture. However, for the same problem, if you look at Finished Goods rationalization in the warehouse and products are stored there pre-labeled, then the 25 entity types would probably make sense. The valuable tool "*Why do we care?*" helps us here.

Step 2: Identify the demand buckets, for example, days in a month or weeks in a year, depending on the drumbeat of the demand. It is necessary to have at least 25 buckets of captured demand to get good measures of the average and particularly variation. The buckets themselves have to be meaningful and often it is useful to take the typical replenishment cycle as the driver. For example, in hospital care units, drug inventories are kept on the floor in electronic vaults with the medications being dispensed directly from them. It is obviously important not to run out of medications, and volume and variation in demand are key to calculating how much to put in there. Therefore, if the vaults were replenished once daily then a day would be a reasonable time bucket for the segmentation.

Step 3: Data is collected from the downstream Customer for different entity types across the time buckets, as in Table 7.16.1. The most common mistake is to look to the Process Planning group for when we *decided* to make the entity, not when it was actually demanded by the Customer. Customer demand rates are typically much smoother than we care to admit, and in fact, their usage rates are even smoother. Internally, we tend to batch entity processing into large lots, which we make on an infrequent basis; so it shows much higher variation in demand than is actually there. For Demand Segmentation, we need to look at *demand* patterns, not our own planned process patterns.

Examples include

- Healthcare floor medication inventories—Consider actual pulls from the vault
- Raw Material segmentation—Production usage of the raw materials
- Production segmentation—Customer demand of manufacturing codes (not product codes)
- Product portfolio—Customer demand of final product codes

Table 7.16.1 Capturing and Analyzing the Data

Product Line X	Week 1	Week 2	Week 3	Week 4	Week 5	Mean	S	CV
Product A	1700.00	9.00	230.00	1.00	10.00	390.00	660.69	1.69
Product B	420.00	333.00	380.00	550.00	390.00	414.60	73.24	0.18
Product C	10.00	1.00	3.00	0.00	2.00	3.20	3.54	1.11
Product D	7.00	5.00	3.00	1.00	4.00	4.00	2.00	0.50
Demand	2137.00	348.00	616.00	552.00	406.00	811.80	739.48	0.91

Step 4: After you have the demand data for each entity type across multiple time buckets, you can calculate the mean of the demand and the coefficient of variation (COV), defined as

$$\text{Coefficient of Variation (COV)} = \frac{\text{Standard Deviation of Demand}}{\text{Mean of Demand}} = \frac{s}{\overline{X}}$$

The COV is used rather than the standard deviation because it is the size of variability *relative* to the total level of demand that is important, rather than the size of the variation itself. For example, if there is demand for an entity type from the Customer with a mean of 1,000,000 units per month, then a standard deviation of 50 units is neither here nor there and the demand is considered to be smooth. However, if the mean of the demand is only 100 units and the standard deviation is the same 50 units, then the demand cannot be considered as smooth. The general rule here is that a COV less than 1 is considered to be very smooth. A COV greater than 2 is highly variable.

The mean and the COV are calculated for each entity type as shown in Table 7.16.1. A Total demand calculation is also shown in the table, which gives an indication of the total variability in demand seen by the process. However, at this stage the important numbers to examine are for the individual entity types.

Step 5: From the data table, create a graph similar to the one shown in Figure 7.16.1, taking the mean of demand as the x-axis and the COV of demand as the y-axis. Each entity type has its own point on the graph.

INTERPRETING THE OUTPUT

Interpreting the Demand Segmentation graph depends heavily on application but can be addressed by breaking the graph into zones, as shown in Figure 7.16.2.

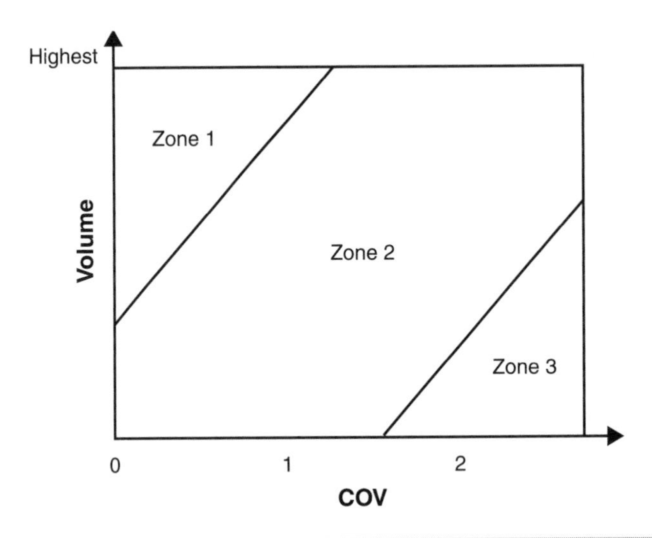

Figure 7.16.2 Demand zones.

Application: Replenishment

Demand Segmentation can be applied to materials management and specifically replenishment as replenishment to a Customer or replenishment of materials internally or from a Supplier—for example, the delivery of product to a Customer, a materials delivery to a line, or a medications delivery to a Care Unit. In this case, the zones would be treated as follows:

- **Zone 1—Deliver direct to line.** High volume, low variation materials usage is so smooth that it allows us to add service value to the Customer by managing their inventory for them. Materials would be delivered on a rate-based system directly to the point of use (POU). There is probably no need to keep anything more than a small buffer of Finished Goods at the end of the supplying process for these types of entities. Payment could be based on a Blanket Order and then Call Offs made as entities are used.
- **Zone 2—Pull System from Customer.** For the middle majority, simple pull triggers from the Customer would allow replenishment when needed (see "Pull Systems and Kanban" in this chapter). If the Global Process Lead Time for entities in this Zone were too long (for example, it is impossible to process the entity from scratch and get it to the Customer POU within acceptable timeframes), then a solution would be to replenish from stock, by keeping a small inventory of Finished Goods at the end of the supplying process. See also "Time—Global Process Lead Time" in this chapter.

- **Zone 3—Make (or Deliver) to order.** High variation, low volume gives such unpredictable demand that it forces the Supplier to deliver only when there is an order. If the Global Process Lead Time for the supplying process is short enough, then there isn't too much of a problem. However, if the Global Process Lead Time is beyond acceptable bounds for delivery Lead Times, then either:
    - The supplying process (or the POU) has to stock inventory, which is probably highly unpalatable
    - Work should be done quickly to reduce the Global Process Lead Time
    - The validity of having the entity types in the portfolio should be questioned (see the next application section)
    - Or the worst case is that the promised delivery time needs to be extended with the Customer

Application: Product Portfolios

Demand Segmentation can be applied to the Product Portfolio to identify opportunities for rationalization and to validate the value in the portfolio. In this case the Zones would be treated as follows:

- **Zone 1—Dedicated Business Unit.** Products in this Zone have high-volume, low variation demand and could be managed independently, either by dedicating a small internal group to manage them or to spin them off as a whole new Business Unit. If the entity types in question have reasonable margin, then these businesses often are known as the "Cash Cows."
- **Zone 2—Majority of portfolio.** Products in this Zone are probably best left as they are. There is always opportunity to rationalize in this majority, but they typically aren't a primary focus.
- **Zone 3—High-value niche products.** Unless they are high-value products, the validity of keeping low volume, highly variable products in the portfolio is questionable. This is the first portfolio area to look to for rationalization or obsolescence opportunities. Other opportunities here might be to combine products in these Zones from multiple operating sites and make them just at one site, thus creating an elevated volume at the single site (and usually a reduced variability too, due to the Central Limit Theorem) and freeing up the other sites from the burden.

Application: Production or Operations Planning

Demand Segmentation when applied to Production or Operations Planning allows processing of the entity types in the different Zones to be planned more effectively, as follows:

- **Zone 1—Repetitive flow rate-based scheduling.** Entity types that have a large, smooth demand do not need to be scheduled individually, they can be rate-based, so that during each time-period a consistent amount is processed with the knowledge that the Customer/Market uses that amount. Slight variation in demand is catered for by using small buffers of inventory, preferably at the POU. For example, in the light bulb manufacturing industry, there is a consistent high level of demand to which the processes are paced. No one in Operations Planning for those companies takes all the orders from DIY and hardware stores on a daily basis and determines what should be made during that shift.
- **Zone 2—Hybrid control Pull System.** For the middle majority of entity types in this Zone, Operations can be successfully governed using internal Pull Systems, to minimize work in process (WIP) inventory.
- **Zone 3—Discrete job order.** For entity types in this Zone, the orders are few and far between and highly variable, so it makes little sense to process them without an order or Customer request.

17: DOE—Introduction

Overview

The Lean Sigma roadmaps are primarily based on understanding the key Xs in a process and understanding how they drive the performance of the process through the equation $Y = f(X_1, X_2,..., X_n)$, where Y is the performance characteristic in question. Early tools (*Process Variables Map, Fishbone Diagram*) in the roadmap identify all the Xs and then subsequent tools narrow the Xs down, first through the application of Team experience (*Cause & Effect Matrix, Failure Mode & Effects Analysis*) and then through the analysis of data sampled passively from the process (*Multi-Vari Study*). At this point, the likely key Xs have been identified and what is required is a tool or series of tools to determine exactly what the equation $Y = f(X_1, X_2,..., X_n)$ is. This is done in Lean Sigma with Design of Experiments (DOE), where the Xs are actively manipulated and the resulting behavior is modeled using mathematical equations to determine which Xs truly are driving the Y or Ys and how. This active manipulation probably includes stopping the line to conduct the DOE, and it could mean that the process generates scrap or lower grade output for some experiments. Clearly the Process Owner(s) need this explanation, and the whole DOE activity requires considerable organizational support.

The biggest mistake Belts make with Designed Experiments is to apply them too early in the project—by far the best approach is to follow the roadmap! By validating the measurement system, identifying all the Xs and then narrowing them down using Team experience and passive data collection analysis (*Multi-Vari*) significantly reduces wasted effort at the experimental stage. Too often Teams sit and brainstorm Xs to try in the Designed Experiment, rather than letting the roadmap guide them to the best Xs—trying to guess important Xs is an almost surefire way to miss an X and expend time and money investigating unimportant Xs!

Designed Experiments are not a single tool, but rather a series of tools falling into three main categories:

- Screening Designs—Used to reduce the set of Xs to a manageable number. These designs are purposely constructed to be highly efficient, but in being so, they give up some (or in some cases a great deal of) information content.
- Characterizing Designs—After a manageable number of Xs are identified (either from earlier Lean Sigma tools or from a Screening Design), these Designs are used to create a full detailed model of the data to understand which Xs contribute what and any interactions that might occur between Xs. Characterizing Designs are less efficient than Screening Designs.
- Optimizing Designs—After the Xs are characterized, the Team might want to optimize the process to gain the best performance. These Designs require many data points, but provide detailed insight into the behavior of the process. Optimization is really an optional element in many projects. Often a solid enough result is gained from a Characterizing Design and tuning to a higher degree is not required or not worth the effort that would be expended.

It is highly recommended that before performing a Designed Experiment of any form, the Belt read this section fully, followed by "DOE—Characterizing," then "DOE—Screening" (Screening Designs are a derivative of Characterizing Designs, so that order makes most sense), and finally "DOE—Optimizing" all in this chapter. This gives enough grounding in the bigger picture of DOE use and ensures that the Belt isn't "dangerous."

DOE is a broad subject area and fortunately many good texts are available covering the theory and some of the practical considerations.[28] Before a Belt or Team performs the first DOE, it is highly recommended that they receive input from an expert, such as a Consultant or Master Black Belt, just to validate their approach.

The key to successful use of DOE in Lean Sigma is to have good planning and to keep the application practically based. As with all tools, some theory does help to understand what the tool does, and more importantly how to apply it—remember, the aim here is to understand the practical implications, so the theory here is minimal. In DOE, the objective is to create a good model of data points collected from the process and then make practical conclusions based on the model. Models are generally of the form

[28] A reference here is *Statistics for Experimenters* by Box, Hunter & Hunter.

$$Y = \beta_0 + (\beta_1 X_1 + \beta_2 X_2 + \beta_3 X_3) + (\beta_{12} X_1 X_2 + \beta_{13} X_1 X_3 + \beta_{23} X_2 X_3) + \beta_{123} X_1 X_2 X_3$$

Where the βs are coefficients in the equation, the Xs are the process Xs, and the Y is the performance characteristic in question. Two Xs appearing together (in fact multiplied together in the equation) represent an interaction in the process between those two Xs. The effect of one X changes the effect of another X. A simple example of this might be time and temperature in an oven. For low temperatures, the time needs to be high, but for high temperature, the time should be low. Interactions between two Xs are extremely common in most industrial processes and common in processes in general. Three Xs appearing together represent an interaction between three Xs. This can be explained as the interaction of two of the Xs changes depending on the value of the other X. Interactions between three Xs are rare, less than one in 500, and if they do occur they tend to be in processes where some chemistry is involved.

The preceding equation might look fairly abstract and probably a little alien to some readers. An example might be for three Xs in an Industrial process:[29]

- Y: Yield
- X_1: Temperature
- X_2: Pressure
- X_3: Volume

The equation would be

$$\text{Yield} = \beta_0 + (\beta_1 \text{Temp} + \beta_2 \text{Press} + \beta_3 \text{Vol}) + (\beta_{12} \text{Temp} \times \text{Press} + \beta_{13} \text{Temp} \times \text{Vol} + \ldots$$

At this point the equation can represent any data points captured for the Y and the Xs. A sample of reality is captured from the process and from the data the βs would be calculated. There are eight βs in the equation, so at least eight data points are required to calculate these "unknowns." In theory any eight data points (within reason) would suffice, but there are configurations of data points that give the best information content for the expended effort. These tend to form common geometric shapes, such as squares and cubes, as shown in Figure 7.17.1. For eight data points, a cube would give an efficient means of getting information on the relative change in three Xs. Two settings would be selected for each X, such as high and low, and all combinations of highs and lows for each X become the eight data points. All the data points must be captured (don't stop when a reasonable result is gained) or the analysis cannot be completed. This is the second big mistake in DOE, stopping before all the data is collected.

---

[29] DOE is equally at home in healthcare, service, and transactional processes.

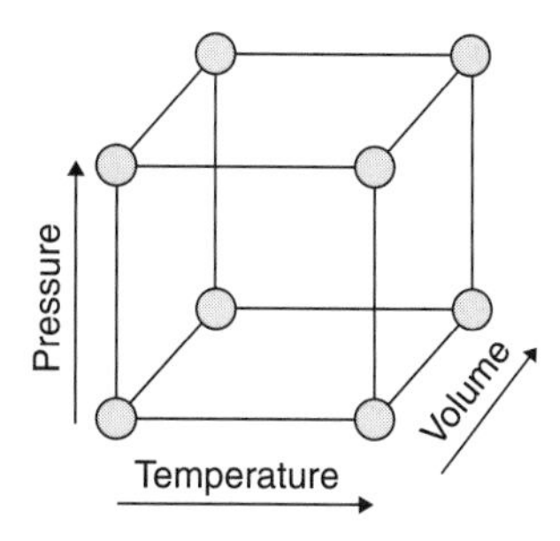

Figure 7.17.1 An example of a data point configuration for three Xs.[30]

If additional data points were captured (above and beyond the eight in the cube) then either:

- A higher degree of confidence is gained in the values of the calculated βs that represent the size of the effects of each X or interaction between Xs.
- A higher order mathematical model (quadratic, cubic, and so on) could be used to represent curvature rather than straight lines. This is covered in detail in "DOE—Optimization" in this chapter.

Some βs might even be zero or so small they are considered negligible. After the βs are calculated and substituted in the equation, it might look something like:

$$\text{Yield} = 67.22 + (8.5 \times \text{Temp} - 3.75 \times \text{Press} + 0.003 \times \text{Vol}) + (0.0054 \times \text{Temp} \times \text{Press} + \ldots$$

Now, if the actual values of temperature, pressure, and volume are known, they can be substituted in the equation and the associated actual yield is predicted.

The equation itself gives an indication of the most influential Xs and interactions, because they have the largest coefficients (βs), negative or positive.

Fortunately, statistical software packages do the hard analysis work and calculate what the βs are and the associated equation. All that the user needs to provide is their selection of design, what the Y and Xs are called, along with the data points collected from the process. The word "all" in the previous sentence could be entirely misleading—that is all that is required to conduct an *analysis*, DOE is much more involved as is described in this section. Another mistake in DOE is to assume that it is merely a case of grabbing a few data points and calculating an equation—rigor is everything.

DOE is thus a systematic (structured design) series of tests (data points) in which various Xs are directly manipulated and the effects on the Ys are observed. This helps determine

[30] This type of point configuration is typical of a Full Factorial Design and is described in "DOE—Characterizing" in this chapter.

- Which Xs most affect Ys
- Which Xs are influential in minimizing the presence of Noise Variables

DOE has its own terminology which can cause some confusion with novice Belts, as follows:

- Ys are known as Responses
- Xs that can be controlled are known as Factors
- Xs that cannot be controlled are known as (Noise) Variables

DOE is not just a single tool, and in fact, it is best not to conduct a single experiment to understand a process. The best approach is to use what is known as sequential experimentation. Simple experiments are conducted first, and after completed, the information they generate is used to plan the next experiment. Table 7.17.1 shows the typical sequence of experimentation.

Sequential experimentation is the most efficient way to understand a process. Belts often make the mistake of trying to answer all of the questions in the first experiment and construct a large complex (and expensive) design. As can be seen in the second step in Table 7.17.1, the design could be in entirely the wrong region, which would make any points above and beyond the simplest design a waste of resources. The general rule here is only expend 25% of the experimental budget in the first experiment.

As mentioned previously, good planning is crucial in DOE. A well-designed experiment eliminates all possible causes except the one being tested. If a change occurs on the Y, then it can be tied directly to the Xs that were directly manipulated. With this in mind, any Noise Variables in the process must be controlled or accounted for so that they don't interfere with understanding the key controllable Factors (Xs). Noise Variables can be managed by

- Randomizing experimental runs—This smears the effect of the Noise Variable across the whole experiment, rather than letting it concentrate on an X that would interfere with understanding.
- Try to keep the Noise Variables constant—Many Noise Variables can be held constant for a short period of time during which the data points are collected (run it on one machine, in a single day or shift and use the same operators throughout).
- Blocking—Make the Noise Variable part of the experiment by letting it be another X in the Design. For example, it is possible to add day, shift, or batch as a Factor to the experiment by replicating the experiment each day or shift.

Table 7.17.1 Order of Sequential Experimentation[31]

Step	Diagram
1) Start with a Screening Design. These are efficient Designs and allow the Belt to eliminate some Xs before examining the remainder in detail. The Screening Designs recommended here are known as Fractional Factorials; instead of conducting all of the combinations of high and low, a partial fraction (a half or quarter) is completed. For more detail see "DOE—Screening" in this chapter.	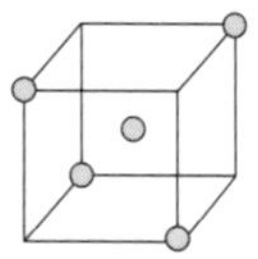
2) Move the study to a new location. Often, the initial study indicates that the experiment was focused in entirely the wrong region. The first study guides the direction to move. Sometimes the second study directs a subsequent region change.	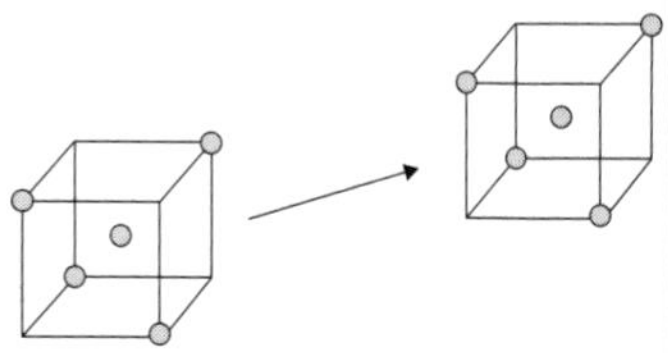
3) Add runs to the Screening Design to isolate ambiguous effects. A Screening Design is efficient and does not give major insight into the Xs, it just guides which Xs to keep. To understand more, the partial fraction should be made up to the whole set of combinations. It becomes known as a Full Factorial. For more detail see "DOE—Characterizing" in this chapter.	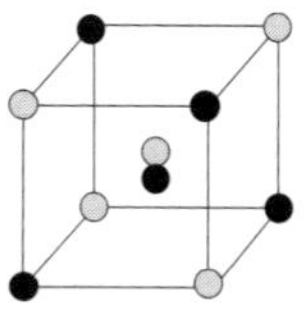
4) Rescale the levels. If the Y does not change significantly across the range of an X, it might be appropriate to stretch the position of the data points (known as Levels) apart.	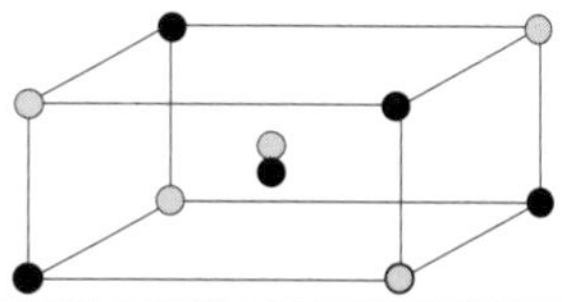
5) Drop and add Factors. This really is not recommended in Lean Sigma, because the Xs should come directly from previous tools. If an X should be considered, it should be included in an earlier Screening Design.	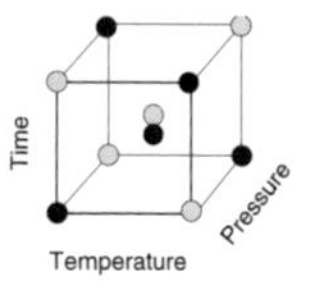 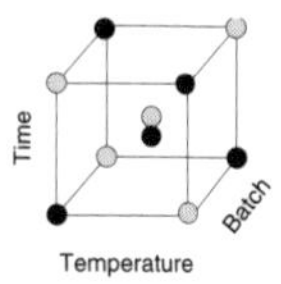 Time, Temperature, Pressure; Time, Temperature, Batch
6) Replicate runs. If a longer term understanding of the process is required, the approach is to replicate the whole experiment or some of its runs at a later time.	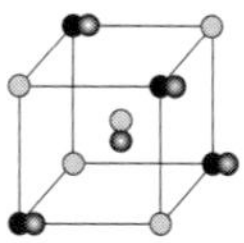
7) Augment designs with additional points. This changes the Design from a Characterizing Design to an Optimizing Design. The additional points allow the investigation of curvature in the solution space to identify peaks and troughs.	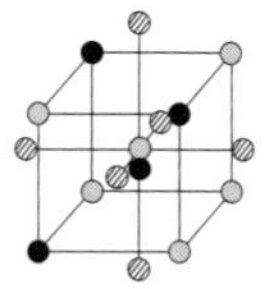

[31] Adapted from SBTI's Lean Sigma training material.

Other techniques used in DOEs are known as Repetition and Replication, which help provide estimates of the natural variability in the experimental system:

- Repetition—Running several samples during one experimental setup run (short-term variability)
- Replication—Replicating the entire experiment in a time sequence (long-term variability)

Both can be used in the same experiment, but both require additional data points to be run and can quickly push up the size and associated cost of the experiment.

In terms of understanding process behavior, both Repetition and Replication push the experiment away from being a snapshot in time to more of a longer term; obviously Replication to a greater degree. The snapshot study forms what is known as Narrow Inference, where the focus is only on one shift, one operator, one machine, one batch, and so on. Narrow inference studies are not as affected by Noise variables.

Broad Inference on the other hand usually addresses entire process (all machines, all shifts, all operators, and so on), and thus, more data is taken during a longer period of time. Broad Inference studies are affected by Noise variables and are the means of taking the effect of Noise Variables into account.

It is better to perform Narrow Inference studies initially until the controllable Xs are understood and then follow these with Broad Inference studies to understand the robustness of the process.

LOGISTICS

Planning, designing the experiment, and data collection require participation of the whole Team. The planning and design can usually be achieved in a 1–2 hour meeting. Data captured depends heavily on the process complexity and time taken to measure the Y. The Belt alone can conduct the analysis, but the results should be fed back as soon as possible to the Team.

ROADMAP

The roadmap to planning, designing, and conducting a sequence of Designed Experiments is as follows:

Step 1: Review initial information from previous studies and Lean Sigma tools. Collect all available information on key Xs and performance data for Ys. A Capability Study for the Ys should be conducted prior to this point.

Step 2: Identify Ys and Xs to be part of the study. These should not just be brainstormed at this point. The Ys should have come from earlier VOC work, specifically from the *Customer Requirements Tree* (see "KPOVs and Data" in this chapter). The Xs should come from earlier Lean Sigma tools, which focus on narrowing them. DOE is an expensive and time-consuming way to examine guessed Xs.

Step 3: Verify measurement systems for the Ys and Xs. The Ys should have been examined before using a Capability Study. The Xs might have been examined earlier, if not then ensure that data on Xs is reliable (see "MSA—Validity," "MSA—Attribute," and "MSA—Continuous" in this chapter).

Step 4: Select the Design type based on the number of Xs. This generally follows the path of sequential experimentation (refer to Table 7.17.1), so in the early stages, perform more efficient screening designs to get an initial idea of the important variables before moving to more complex detailed designs. Don't try to answer all the questions in one study, rely on a series of studies. Two-level designs are efficient designs and are well suited for Screening and Characterizing in Lean Sigma projects.

Current baseline conditions or the best setup should be included in the experiment by the use of Centerpoints, for example, points in the body-center of the cube for a three Factor Design.

For each experiment, it is important to either incorporate or exclude Noise Variables. This can be done with randomization or by blocking the Noise Variable into the experiment (as explained in "Overview" in this section).

The Design of the experiment is heavily affected by its purpose. Determine up front what inferences are required based on the results. A certain amount of redundant data points help provide a better estimate for error in the analysis and give greater confidence in the inferences. To gain redundancy, it is necessary to either repeat each run or replicate the experiment. Repetition allows some short-term variation to creep in; Replication allows longer-term variation. Sample size is a key consideration in determining how many redundant data points are included. A good approach is to include about 25% Repeated or Replicated points. It is usually best to keep redundancy reasonably low until the design appears to at least be in the right region of the solution space.

Step 5: Create and submit a DOE Proposal to gain signoff to proceed. DOE runs require time and money to conduct, and the decision to run them should be made in the same way any other business spending is considered. The Champion and Process Owners want to understand

- The measurable objective
- How the DOE results tie into Business Metrics
- The experimentation cost
- Involvement of internal and external Customers

- The time needed to conduct the runs
- What analysis is planned
- What was learned and subsequently remedied from the pilot runs

The simplest way to communicate this information to the appropriate parties is to create a DOE Proposal that contains the following:

- Problem Statement
- Links to a major business metric or metrics (sometimes known as a strategic driver or business pillar)
- Experiment objective(s)
- Ys
- Measurement System Analysis results for Ys
    - Accuracy
    - Repeatability and Reproducibility
- Xs
    - Levels for Controllable Factors
    - Noise Variables and how they are to be controlled
- Overview of the Design selected and the data collection plan
- Budget and deadlines
- Team Members involved

Step 6: Assign clear responsibilities for proper data collection.

Step 7: Perform a pilot run to verify and improve data collection procedures.

Step 8: Capture data using the improved data collection procedures. Identify and record any extraneous sources of variation or changes in condition.

Step 9: Analyze the collected data promptly and thoroughly, for more detail see "DOE—Screening," "DOE—Characterizing," and "DOE—Optimizing" in this chapter. Analysis generally follows the following path:

- Graphical
- Descriptive
- Inferential

Typically, analysis initially involves considering all Factors and Interactions and then eliminating unimportant elements and rerunning the reduced model. This procedure is repeated until an acceptable model is developed.

Step 10: Always run one or more verification runs to confirm the results. It is often useful to go from a Narrow to Broad inference study (short-term to longer-term). More noise is apparent in the longer-term study so be ready to accept changes.

Step 11: Create a DOE final report, which typically contains

- Executive Summary or Abstract
- Problem Statement and Background
- Study Objectives
- Output Variables (Ys)
- Input Variables (Xs)
- Study Design
- Procedures
- Results and Data Analysis
- Conclusions
- Appendices
  - Detailed data analysis
  - Original data if practical
  - Details on instrumentation or procedures

Step 12: Plan the next experiment based on the learning from this experiment. The best time to design an experiment is after the previous one is finished. The roadmap does not have to be completely restarted from Step 1, because the objectives and so on remain the same. Go to Step 4 and repeat Step 4 through Step 12 until the Team thinks the solution is in hand. Sequential experimentation steps might include (as per Table 7.17.1)

- Adding runs to isolate any ambiguous (confounded) effects
- Folding over to complete the next runs of a factorial
- Constructing the full factorial
- Moving the factorial toward the optimum
- Performing a response surface to identify the optimum process window

In general, the approach is to move from simple, efficient designs to more detailed designs involving more redundant points and from first order to second order models.

OTHER CONSIDERATIONS

Management plays a key role in any DOE work, but the DOE approach often presents difficulties to traditional management styles. As a Team, it is important to help managers understand

- Focus should be on the DOE process, which might not deliver the solution after the first experiment.
- Progress depends on previous experiments and the approach is not linear. Improvements cannot be made until the whole path has been followed.
- It is difficult to set deadlines for results when the exact path for the sequential experimentation will not be known from the outset.
- Focus should be on the timely completion of sequential studies and trust that the results will come.
- Systematic experimentation should be encouraged and expected.
- Experiments increase knowledge but do not always solve the problem.
- Experiments stimulate unconventional thinking.
- Experimentation is a catalyst for engineering, not a replacement.

Given the complexity of the tool, it is no surprise, that there are a number of ways that it can go wrong:

- Factor levels are put too close together or too far apart. If the high and low values for the Xs are too close together then no effect might be seen for that X and its importance could be incorrectly discounted. If the levels are chosen too far apart, then it is possible that a significant area of interest is not examined between the levels and again the X might be incorrectly discounted. Choosing the levels is discussed more in "DOE—Screening," "DOE—Characterizing," and "DOE—Optimizing" in this chapter.
- Nonrandom experiments produce spurious results. Randomization spreads the effect of Noise Variables across the whole experiment and prevent them being associated with just one X.
- The Sample Size is too small. The recommendation is to replicate at least 25% of the data points to gain a good understanding of experimental error and more confidence in the significance of the Xs.
- Measurement Systems are not adequate. This is typically the biggest mistake Belts make. If a Measurement System has too much noise in it, it can completely overshadow important X effects. All Measurement Systems (specifically on the Ys) need to be validated as capable before any DOE work commences.

- Measurement systems do not measure what the Team thinks they do.
- No pilot run is done and experimental runs are flawed because of lack of discipline.
- Confirmation runs are not done to verify results. Even the best Belt can analyze data incorrectly. Small-scale confirmation runs ensure no full-scale mishaps occur.
- Data or experimental entities are lost. Be sure to keep track of all data and samples and get any data into electronic form and backed up as soon as possible.

18: DOE—SCREENING

OVERVIEW

Screening a process using Designed experiments involves

- Determining which Xs (both controlled and uncontrolled) affect the Ys
- Eliminating Xs that do not have a large effect from subsequent investigation

Before reading this section it is important that the reader has read and understood "DOE—Introduction" and "DOE—Characterizing" in this chapter. This section covers only designing, analyzing, and interpreting a Screening Design, not the full DOE roadmap.

Screening Designs in Lean Sigma are based on the same set of experimental designs as used in Characterizing Designs known as 2-Level Factorials. These are chosen because they

- Are good for early investigations because they can look at a large number of factors with relatively few runs
- Can be the basis for more complex designs and lend themselves well to sequential studies
- Are fairly straightforward to analyze

Full-Factorial Designs are efficient designs for a small to moderate number of Factors; however, they quickly become inefficient for more than five Factors, as shown in Table 7.18.1.

The solution here is to consider the model underpinning any experimental design. For a 3-Factor design the model is

$$Y = \beta_0 + (\beta_1 X_1 + \beta_2 X_2 + \beta_3 X_3) + (\beta_{12} X_1 X_2 + \beta_{13} X_1 X_3 + \beta_{23} X_2 X_3) + \beta_{123} X_1 X_2 X_3$$

Table 7.18.1 Relationship between Number of Factors and Number of Experimental Runs

Factors (k)	Runs (2^k)
2	4
3	8
4	16
5	32
6	64
7	128
8	256
9	512
10	1024
15	32768

For more factors the model would include 4-Factor Interaction terms and higher. 3-Factor Interactions are incredibly rare and 4-Factor Interactions and higher are essentially non-existent. This effectively means that a large number of runs (data points) in the experiment are being used to calculate terms in the model that really have no practical value. A Screening Design takes advantage of this by assuming higher order interactions are negligible, and it is possible to do a fraction of the runs from a Full Factorial and still get good estimates of low-order interactions. By doing this, a relatively large number of Factors can be evaluated in a relatively small number of runs.

Fractional Factorials, as they are known, are based on the following:

- **The Sparsity of Effects Principle.** Systems are usually driven by Main Effects and low-order interactions.
- **The Projective Property.** Fractional Factorials can represent full-factorials after some effects demonstrate weakness and their terms are eliminated from the model.
- **Sequential Experimentation.** Fractional Factorials can be combined into more powerful designs; for example, Half-Fractions can be "folded over" into a Full Factorial.

The key to all Designed Experiments is the relationship between the data points chosen and the model generated, effectively the structure of the design. Fractional Factorials are no exception to that. If only a fraction of the runs are needed, which is the best fraction to consider? This is best demonstrated through example.

Suppose you want to investigate four Input Variables (typically 16 runs as a Full factorial) but cannot afford to do any more than eight runs, which is half as many runs. The Full Factorial associated with eight runs contains three Factors, so in some way you have to insert the fourth Factor without compromising the structure of the Design. The Design Matrix for three factors is shown in its entirety in Figure 7.18.1, including all the interaction terms. Full Factorials are powerful and efficient because it is possible to separate the effect of every X and all the interactions from one another with the Design Matrix shown. If the data shows a result due to an X or Interaction, then it can be attributed directly to that X or Interaction. The property that allows this to occur is known as orthogonality—every column of the matrix is orthogonal to every other column.[32] The fourth Factor needs to be added to a column that is orthogonal to all the others. Unfortunately, given the super-efficiency of Full factorials, there are no other orthogonal combinations of +1s and −1s. To add the fourth Factor it has to be put into a column that is already in use. In theory, higher order Interactions are rarer than low order Interactions and Factors, so the fourth Factor is placed in the column representing the highest order interaction, the 3-Factor Interaction AxBxC.

A	B	C	A × B	A × C	B × C	A × B × C
−1	−1	−1	1	1	1	−1
1	−1	−1	−1	−1	1	1
−1	1	−1	−1	1	−1	1
1	1	−1	1	−1	−1	−1
−1	−1	1	1	−1	−1	1
1	−1	1	−1	1	−1	−1
−1	1	1	−1	−1	1	−1
1	1	1	1	1	1	1

Figure 7.18.1 Design Matrix for 3-Factor Full Factorial.

In this case, when the AxBxC Interaction is replaced with Factor D, the expression is that AxBxC is aliased with D or confounded with D. If the data shows that D is an important Factor, it is impossible to know whether it is D that is important or the 3-Factor

[32] No single column can be expressed as multiples of sums of other columns. In effect the columns are at "90 degrees" to one another. Common graphical axes, X, Y, and Z (known as Cartesian coordinates), are orthogonal because there are no multiples of lines in the X or Y directions that add up to a line in the Z direction.

interaction AxBxC. As mentioned before, 3-Factor Interactions are extremely rare; therefore, the assumption is the effect would be due to Factor D.

This is a half-fraction of a 4-Factor design because instead of using $2^4 = 16$ runs, the Design uses only eight runs to evaluate four factors. This improvement in efficiency is incredibly useful, but there is a cost—the higher order interaction is no longer available. When assessing what has been lost, the measure is known as Resolution:

- Resolution III Designs
  - No Factors are aliased with other Factors
  - Factors are aliased with 2-Factor Interactions
- Resolution IV Designs
  - Factors are aliased with 3-Factor Interactions
  - No Factors are aliased with other Factors or with 2-Factor Interactions
  - 2-Factor Interactions are aliased with other 2-Factor Interactions
- Resolution V Designs
  - Factors are aliased with 4-Factor Interactions (considered to be non-existent)
  - 2-Factor Interactions are aliased with 3-Factor Interactions

The preceding half-fraction 2^4 Design is considered a Resolution IV design. Any Screening Design that is Resolution V and above is a good reliable Design and does not cause any problems during analysis. Some care must be taken when using Resolution III and IV Designs.

After the Design has been chosen, the analysis is almost indistinguishable from the analysis of a Full Factorial Design.

ROADMAP

The roadmap to designing, analyzing, and interpreting a Screening DOE is as follows:

Step 1: Identify the Factors (Xs) and Responses (Ys) to be included in the experiment. Each Y should have a verified Measurement System. Factors should come from earlier narrowing down tools in the Lean Sigma roadmap and should not have been only brainstormed at this point.

Step 2: Determine the levels for each Factor, which are the Low and High values for the 2-Level Factorial. For example, for temperature it could be 60° C and 80°C. Choice of these levels is important because if they are too close together, then the Response might not change enough across the levels to be detectable. If they are too far apart, then all of the action could have been overlooked in the center of the design. The phrase that best sums up the choice of levels is "to be bold, but not reckless." Push the levels outside regular operating conditions, but not so far to make the process completely fail.

Step 3: Determine the Design to use. This depends on a trade-off between the Resolution of the Design and the budget available for the experimentation. The standard approach is to also add two Center Points in the middle of the Design at point (0,0,..., 0) to be "redundant" points to get a good estimate of error. This also somewhat alleviates the problem of pushing the Levels too far apart as described in Step 2. More than two Center Points does not make sense because the focus here is efficiency of Design.

Step 4: Create the Design Matrix in the software with columns for the Response and each Factor. An example is shown in Figure 7.18.2 using 16 runs for five Factors and a single Response, Tensile Strength. The software created the additional columns (C1-C4) in this case to form part of its analysis. This is a Resolution V Design.

Tensile Strength ***

↓	C1	C2	C3	C4	C5	C6-T	C7	C8	C9	C10
	StdOrder	RunOrder	CenterPt	Blocks	Dispense	Supplier	Force	Temp	Conc	Tensile Strength
1	1	15	1	1	10	A	100	140	6	
2	2	16	1	1	15	A	100	140	3	
3	3	6	1	1	10	B	100	140	3	
4	4	7	1	1	15	B	100	140	6	
5	5	11	1	1	10	A	120	140	3	
6	6	12	1	1	15	A	120	140	6	
7	7	13	1	1	10	B	120	140	6	
8	8	5	1	1	15	B	120	140	3	
9	9	9	1	1	10	A	100	180	3	
10	10	10	1	1	15	A	100	180	6	
11	11	8	1	1	10	B	100	180	6	
12	12	1	1	1	15	B	100	180	3	
13	13	14	1	1	10	A	120	180	6	
14	14	3	1	1	15	A	120	180	3	
15	15	2	1	1	10	B	120	180	3	
16	16	4	1	1	15	B	120	180	6	

Figure 7.18.2 Example data entry for a Screening DOE (output from Minitab v14).

Step 5: Collect the data from the process as per the runs in the Design Matrix in Step 4. The runs should be conducted in random order to ensure that external Noise Variables don't affect the results. The corresponding Y value(s) for each run should be entered directly into the Design Matrix in the software. Be sure to keep a backup copy of all data.

Step 6: Run the DOE analysis in the software, including the terms in the model as follows:

- Resolution III or IV—Just the Factors, no Interactions
- Resolution V or VI—Factors and 2-Factor Interactions (no 3-Factors or higher)
- Resolution VII and higher—Factors, 2-Factor, and 3-Factor interactions (no 4-Factors or higher)

For Resolution VI and above it might be well worth considering a smaller Fraction.

Step 7: Most software packages give a visual representation of the effects of each of the Factors and Interactions, which can be invaluable in reducing the model (eliminating terms in the model with negligibly small coefficients). For the preceding 5-Factor example, the Pareto Chart of the effects is shown in Figure 7.18.3. The Design is a Resolution V Design, so terms for the 3-Factor Interactions and higher have been eliminated (pooled into the Error term).

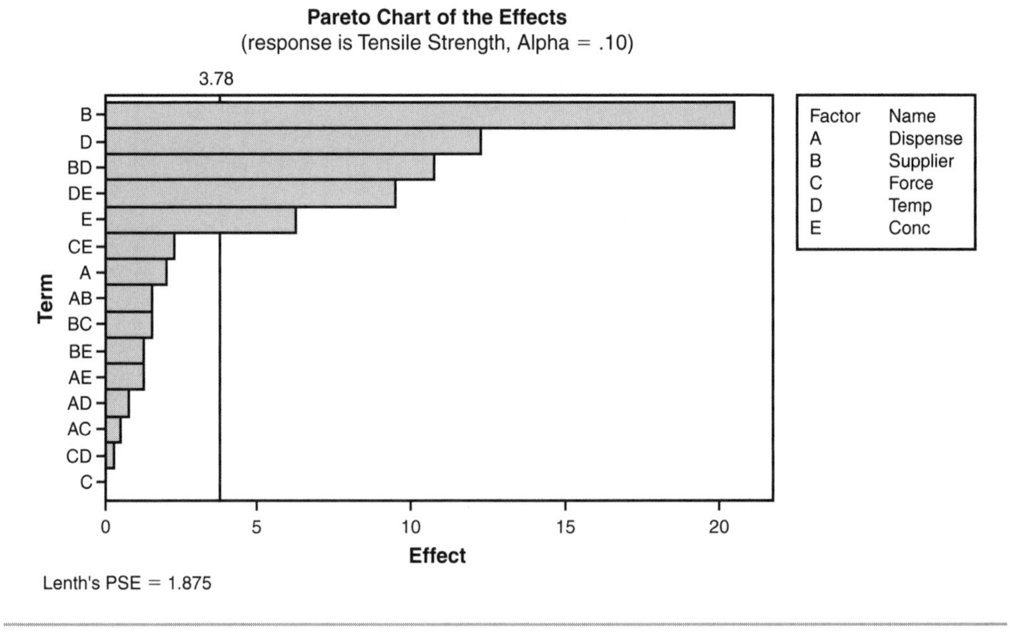

Figure 7.18.3 An example of a Pareto Chart of Effects for the 5-Factor Fractional Factorial (output from Minitab v14).

The focus should be on the highest order interaction(s) first. The vertical line at 3.78 in the figure represents the p-value cutoff point, in this case of 0.10. The higher value of 0.1 versus 0.05 is used to avoid eliminating any important terms accidentally. Usually it is best to eliminate a few of the bottom terms first and rerun the model advancing cautiously, a few terms dropped at a time, each time focusing on the terms to the left of the cutoff line. If this is done, the terms to remove from the model are all the terms including and below CE in the Figure.

In some analyses, the software does not allow the term for a Factor, such as C, to be removed from the model because the Factor in question appears in a higher order Interaction, such as AxC, that is not eliminated from the model. This is known as hierarchy. In this case, keep the Factor in place and it could be eliminated later or it might have to stay in the final model even if it does not contribute much by itself, only in the Interaction form.

For designs using Center Points, inspect the F-test for Curvature. If the p-value here is large, then there is no curvature effect to analyze and these can be pooled in to calculate the error term.

Step 8: Based on the preceding results, rerun the reduced model with only the significant effects. For the 5-Factor example, the results are shown in Figure 7.18.4. The analytical results are shown in Figure 7.18.5.

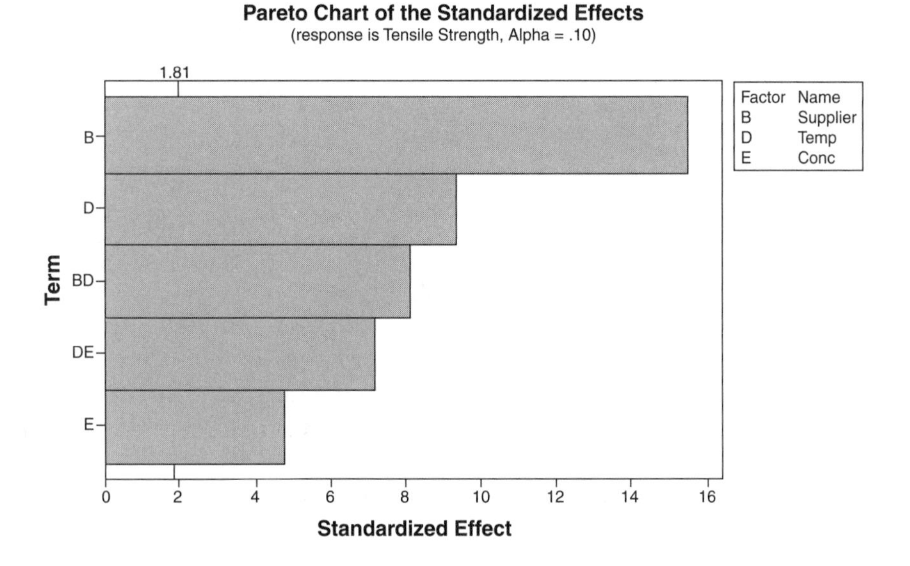

Figure 7.18.4 An example of a Pareto Chart of Effects for the reduced 5-Factor Fractional Factorial (output from Minitab v14).

Factorial Fit: Tensile Strength versus Supplier, Temp, Conc

Estimated Effects and Coefficients for Tensile Strength (coded units)

Term	Effect	Coef	SE Coef	T	P
Constant		65.250	0.6626	98.47	0.000
Supplier	20.500	10.250	0.6626	15.47	0.000
Temp	12.250	6.125	0.6626	9.24	0.000
Conc	−6.250	−3.125	0.6626	−4.72	0.001
Supplier*Temp	10.750	5.375	0.6626	8.11	0.000
Temp*Conc	−9.500	−4.750	0.6626	−7.17	0.000

S = 2.65047   R − Sq = 97.89%   R − Sq (adj) = 96.84%

Analysis of Variance for Tensile Strength (coded units)

Source	DF	Seq SS	Adj SS	Adj MS	F	P
Main Effects	3	2437.50	2437.50	812.500	115.66	0.000
2-Way Interactions	2	823.25	823.25	411.625	58.59	0.000
Residual Error	10	70.25	70.25	7.025		
Lack of Fit	2	7.25	7.25	3.625	0.46	0.647
Pure Error	8	63.00	63.00	7.875		
Total	15	3331.00				

Figure 7.18.5 Numerical analysis for the reduced model of 5-Factor Fractional Factorial (output from Minitab v14).

At this point, the estimate of error should be good because ten pieces of information have been pooled to create it (DF = 10 for the Residual Error). According to the Lean Sigma analysis rule for p-values less than 0.05, this is as far as the analysis can go because all the p-values are below 0.05; so the associated Factors or Interactions are significant. However, it seems unlikely that all five remaining terms contribute to the big process picture. To understand their contribution, the Epsilon^2 value should be calculated.

Step 9: Calculate Epsilon^2 for each significant effect in the reduced model. In Minitab this is done using a separate function called the Balanced ANOVA—most other statistical software packages provide this. The results of this can be seen in Figure 7.18.6.

It is clear from Figure 7.18.6 that all the remaining terms give meaningful contribution and should be kept in place for subsequent studies. If there were small contribution terms, a further reduction in the model would be conducted.

Analysis of Variance for Tensile Strength

Source	DF	SS	Eps-Sq
Supplier	1	1681.00	50.47
Temp	1	600.25	18.02
Conc	1	156.25	4.69
Supplier*Temp	1	462.25	13.88
Temp*Conc	1	361.00	10.84
Error	10	70.25	2.11
Total	15	3331.00	

Figure 7.18.6 Epsilon^2 Calculation for 5-Factor example (output from Minitab v14 with Epsilon^2 hand calculated and typed in).

The final ANOVA table is the one already calculated and shown in Figure 7.18.5. The model explains 97.89% of the variation in the run data (R-Sq). The R-Sq(adj) value is close to the R-Sq value, which indicates that there aren't any redundant terms in the model. The Lack of Fit of the model is insignificant (the p-value is well above 0.05).

By closely controlling the three Factors in question (the Interactions are also taken care of if the Factors are controlled because they are comprised of the Factors), some 97.89% of the variability in the process is contained.

Step 10: Formulate conclusions and recommendations. The output of a Screening Design is not the optimum settings for all of the Factors identified; it merely identifies the process Factors that should be carried over to Characterization. Here the model is reduced significantly to the point that with so many terms pooled into the error, it might be possible to skip a Characterizing Design and go straight to an Optimizing Design.

Recommendations here would certainly include better control on the three identified Factors, but quite possibly a subsequent Optimizing DOE to determine the optimum settings for the three factors.

OTHER OPTIONS

Some potential elements that were not considered here include

- Attribute Y data—In the preceding analyses, all of the Ys were Continuous data, which made the statistics work more readily. If the Y is Attribute data, then quite often a key assumption of equal variance across the design is violated. To remove this issue, Attribute data has to be transformed, which is considered well beyond the scope of this book.

- Blocking on a Noise Variable—The preceding examples did not include any Blocked Variables. Blocks are added if the Team is not sure that a potential Noise Variable could have an effect. The Noise Variable is tracked for the experiment and effectively entered as an additional X in the software. After the initial analysis run, when the first reductions are made the effect of Block would appear with its own p-value. If the p-value is above 0.1 then eliminate the Block from the model, because it has no effect. If the Block is important then it should be included in the model and subsequent analysis and experimental designs.
- Center Points and Curvature—Center Points are useful to find good estimates of Error, but they also allow investigation of Curvature in the model. When the initial analysis is run, the Curvature should have its own p-value. Again if the p-value is high, then there is no significant effect from Curvature and the Curvature term can be eliminated from the reduced model going forward. Curvature is examined in greater detail in "DOE—Optimizing" in this chapter.

19: DOE—CHARACTERIZING

OVERVIEW

Characterizing a process using Designed experiments involves

- Determining which Xs (both controlled and uncontrolled) most affect the Ys
- Identifying critical process and noise variables
- Identifying those variables that need to be carefully controlled in the process.

Before reading this section it is important that the reader has read and understood "DOE—Introduction" also in this chapter. This section covers only designing, analyzing, and interpreting a Characterizing Design, not the full DOE roadmap.

Characterizing Designs[33] in Lean Sigma are based on a set of experimental designs known as Full Factorials and specifically the most efficient of those known as 2^k Factorials. These are chosen because they

- Are good for early investigations because they can look at a large number of factors with relatively few runs
- Can be the basis for more complex designs and thus lend themselves well to sequential studies
- Are fairly straightforward to analyze.

[33] For more detail see *Statistics for Experimenters* by Box, Hunter and Hunter.

A 2^k factorial refers to k factors (Xs), each with two levels (a High and a Low value for the X). Thus, a 2-Factor Design is known as a 2^2 factorial and has two factors, each with two levels, and can be done in $2^2 = 4$ runs, as shown in Figure 7.19.1a. Likewise a 2^3 factorial has three factors, each with two levels and can be done in $2^3 = 8$ runs, as shown graphically in Figure 7.19.1b.

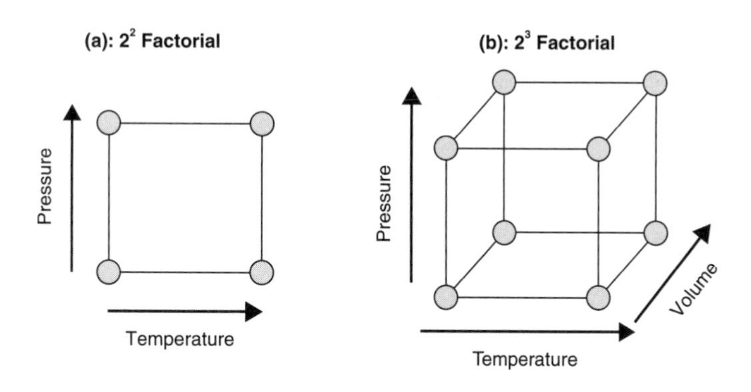

Figure 7.19.1 2-Factor and 3-Factor Full Factorial Designs.

Thus, to conduct a 2-Factor Full Factorial with two levels for each X (such as high and low), there would be four runs required, as follows:

- X_1 at the low level, X_2 at the low level
- X_1 at the low level, X_2 at the high level
- X_1 at the high level, X_2 at the low level
- X_1 at the high level, X_2 at the high level

Writing it this way looks messy, so designs for 2^k factorials are usually written as matrices shown in a specific standard order.[34] Also, for convention (and to enable the mathematics in the analysis itself), the low level of a factor is renamed as the "−1" level and the high level is renamed as the "+1" level. This is known as using coded units. An additional benefit here is that Attribute Xs can also be included in the design, for example Supplier A might be the −1 level and Supplier B the +1 level. Many Belts worry unnecessarily about coding and its implications. A Belt should know it exists and looks to the software to show which type of units (coded or actual) is displayed. Example design matrices for a 2^2 and a 2^3 factorial are shown in Figure 7.19.2.

[34] Known as Yates' Order.

Temp	Conc
−1	−1
1	−1
−1	1
1	1

Temp	Conc	Catalyst
−1	−1	−1
1	−1	−1
−1	1	−1
1	1	−1
−1	−1	1
1	−1	1
−1	1	1
1	1	1

Figure 7.19.2 Design matrices for a 2^2 and a 2^3 Full Factorial Design.

Notice that the 2^3 Factorial matrix actually contains the 2^2 Factorial matrix. This is a key property in the use of these types of Designs for sequential experimentation. If the smaller 2-Factor Design for the Xs Temperature and Concentration had already been run and it was important to add a third Factor Catalyst, the third Factor could be added easily.

After the Design has been constructed, it is simply a matter of running the experiments depicted by each row in the matrix, so for the 2^2 Factorial:

- −1 (low) level for Temperature, −1 (low) level for Concentration
- +1 (high) level for Temperature, −1 (low) level for Concentration
- And so on

In reality, the runs would be done in random order (to prevent external Noise Variables affecting the results) and most DOE software packages generate the random order automatically. For each run, the associated value for the Y or Ys are recorded and the values become additional columns in the matrix, as in Figure 7.19.3. From the first run in the Figure when the process was run with a low (−1) value of Temperature and a low (−1) value of Concentration and all other Xs were maintained as constant as possible, the resulting Y value (perhaps a Yield or similar) was 23(%).

The purpose of DOE is to understand the effect of the Xs. To do this, the data must be analyzed to determine what contribution each X makes to the changes in the Y. From Figure 7.19.3 there are some simple calculations that can be made:

- The average Y value at the low (−1) Temperature level is the average of runs 1 and 3. This equates to $(23 + 28) \div 2 = 25.5$
- The average Y value at the high (+1) Temperature level is the average of runs 2 and 4. This equates to $(5 + 9) \div 2 = 7$

- The effect of going from the low (−1) level to the high (+1) level of Temperature is (7−25.5) = −18.5 units of Y. Thus, as the Temperature increases from the low level to the high level, Y goes down by 18.5 units.
- By a similar calculation, the effect of going from the low (−1) level to the high (+1) level of Concentration is (28 + 9) ÷ 2 − (23 + 5) ÷ 2 = + 4.5 units of Y. Thus, as Concentration increases from the low level to the high level, Y goes up by 4.5 units.

Temp	Conc	Y
−1	−1	23
1	−1	5
−1	1	28
1	1	9

Figure 7.19.3 Design Matrix after the associated response data has been added.

Temperature seems to have the largest effect on the Y, although it is negative. If Temperature in this process is not controlled well, then it is likely that there would be large swings in Y associated with the changes in Temperature. Concentration seems to have an effect, but a much smaller one.

Figure 7.19.4 shows a graphical representation of an interaction between 2 Xs. A 2-Factor interaction is when the effect of one X changes depending on the level of the second X. A simple example of this is when cooking entities in an oven with two factors, such as time and temperature. For low time, a high temperature is required, but for high time a low temperature is required.

Figure 7.19.4 depicts a situation somewhat similar to this. From the figure, it is easy to see that two diagonally opposite corners in the plot are lowered and the other two are raised, forming a "twist" in the solution landscape. Thus, to detect an interaction, the effect is calculated from the differences in the averages of the diagonal corners. This is mathematically analogous to calculating a third column in the Design Matrix that is the multiple of the levels of the 2 Xs, as shown in Figure 7.19.5.

The corners of the Design space at [−1,−1] (run 1) and [+1, +1] (run 4) are diagonally opposite from one another and have an Interaction Level of +1. Similarly, the two other corners (runs 2 and 3) are diagonally opposite to each other and have an Interaction Level of −1. To calculate the size of the interaction for the example in Figure 7.19.3, the equation takes the average of the runs at the −1 level from the average of the runs at the +1 level for the interaction:

(Average of runs 1 & 4) − (Average of runs 2 & 3) = (23 + 9) ÷ 2 − (5 + 28) ÷ 2 = 0.5

For this example, there is almost no interaction at all shown by the sample of data collected.

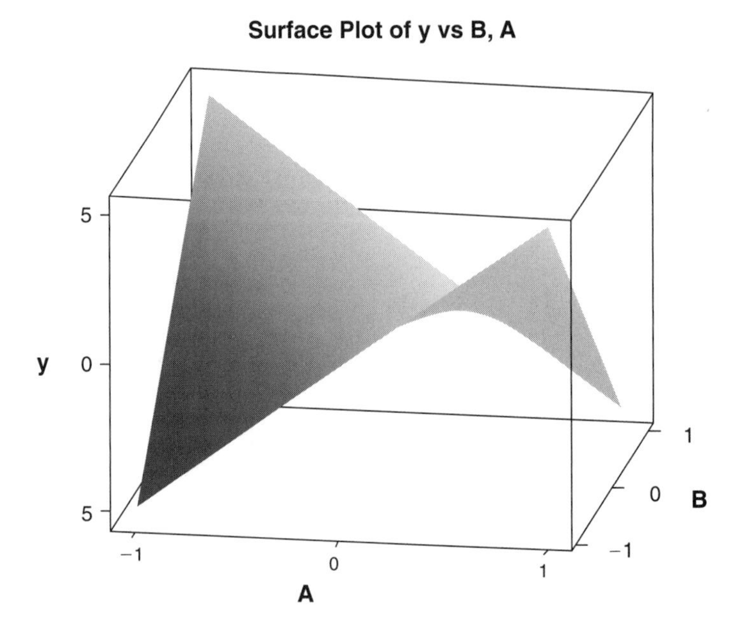

Figure 7.19.4 Surface plot of an Interaction between 2 Xs (output from Minitab v14).

Temp	Conc	Interaction
−1	−1	1
1	−1	−1
−1	1	−1
1	1	1

Figure 7.19.5 Design Matrix including the interaction between temperature and concentration.

In "DOE—Introduction" in this chapter, it was stated that DOE is about modeling the data. At this point, there is no model, but the size of the effects form the model. A generic 2-Factor model is

$$Y = \beta_0 + \beta_1 X_1 + \beta_2 X_2 + \beta_{12} X_1 X_2$$

The term β_0 is actually the average value of Y across all the data points, that is $(23 + 9 + 5 + 28) \div 4 = 16.25$. The other terms are half the size of the effects just calculated:[35]

[35] An Effect is the size of change associated in going from –1 to + 1, which means 2 units. A coefficient is the size of change incurred for 1 unit of change in the X and is half the size of the Effect.

- β_1 = Temperature Effect / 2 = –18.5 ÷ 2 = –9.25
- β_2 = Concentration Effect / 2 = + 4.5 / 2 = 2.25
- β_{12} = Interaction Effect / 2 = 0.5 / 2 = 0.25

The final equation in coded units (-1s and +1s) for this data is thus:

$$\text{Yield} = 16.25 - (9.25 \times \text{Temp}) + (2.25 \times \text{Conc}) + (0.25 \times \text{Temp} \times \text{Conc})$$

For any known values of Temperature and Concentration, if they were expressed in coded units, the associated Yield could be calculated.

All of this theory is important to at least follow, but in the practical world of being a Belt, it is a lengthy process to do this by hand. Statistical software packages make this incredibly simple to generate. Figure 7.19.6 shows the output of a Full Factorial analysis for the previous example. As you can see from the top part of the output, the equation calculated by hand seems to have been done correctly, which is always a relief.

Factorial Fit: y versus A, B

Estimated Effects and Coefficients for y (code units)

Term	Effect	Coef
Constant		16.250
A	−18.500	−9.250
B	4.500	2.250
A*B	−0.500	−0.250

S = *

Analysis of Variance for y (coded units)

Source	DF	Seq SS	Adj SS	Adj MS	F	P
Main Effects	2	362.500	362.500	181.250	*	*
2-Way Interactions	1	0.250	0.250	0.250	*	*
Residual Error	0	*	*	*		
Total	3	362.750				

Figure 7.19.6 Full factorial analysis for example data listed in Figure 7.19.3 (output from Minitab v14).

The bottom part of Figure 7.19.6 shows an ANOVA table (for more detail see "ANOVA" in this chapter). ANOVA divides the total variability in the data into its component pieces, in this case the variation due to

- Main Effects (Temperature and Concentration)
- Temperature-Concentration interaction
- Error (the part not explained by the preceding three components)

The first place to look in an ANOVA is at the p-values. The p-values here are associated with the signal-to-noise ratio of the size of the effects versus the background noise and how unlikely a ratio of that size is to occur purely by random chance. Thus, the hypotheses that the p-values relate to are

- H_0: Factor has no effect
- H_a: Factor has an effect

If the p-value is low (less than 0.05), then it is unlikely that an effect of that magnitude relative to background noise could have occurred purely by random chance; thus, the Null Hypothesis H_0 is probably incorrect.

What is unusual about the ANOVA table shown in Figure 7.19.6 is that the p-values don't exist because there is no Error value for the noise element of the signal-to-noise ratio. The issue here is the sheer efficiency of the design. The model has four coefficients, namely β_0, β_1, β_2, and β_{12} and there were only four data points run; four data points for four unknowns doesn't give redundancy to give an estimate of background variation (noise).

There are two options available to create the p-values, either add more runs or reduce the number of coefficients. In "DOE—Introduction" in this chapter, it recommends using at least 25% additional runs, in this case a minimum of one extra run. This run should be done as the Center Point at (0,0) and would immediately provide the much-needed p-values.

However, in this example there is the latter option of reducing the number of coefficients. The first analysis showed that the Interaction has a small coefficient relative to the size of the other coefficients. Ignoring this coefficient completely, there are now four runs for three coefficients, namely β_0, β_1, and β_2. The analysis can be rerun and the data point that was used to calculate the β_{12} coefficient can be diverted to help understand the size of the background noise (in a limited way, but it's better than nothing). This is known as the "reduced model" and the associated analysis results are shown in Figure 7.19.7.

By simply "pooling" the Interaction coefficient β_{12} into the error term, there are now a number of statistics available:

- The initial place to look is the p-value, which in this case indicates that the Main Effects are significant; the size of the Main Effects are 725 times greater than the background noise with a likelihood of occurrence of $p = 0.026$. This is less than the standard cutoff of 0.05.[36]

- The R-Sq value of 99.93% means that of all the variation in the data, 99.93% of it is explained by the model.
- R-Sq(adj) is close to R-Sq which means there probably aren't any redundant terms in the model. If R-Sq(adj) drops well below R-Sq, then there are terms in the model that don't contribute anything.
- $S = 0.5$ is the standard deviation of the background noise, which is small compared with the size of the Main Effects.

Factorial Fit: y versus A, B

Estimated Effects and Coefficients for y (code units)

Term	Effect	Coef	SE Coef	T	P
Constant		16.250	0.2500	65.00	0.010
A	−18.500	−9.250	0.2500	−37.00	0.017
B	4.500	2.250	0.2500	9.00	0.070

S = 0.5 R − Sq = 99.93% R − Sq (adj) = 99.79%

Analysis of Variance for y (coded units)

Source	DF	Seq SS	Adj SS	Adj MS	F	P
Main Effects	2	362.500	362.500	181.250	725.00	0.026
Residual Error	1	0.250	0.250	0.250		
Total	3	362.750				

Figure 7.19.7 The reduced analysis results (output from Minitab v14).

The output in Figure 7.19.7 doesn't show a key piece of information called Percentage Contribution or otherwise known as Epsilon^2. In simple terms, this is the amount of variation each Factor explains out of the total variation seen in the data. Figure 7.19.8 shows a separate ANOVA analysis run in the same software to generate the numbers to calculate this.

Epsilon^2 is defined as follows:

[36] In fact, in DOE the cutoff used for the p-value in the initial stages of analysis is $p = 0.1$ so that no important coefficients are eliminated accidentally. DOEs are efficient with data, so a more cautious approach is used.

$$\text{Epsilon}^2(\text{Factor 1}) = \frac{\text{SS (Factor 1)}}{\text{SS (Total)}} \times 100$$

Analysis of Variance for Y

Source	DF	SS	MS	F	P
Temp	1	342.25	342.25	1369.00	0.017
Conc	1	20.25	20.25	81.00	0.070
Error	1	0.25	0.25		
Total	3	362.75			

S = 0.5   R – Sq = 99.93%   R – Sq (adj) = 99.79%

Figure 7.19.8 Balanced ANOVA analysis to generate Epsilon² (output from Minitab v14).

For the example shown in Figure 7.19.8, the Epsilon² values are

- Epsilon² (Temp) = {SS(Temp) / SS(Total)} × 100 = {342.25 / 362.75} ×100 = 94.35%
- Epsilon² (Conc) = {SS(Conc) / SS(Total)} × 100 = {20.25 / 362.75} ×100 = 5.58%

This can be seen graphically in Figure 7.19.9, which really demonstrates how large a contribution is made by Temperature (interestingly the Error term is so small it is invisible on the graphic).

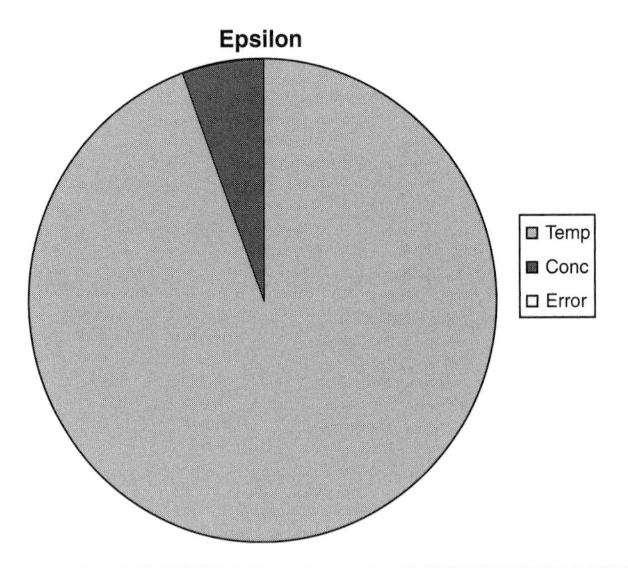

Figure 7.19.9 Graphical representation of Epsilon².

Temperature has such a large effect on the Yield that (unless an X was missed along the way) controlling Temperature alone would go a long way to keeping this process under control.

The next question to be answered at this point would be "at what value should Temperature be set to gain the maximum Yield?" That could be done by taking Temperature (and possibly Concentration) as the X Variable in a *Regression Study*, or taking Concentration and Temperature forward into an Optimizing DOE (see "DOE—Optimizing" in this chapter).

ROADMAP

The roadmap to designing, analyzing, and interpreting a Characterizing DOE is as follows:

Step 1: Identify the Factors (Xs) and Responses (Ys) to be included in the experiment. Each Y should have a verified Measurement System. Factors should come from earlier narrowing down tools in the Lean Sigma roadmap and should not have been only brainstormed at this point.

Step 2: Determine the levels for each Factor, which are the Low and High values for the 2-Level Factorial. For example, for Temperature it could be 60°C and 80°C. Choice of these levels is important in that if they are too close together then the Response might not change enough across the levels to be detectable. If they are too far apart then all of the action could have been overlooked in the center of the design. The phrase that best sums up the choice of levels is "to be bold, but not reckless." Push the levels outside regular operating conditions, but not so far as to make the process completely fail or to make the process unsafe in any way.

Step 3: Determine the Design to use. For 2-Level Full Factorials this is simple because there is only one Design available for a given number of Factors. A choice has to be made about how to add additional "redundant" points to get a good estimate of error. The standard approach is to add two to four Center Points in the middle of the Design at point (0,0,..., 0). This somewhat alleviates the problem of pushing the Levels too far apart as described in Step 2. More than four Center Points doesn't provide much additional information for the investment needed. For two or three Factors, two Center Points usually is fine.

Step 4: Create the Design Matrix in the software with columns for the Response and each Factor. An example is shown in Figure 7.19.10 for three Factors and two Responses, Yield and Strength. The software created the additional columns (C1-C4) in this case to form part of its analysis.

Step 5: Collect the data from the process as per the runs in the Design Matrix in Step 4. The runs should be conducted in random order to ensure that external Noise Variables don't affect the results. The corresponding Y value(s) for each run should be entered directly into the Design Matrix in the software. Be sure to keep a backup copy of all data.

Worksheet 3 ***

↓	C1	C2	C3	C4	C5	C6	C7	C8	C9
	StdOrder	RunOrder	CenterPt	Blocks	Temperature	Pressure	Concentration	Yield	Strength
1	1	2	1	1	60	450	0.2		
2	2	10	1	1	80	450	0.2		
3	3	5	1	1	60	500	0.2		
4	4	4	1	1	80	500	0.2		
5	5	9	1	1	60	450	0.4		
6	6	6	1	1	80	450	0.4		
7	7	3	1	1	60	500	0.4		
8	8	1	1	1	80	500	0.4		
9	9	7	0	1	70	475	0.3		
10	10	8	0	1	70	475	0.3		
11									

Figure 7.19.10 Example data entry for a Characterizing DOE (output from Minitab v14).

Step 6: Run the DOE analysis in the software including all terms in the model (all Factors and all Interactions). If the Design has more than three factors, run the model showing only 3-way and 2-way Interactions. Interactions in four or more variables do not occur.

Step 7: Most software packages give a visual representation of the effects of each of the Factors and Interactions, which can be invaluable in reducing the model (eliminating terms in the model with negligibly small coefficients). An example of a 4-Factor Full Factorial is shown in Figure 7.19.11. The focus should be on the highest order interaction(s) first. The vertical line at 6.3 in the figure represents the p-value cutoff point, in this case 0.10. The higher value of 0.1 versus 0.05 is used to avoid eliminating any important effects accidentally. Focusing on the highest order Interactions, it is clear that all four 3-way Interactions are below the cutoff line and should be eliminated from the model. Also, any 2-way Interactions with effects less than the 3-way Interactions being eliminated, should also be removed, in this case AC, CD, AD.

The numerical p-values for each of the Factors and Interactions should be available in the software. Here they are shown in Figure 7.19.12. It is clear that the elimination of the 3-way Interactions makes sense because their p-values are all well above 0.1.

For designs using Center Points, inspect the F-test for Curvature. If the p-value here is large, then there is no curvature effect to analyze and these too can be pooled in to calculate the error term.

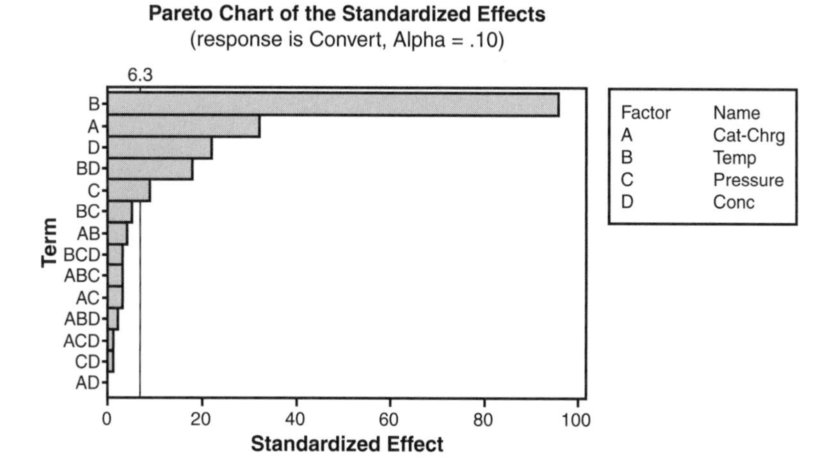

Figure 7.19.11 Example Pareto Chart of Effects for a 4-Factor Full Factorial (output from Minitab v14).

Factorial Fit: Convert versus Cat-Chrg, Temp, Pressure, Conc

Estimated Effects and Coefficents for Convert (coded units)

Term	Effect	Coef	SE Coef	T	P
Constant		72.250	0.1250	578.00	0.001
Cat-Chrg	-8.000	-4.000	0.1250	-32.00	0.020
Temp	24.000	12.000	0.1250	96.00	0.007
Pressure	-2.250	-1.125	0.1250	-9.00	0.070
Conc	-5.500	-2.750	0.1250	-22.00	0.029
Cat-Chrg*Temp	1.000	0.500	0.1250	4.00	0.156
Cat-Chrg*Pressure	0.750	0.375	0.1250	3.00	0.205
Cat-Chrg*Conc	-0.000	-0.000	0.1250	-0.00	1.000
Temp*Pressure	-1.250	-0.625	0.1250	-5.00	0.126
Temp*Conc	4.500	2.250	0.1250	18.00	0.035
Pressure*Conc	-0.250	-0.125	0.1250	-1.00	0.500
Cat-Chrg*Temp*Pressure	-0.750	-0.375	0.1250	-3.00	0.205
Cat-Chrg*Temp*Conc	0.500	0.250	0.1250	2.00	0.295
Cat-Chrg*Pressure*Conc	-0.250	-0.125	0.1250	-1.00	0.500
Temp*Pressure*Conc	-0.750	-0.375	0.1250	-3.00	0.205

Figure 7.19.12 Numerical analysis of 4-Factor Full Factorial (output from Minitab v14).

Step 8: Based on the preceding results, rerun the reduced model with only the significant effects.

In some analyses the software does not allow the term for a Factor, such as C, to be removed from the model because the Factor in question appears in a higher order

Interaction, such as AxC, that is not being eliminated from the model. This is known as hierarchy. In this case, just keep the Factor in place and it could be eliminated later or it might have to stay in the final model even if it does not contribute much by itself, only in the Interaction form.

For the 4-Factor example, the results are shown in Figure 7.19.13. At this point, the estimate of error should be good because eight pieces of information have been pooled to create it (DF = 8 for the Residual Error). According to the Lean Sigma analysis rule for p-values less than 0.05, this is as far as the analysis can go because all the p-vales are below 0.05; so the associated Factors or Interactions are significant. However, it seems unlikely that all the eight remaining terms really contribute to the big process picture. To understand their contribution, the Epsilon^2 value should be calculated.

Factorial Fit: Convert Versus Cat-Chrg, Temp, Pressure, Conc

Estimated Effects and Coefficients for convert (coded units)

Term	Effect	Coef	SE Coef	T	P
Constant		72.50	0.2577	280.37	0.000
Cat-Chrg	−8.000	−4.000	0.2577	−15.52	0.000
Temp	24.000	12.000	0.2577	46.57	0.000
Pressure	−2.250	−1.125	0.2577	−4.37	0.002
Conc	−5.500	−2.750	0.2577	−10.67	0.000
Cat-Chrg*Temp	1.000	0.500	0.2577	1.94	0.088
Temp*Pressure	−1.250	−0.625	0.2577	−2.43	0.042
Temp*Conc	4.500	2.250	0.2577	8.73	0.000

S = 1.03078 R − Sq = 99.70% R − Sq (adj) = 99.43%

Analysis of Variance for Convert (coded units)

Source	DF	Seq SS	Adj SS	Adj MS	F	P
Main Effects	4	2701.25	2701.25	675.313	635.59	0.000
2-Way Interactions	3	91.25	91.25	30.417	28.63	0.000
Residual Error	8	8.50	8.50	1.063		
Total	15	2801.00				

Figure 7.19.13 Numerical analysis for the reduced model of 4-Factor Full Factorial (output from Minitab v14).

Step 9: Calculate Epsilon^2 for each significant effect in the reduced model. To do this, the Sequential Sum of Squares (Seq SS) for each effect is represented as a percentage of the Total Seq SS (2801.00 in this case). To list the Seq SS for each effect separately in Minitab requires another function called Balanced ANOVA—most other statistical software packages provide this in the ANOVA function. The results of this can be seen in Figure 7.19.14.

It is clear from the figure that only the effects from the following give meaningful contribution, the other effects are negligible.

- Catalyst-Charge
- Temperature
- Concentration
- Temperature-Concentration Interaction (even this is debatable because it is so small)

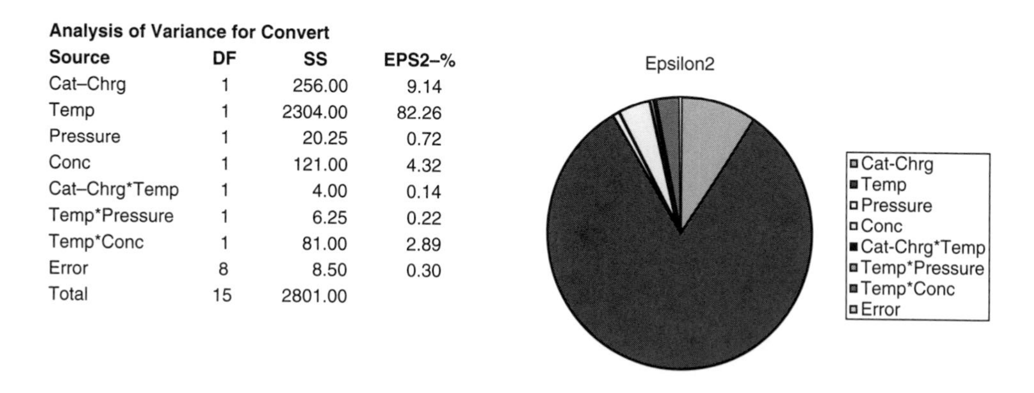

Analysis of Variance for Convert

Source	DF	SS	EPS2–%
Cat–Chrg	1	256.00	9.14
Temp	1	2304.00	82.26
Pressure	1	20.25	0.72
Conc	1	121.00	4.32
Cat–Chrg*Temp	1	4.00	0.14
Temp*Pressure	1	6.25	0.22
Temp*Conc	1	81.00	2.89
Error	8	8.50	0.30
Total	15	2801.00	

Figure 7.19.14 Epsilon² Calculation for 4-Factor example (output from Minitab v14 with Epsilon² hand calculated and typed in).

Step 10: Run the final reduced model for the effects identified in Step 9. The final ANOVA table is shown in Figure 7.19.15. The model explains 98.61% of the variation in the run data (R-Sq). The R-Sq(adj) value is close to the R-Sq value, which indicates that there aren't any redundant terms in the model. The Lack of Fit of the model is insignificant (the p-value is well above 0.05). By closely controlling the three Factors in question (the Interaction is also taken care of if the Factors are controlled because it comprises two of the remaining Factors), about 98.61% of the variability in the process is contained.

Step 11: Formulate conclusions and recommendations. The output of a Characterizing Design is not the optimum settings for all of the Factors identified; it merely identifies the critical process Factors. Recommendations here would certainly include better control on the three identified Factors, but quite possibly a subsequent Optimizing DOE to determine the optimum settings for the three factors.

Factorial Fit: Convert versus Cat-Chrg, Temp, Conc

Estimated Effects and Coefficients for Convert (coded units)

Team	Effect	Coef	SE Coef	T	P
Constant		72.250	0.4707	153.48	0.000
Cat-Chrg	−8.000	−4.000	0.4707	−8.50	0.000
Temp	24.000	12.000	0.4707	25.49	0.000
Conc	−5.500	−2.750	0.4707	−5.84	0.000
Temp*Conc	4.500	2.250	0.4707	4.78	0.001

S = 1.88294 R − Sq = 98.61% R−Sq (adj) = 98.10%

Analysis of Variance for Convert (coded units)

Source	DF	Seq SS	Adj SS	Adj MS	F	P
Main Effects	3	2681.00	2681.00	893.667	252.06	0.000
2-Way Interactions	1	81.00	81.00	81.000	22.85	0.001
Residual Error	11	39.00	39.00	3.545		
Lack of Fit	3	5.00	5.00	1.667	0.39	0.762
Pure Error	8	34.00	34.00	4.250		
Total	15	2801.00				

Figure 7.19.15 Final analysis results for 4-Factor example (output from Minitab v14).

OTHER OPTIONS

Some potential elements that were not considered

- Attribute Y data—In the preceding analyses, all of the Ys were Continuous data, which made the statistics work more readily. If the Y is Attribute data, then often a key assumption of equal variance across the design is violated. To remove this issue, Attribute data has to be transformed, which is considered well beyond the scope of this book.
- Blocking on a Noise Variable—The preceding examples did not include any Blocked Variables. Blocks are added if the Team is not certain that a potential Noise Variable could have an effect. The Noise Variable is tracked for the experiment and effectively entered as an additional X in the software. After the initial analysis run, when the first reductions are made, the effect of Block would appear with its own p-value. If the p-value is above 0.1 then eliminate the Block from the model, as it has no effect. If the Block is important, then it should be included in the model and subsequent analysis and experimental designs.

- Center Points and Curvature—Center Points are useful to find good estimates of Error, but they also allow investigation of Curvature in the model. When the initial analysis is run, the Curvature should have its own p-value. Again if the p-value is high, then there is no significant effect from Curvature and the Curvature term can be eliminated from the reduced model going forward. Curvature is examined in greater detail in "DOE—Optimizing" in this chapter.

20: DOE—OPTIMIZING

OVERVIEW

Optimizing a process using Designed experiments involves

- Determining at what level the critical Inputs should be set
- Determining real specification limits

Before reading this section it is important that the reader has read and understood "DOE—Introduction," "DOE—Characterizing," and "DOE—Screening" in this chapter. This section covers only designing, analyzing, and interpreting an Optimizing Design, not the full DOE roadmap.

Optimizing Designs (sometimes known as Response Surface Methodology[37]) in Lean Sigma are based on the set of experimental design techniques known as sequential experimentation and rely on Screening Designs as well as Characterizing Designs. Screening Designs allow a narrowing of the Xs down to the few potential candidates. Characterizing Designs determine for the shortened list of Xs how much each X contributes and how it contributes. The result of Screening and Characterizing Designs is a shortlist of critical Xs that need to be controlled to maintain a consistent level of the Y. What they do not provide is an optimized Y.

To explain this more readily, it is best to follow an example. Assume the first DOE work was a Screening Design that narrowed the Xs down to two key Xs. The associated Y is the Yield of the process. A subsequent Full Factorial on the two Xs yielded the results shown in Figure 7.20.1. The model explains 91.06% of the variation in the data and both Xs are significant.

To take this analysis further, an understanding of the solution space needs to be gained. Because there are only two Factors involved here, it is possible to represent the solution space graphically with a Surface Map of the model, as shown in Figure 7.20.2.

[37] For more detail see *Introduction to Linear Regression Analysis* by Douglas Montgomery and Elizabeth Beck.

Results for: RSM 1

Factorial Fit: Yield versus Time, Temp

Estimated Effects and Coefficients for Yield (coded units)

Term	Effect	Coef	SE Coef	T	P
Constant		62.014	0.6011	103.16	0.000
Time	4.700	2.350	0.7952	2.96	0.042
Temp	9.000	4.500	0.7952	5.66	0.005

S = 1.59049 R − Sq = 91.06% R − Sq(adj) = 86.59%

Figure 7.20.1 Example Characterizing Design analysis results for two Factors (output from Minitab v14).

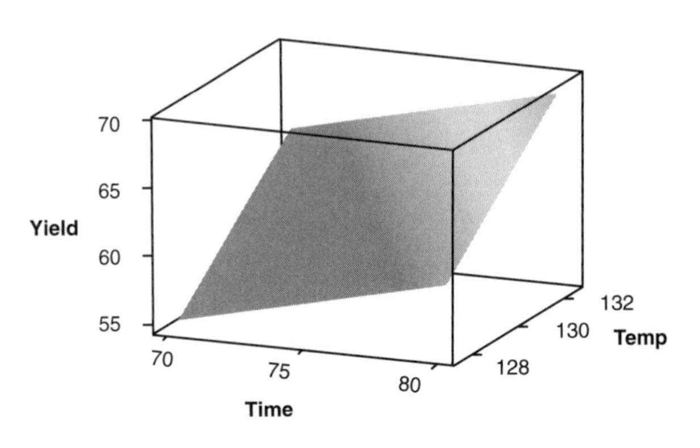

Figure 7.20.2 Surface Plot of Yield versus Time and Temperature (output Minitab v14).

Clearly to improve the Yield, the Time and Temperature need to be changed to levels outside the Design space, preferably toward the top right-hand corner; the question is where? The Surface Map gives a good visual representation of the direction to travel, but the Contour Plot, as shown in Figure 7.20.3, really shows the best direction to travel in terms of Time and Temperature to get an improved Yield.

If the journey is commenced from the Center Point of the Plot, the fastest way to move in the direction of increased Yield is known as the Path of Steepest Ascent and is perpendicular to the contour lines, as drawn on Figure 7.20.3. To proceed, additional data points (experimental runs) should be taken at regular reasonable intervals up the

Path of Steepest Ascent until a maximum Yield is reached. This direction can be approximated by eye if there are only two Factors (as in this case) or calculated from the analysis results in Figure 7.20.1 irrespective of the number of Factors involved. Either way is appropriate in this case where there are only two Factors; however, Belts can become a little confused with the latter approach so the former is recommended here.

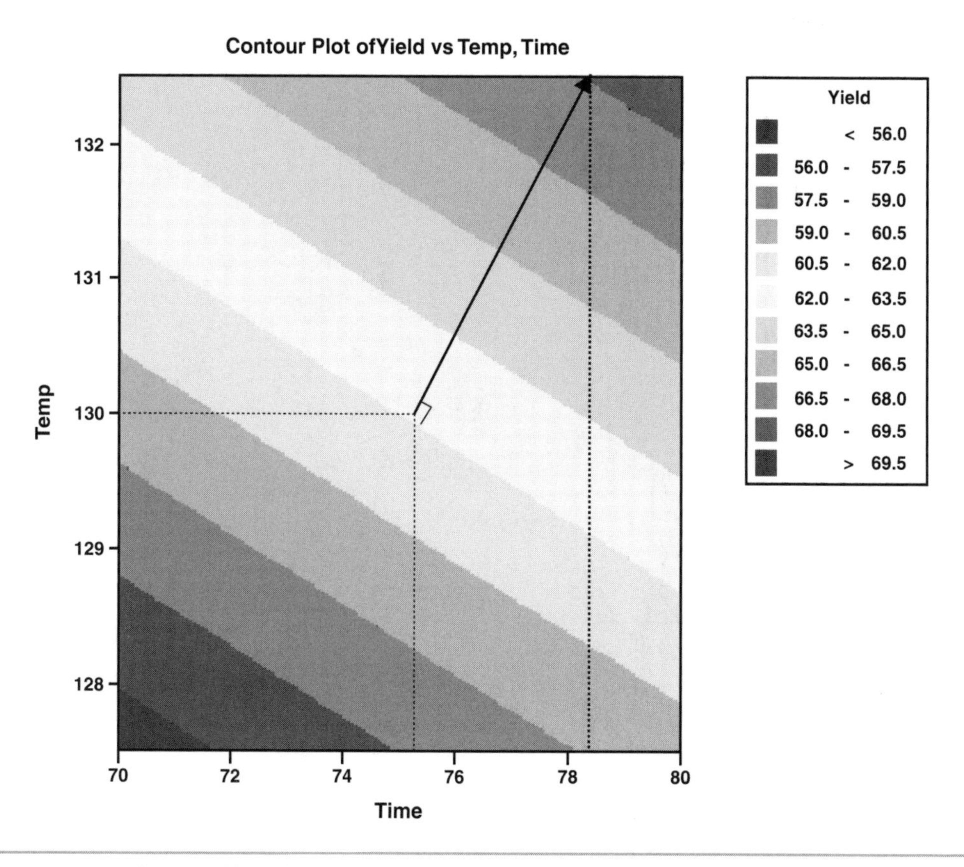

Figure 7.20.3 Contour Plot of Yield versus Time and Temperature (output Minitab v14).

To approximate the Path of Steepest Ascent, use the Contour Plot as shown in Figure 7.20.1. Extend the arrow out from the Center Point perpendicular to the contours until it reaches the boundary of the square. Read off the coordinates of the point where the arrow touches the square. In this case Time ≈ 78.2, Temperature = 132.5.

As Temperature goes up 2.5 units from the Center Point, Time needs to go up approximately 3.2 units to stay on the Path of Steepest Ascent. From this, a simple table can be created and populated as shown in Table 7.20.1. Just a few of the steps were run with the associated Yield value in the table.

Table 7.20.1 Calculating the Path of Steepest Ascent

	Time	Temp	Yield
Origin (Center Point)	75	130	62.3
Step Size	3.2	2.5	
Origin + 1 Step	78.2	132.5	
Origin + 2 Steps	*81.4*	*135*	71.0
Origin + 3 Steps	*84.6*	*137.5*	
Origin + 4 Steps	*87.8*	*140*	
Origin + 5 Steps	*91*	*142.5*	85.1
Origin + 6 Steps	*94.2*	*145*	
Origin + 7 Steps	*97.4*	*147.5*	
Origin + 8 Steps	*100.6*	*150*	56.7
Origin + 9 Steps	*103.8*	*152.5*	

Clearly the Yield reaches a maximum value somewhere around "Origin + 5 Steps" at [91, 142], so no further runs are conducted after "Origin + 8 Steps." The experimental sequence could be halted at this point with a new, much improved Yield of 85%, but there might be an even better Yield value somewhere in the solution space.

The approach at this point would be to start the process again with a new Full Factorial centered somewhere around the highest point found from the Path of Steepest Ascent. Based on attainable settings for the Xs, the Team, in this case, decides to center the experiment on [90,145] and stretch the levels of the Xs to cover a broader area, as shown in Figure 7.20.4.

The Team chose new levels:

- Time: Low = 80, High = 100
- Temperature: Low = 140, High = 150

The Team designed a Full Factorial, shown as Experiment 3 in Figure 7.20.4. The runs in Experiment 3 were conducted on the process (including two Center Points) and the associated Yield recorded. Analysis of the runs gave the results shown in Figure 7.20.5.

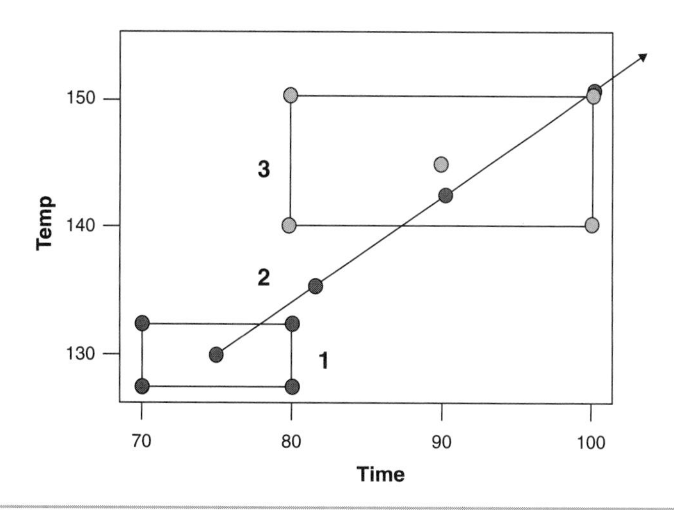

Figure 7.20.4 Sequence of experimentation for the Time-Temperature example.

Factorial Fit: Yield versus Time, Temp

Estimated Effects and Coefficients for Yield (coded units)

Term	Effect	Coef	SE Coef	T	P
Constant		82.975	0.7654	108.41	0.000
Time	−4.050	−2.025	0.7654	−2.65	0.118
Temp	2.650	1.325	0.7654	1.73	0.226
Time*Temp	−9.750	−4.875	0.7654	−6.37	0.024
Ct Pt		4.992	1.1692	4.27	0.051

Figure 7.20.5 Analysis results for experiment 3 in the Time-Temperature example (output from Minitab v14).

At first glance, the results seem a little strange; the linear model is no longer correct (high p-values) and there are now interaction and curvature effects (both with low p-values). This makes complete sense though upon further consideration. The previous experiment showed liner effects representing the side of a hill. Here it looks as though the experiment is sitting on the top of the same hill. There are no linear effects, but there is curvature, which are both indicative of a peak in the response surface. It could be that the experimental region is close to the maximum value for Yield. If there is curvature present then a linear model is no longer of any use, a quadratic model would be better.

To see curvature in a given direction, at least three points are needed in a line. To do this here, the model is expanded by adding what are known as "Star Points" or "Axial Points" to create a new design, known as a Central Composite Design, as shown in Figure 7.20.6.

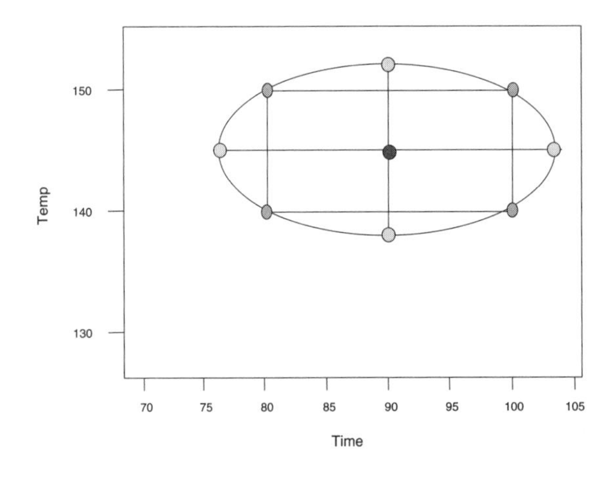

Figure 7.20.6 Adding a Central Composite Design to the time-temperature example.

This design is efficient in representing curvature. As can be seen from the figure, the addition of the Star Points creates a line of three points horizontally, vertically, and on both diagonals. Only the four Star Points are required as extra runs to create this design from the Full Factorial. Two additional Center Points are usually added to make sure that conditions haven't changed from when the Full Factorial was run to when the additional points were run. All the Full Factorial runs, along with the Star Points and all four Center Points, are analyzed together in the Central Composite Design. A Block is used to distinguish the first set of runs (Full Factorial) from the second set (Star Points) and to check that no conditions have changed. The results are shown in Figure 7.20.7.

From the Figure it is clear that the Block has no effect (a p-value above 0.05); so conditions appear not to have changed from the Full factorial to the addition of the Star Points. The analysis can be rerun eliminating the Block term and the results are shown in Figure 7.20.8.

Response Surface Regression: Yield versus Block, Time, Temp

The analysis was done using coded units.

Estimated Regression Coefficients for Yield

Term	Coef	SE Coef	T	P
Constant	87.5141	0.6617	132.252	0.000
Block	0.7025	0.4332	1.621	0.149
Time	−1.3939	0.5760	−2.420	0.046
Temp	0.3687	0.5760	0.640	0.542
Time*Time	−2.2298	0.6053	−3.684	0.008
Temp*Temp	−3.1992	0.6053	−5.285	0.001
Time*Temp	−4.8750	0.8105	−6.015	0.001

S = 1.621   R − Sq = 92.3%   R − Sq (adj) = 85.7%

Figure 7.20.7 Analysis results of the Central Composite Design for the time-temperature example (output from Minitab v14).

The analysis was done using coded units.

Estimated Regression Coefficients for Yield

Term	Coef	SE Coef	T	P
Constant	87.5047	0.7259	120.539	0.000
Time	−1.3939	0.6319	−2.206	0.058
Temp	0.3687	0.6319	0.583	0.576
Time*Time	−2.2214	0.6641	−3.345	0.010
Temp*Temp	−3.1908	0.6641	−4.805	0.001
Time*Temp	−4.8750	0.8892	−5.483	0.001

S = 1.778   R − Sq = 89.4%   R − Sq (adj) = 82.8%

Figure 7.20.8 Analysis results of the Central Composite Design excluding the Block (output from Minitab v14).

The model seems reasonable in that it explains 89.4% of the variation seen in the data (from the R-Sq value). The R-Sq(adj) value is a little lower at 82.8% indicating some redundant terms in the model, but that can be explained by the inclusion of the Factors Time and Temperature, even though their p-values indicate they aren't significant. The Factors cannot be removed for the sake of hierarchy. The equation for Yield can be determined from the "Coef" column, which lists the coefficients:

$$\text{Yield} = 87.5047 - 1.3939\,(\text{Temp}) - 2.2214\,(\text{Time}^2) - 3.1908\,(\text{Temp}^2) - 4.8750\,(\text{Temp*Time})$$

Remember these are in coded units; most software packages give the equation in actual units as well. Figure 7.20.9 shows the graphical representation of the response surface at this point.

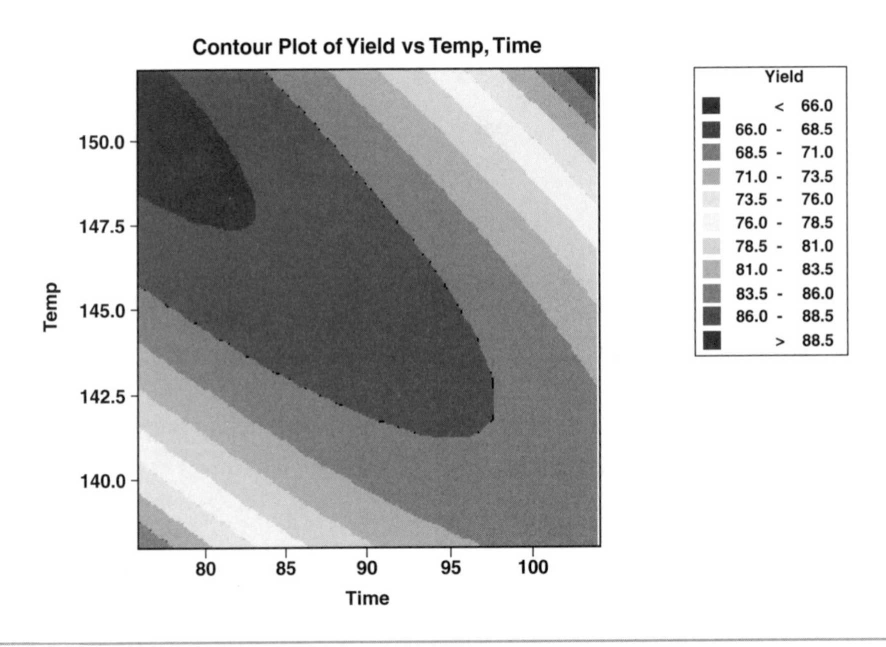

Figure 7.20.9 Contour Plot of the response surface from the Central Composite Design (output from Minitab v14).

From the Contour Plot it seems that there is another area of even higher Yield off to the top left corner. To proceed, the approach would be to determine the Path of Steepest Ascent in that direction and conduct experiments along the Path until a maximum value is reached and then center a further experiment in that region. For the example in question, an expanded response surface is shown in Figure 7.20.10, which clearly shows the progression from the original experiment up to the top of a saddle and then the potential for further experiments leading to a higher Yield in the top left corner.

Most DOE software packages provide an optimization algorithm, which can be applied to the model to determine the maxima or minima. Many Belts fall into the mode of optimizing beyond practicality. Remember that this is an imperfect model that is based on just a sample of data from a varying process with inherent noise in the Measurement System. To optimize to anything more then two significant figures is fairly meaningless. Use the optimizer instead to identify a region of operation. If there is a choice between operating regions then choose the flattest one, because this creates a more robust process (variation in the Xs has little effect in the Y).

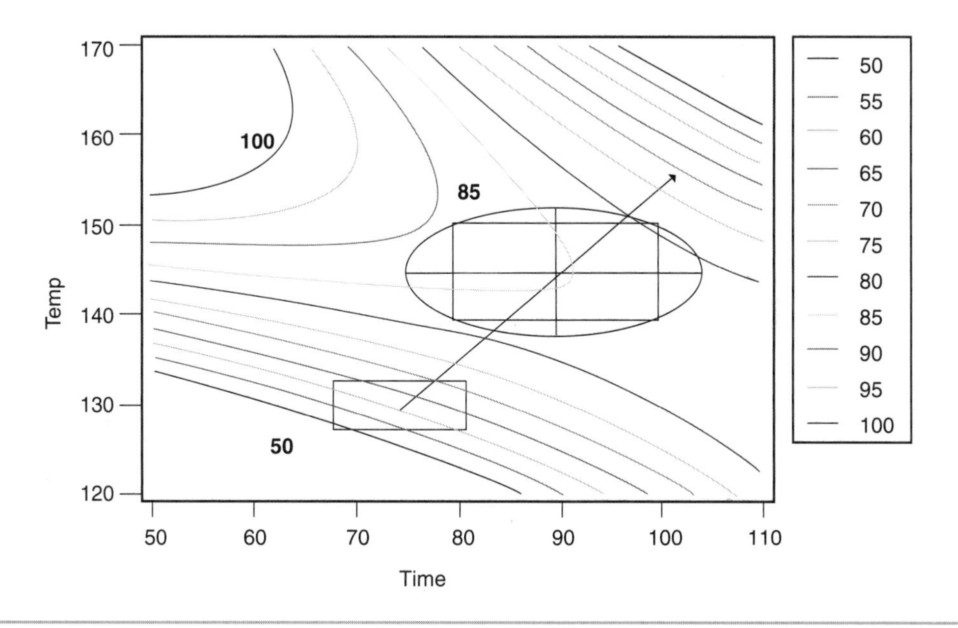

Figure 7.20.10 The "big picture view" of Yield (output modified from Minitab v13).

The final step in the process is to set the specifications for the Xs based on contour lines in the Y. If the Yield has to be kept above a certain value, then simply reading the associated value of the Xs from the Contour Plot can give specifications for the Xs. This is shown graphically in Figure 7.20.11; in this example, to maintain a Yield above 86%, the Xs should be controlled within the following specifications:

- Time: LSL[38] = 87, USL = 93
- Temperature: LSL = 143°, USL = 147°

[38] Lower Specification Limit and Upper Specification Limit.

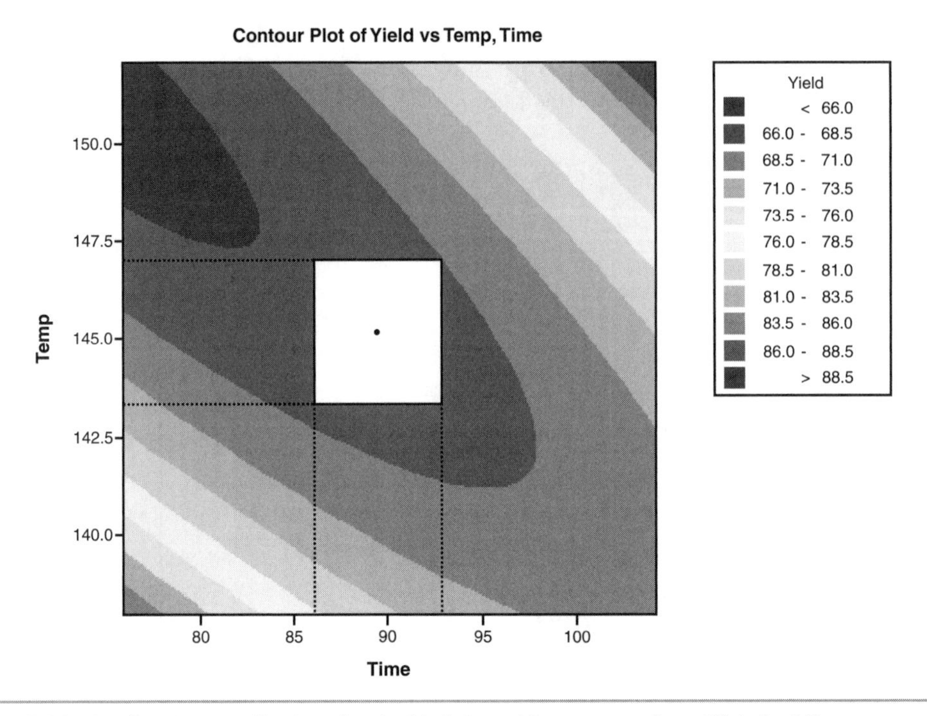

Figure 7.20.11 Creating specifications for the Xs (adapted from output from Minitab v14).

ROADMAP

The roadmap to designing, analyzing and interpreting an Optimizing DOE is as follows:

Step 1: Conduct a Screening DOE to narrow down the Xs (see "DOE—Screening" in this chapter).

Step 2: For the reduced set of Xs conduct a Characterizing DOE to determine the critical Xs and the shape of the response surface at this point using a Contour Plot and Surface Plot. Include two Center Points in the Full Factorial Design. For more detail see "DOE—Characterizing" in this chapter.

Step 3: Determine the Path of Steepest Ascent if the Y is to be maximized, the Path of Steepest Descent if the Y is to be minimized. This can be done using the graphical method described in "Overview" in this section (note this is applicable only to 2-Factor experiments) or using the analytical method shown in "Other Options" later in this section (applicable to two or more Factors).

Step 4: Collect the data from the process as per the runs on the Path of Steepest Ascent (or Descent) in Step 3. After a turning point is reached along the Path, stop.

Step 5: Center another Full Factorial around the highest (lowest) point found on the Path of Steepest Ascent (Descent). Include two Center Points in the Design. Determine the shape of the response surface at this point using a Contour Plot and Surface Plot. For more detail see "DOE—Characterizing" in this chapter. If the Design is still on the side of a hill, repeat Steps 3–5 until the surface levels out.

Step 6: Add Star Points and two additional Center Points to the final Full Factorial from Step 5 to create a Central Composite Design Matrix. These should be added as a second Block to ensure that conditions haven't changed from the Full Factorial runs.

Step 7: Capture data from the process as per the Design Matrix. Complete all of the runs in the Design Matrix.

Step 8: Analyze the data from the Central Composite Design runs. If a potentially higher area of Y is seen outside of the current Design space, then repeat Steps 3–8. If the optimum value of the Y appears to be within the current design space, then use the optimizer algorithm in the software or read the best value by eye from the Contour Plot to identify the best operating conditions.

Step 9: Conduct a confirmatory run at the optimum process settings.

Step 10: Determine specifications for the Xs to maintain the Y at the optimum value, as described in "Overview."

Step 11: Formulate conclusions and recommendations. Process settings are generally implemented at this point and recommendations for improvements to control the Xs are submitted or put in place.

Other Options

Calculating the Path of Steepest Ascent for Two or more Factors

To calculate the exact Path of Steepest Ascent for two or more Factors, the model equation in coded units (–1s and +1s) from the analysis is used. Using the example from "Overview" in this section:

$$\text{Yield} = 62.0 + 2.35 \times \text{Time} + 4.50 \times \text{Temp}$$

For any model equation, the Path can be calculated by choosing the step size in one of the process variables, for example, $\Delta X_1 = 1$ (usually a step size of 1 or 2 at most). The step size for the other variable(s) is

$$\Delta X_k = \frac{\beta_k}{\beta_1} \bullet \Delta X_1$$

where:

- ΔX_1 is the step size in variable X_1, which is Time in this example. The step size taken is 1 or 2 coded units.
- β_1 is the coefficient of the Factor X_1 (Time) from the model, which is 2.35 in this example.
- β_k is the coefficient from the model of the Factor in question. In this case, the Factor is X_2 (temperature) and its coefficient is 4.5.

So for the preceding example, for every increase of 1 coded unit of Time, the associated step size for temperature to follow the Path of Steepest Ascent is

$$\Delta X_k = \frac{\beta_k}{\beta_1} \bullet \Delta X_1 = \frac{4.5}{2.35} \times 1.0 = 1.91 \text{ units}$$

It should be quite straightforward to create a table using a spreadsheet to represent this, as shown in Table 7.20.2. Remember that the listed equation for Time and Temperature from the analysis is in coded units, so before the experiment can be run, the coded units have to be converted back to real life or "Natural Units," as shown in the table in the columns "Nat Time" and "Nat Temp."

Table 7.20.2 Calculating the Path of Steepest Ascent

Steps	Coded Time	Coded Temp	Nat. Time	Nat. Temp	Run	Yield
Origin	0	0	75	130	5,6,7	62.3 (Mean)
Δ	1	1.91	5	4.8		
Origin + 1Δ	1	1.91	80	134.8	8	73.3
Origin + 2Δ	2	3.82	85	139.6		
Origin + 3Δ	3	5.73	90	144.4	9	86.8
Origin + 4Δ	4	7.64	95	149.2		
Origin + 5Δ	5	9.55	100	154	10	58.2

This is done with the equation:

$$\begin{matrix}\text{Value in} \\ \text{Natural Unit}\end{matrix} = \left\{ \frac{\text{Natural Range}}{2} \times \begin{matrix}\text{Value in} \\ \text{Coded Units}\end{matrix} \right\} + \text{Natural Centre Point}$$

For the example here:

$$\begin{matrix}\text{Time Value in}\\ \text{Natural Unit}\end{matrix} = \left\{\frac{80-70}{2} \times \begin{matrix}\text{Time Value in}\\ \text{Coded Units}\end{matrix}\right\} + 75 = \left\{5 \times \begin{matrix}\text{Temp Value in}\\ \text{Coded Units}\end{matrix}\right\} + 75$$

and

$$\begin{matrix}\text{Time Value in}\\ \text{Natural Unit}\end{matrix} = \left\{\frac{132.5-127.5}{2} \times \begin{matrix}\text{Temp Value in}\\ \text{Coded Units}\end{matrix}\right\} + 130 = \left\{2.5 \times \begin{matrix}\text{Temp Value in}\\ \text{Coded Units}\end{matrix}\right\} + 130$$

The next steps are the same as in "Overview" in this section; the runs are conducted sequentially on the process until a highest value is determined. At this point a Full Factorial Design is centered on the highest point from the Path of Steepest Ascent.

21: Fishbone Diagram

Overview

The ubiquitous Fishbone Diagram (also known as a Cause & Effect Diagram) is a well-known quality tool but was dropped from the Six Sigma Process Improvement Roadmap a number of years ago, replaced completely by the *Process Variables Map* and *Cause & Effect Matrix* combination. It does however have a place in a few Lean Sigma problem categories that aren't related to a single process.

The Fishbone Diagram (obviously named for its looks) is a Team brainstorming tool to help identify potential root causes to problems, or in Lean Sigma terms Key Process Input Variables (KPIVs) or Xs. The problem is tackled by examining six major process related areas (Branches):

- Man (People)
- Machine
- Material
- Measurement
- Method
- Mother Nature (Environment)

These are then broken down further into Sub-Branches, effectively following the *5 Whys* principal until some potential root causes (Xs) are identified. Therefore, the tool is another means to generate process Xs.

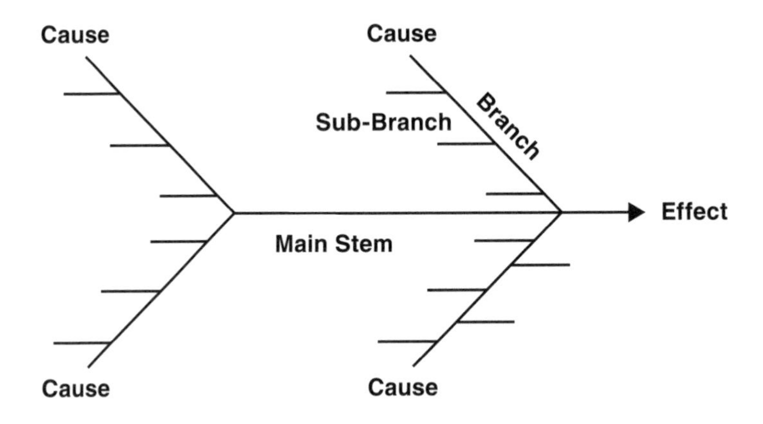

Figure 7.21.1 Structure of a Fishbone Diagram.

An example Fishbone Diagram is shown in Figure 7.21.2.

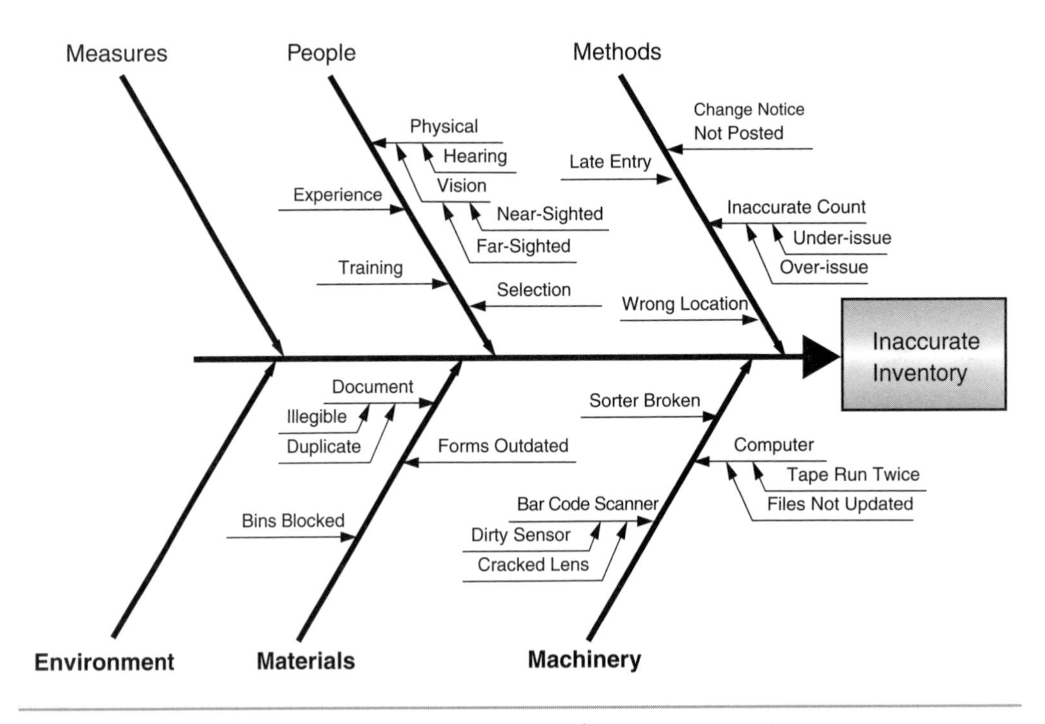

Figure 7.21.2 Example Fishbone Diagram relating to warehouse inventory variances.

Logistics

The Fishbone Diagram is absolutely a Team endeavor. It typically takes about one to two hours to complete in a brainstorming session with Team members and other key process experts. It is best done in one sitting, so if it can't be completed during the time available postpone until a more suitable time.

Belts are strongly advised not to create a straw man of the Diagram prior to the meeting; this approach usually causes a lack of buy-in to the result and also potential missed Xs.

Roadmap

The Roadmap is relatively straightforward:

Step 1: Create the Fishbone structure on a flipchart-sized sheet of blank paper. Place on the wall in landscape orientation. Draw a horizontal line with a box connected at the far right and write the problem or effect in the box. Draw six branches off the main stem and categorize them as People, Material, Method, Machine, Measurement, and Environment.

Step 2: Brainstorm Xs in each of the categories and capture them onto a branch or sub-branch under the correct category. It is often useful to use a sticky note for each X so they can be moved around freely. Use the *5 Whys* to drill down each cause category until an X is reached. Similar to the *Process Variables Map*, Xs need to be actionable or tangible items. For example, an X "Operator" isn't really an X, it is more of a group of Xs—what specifically is it about the Operator?:

- Height
- Weight
- Handedness
- Availability
- Hair color

Table 7.21.1 can help break down the groups and also identify those trickier Xs.

Interpreting the Output

The biggest mistake Belts make is reverting to the traditional use of the Fishbone Diagram; Process Improvement Teams would look at all the Xs generated and from gut-feel decide which Xs to pursue. This is *incorrect* in Lean Sigma and in fact, could completely derail the project.

Table 7.21.1 Fishbone Diagram Checklist[39]

Measurement	Methods	Machine
Poor repeatability	Incorrect definition	Machine maintenance or calibration
Poor reproducibility	Incorrect sequence	Machine controls or lack of controls
Poor accuracy	Missing definitions, implicit rules	Machine fault or defect
Poor stability	Poor process controls	Software or network fault
Poor linearity	Poor measurement controls	Machine related contamination
Invalid measurement or test method	Lack of critical information	Machine tooling or fixtures
Excessive test or measurement	Incorrect information	Incorrect machine or tester
	Excessive queues or outtime	
	Handling	
	Orientation	
	Poor management of change	
	Incorrect revision	

Materials	Environment	People
Defective	Physical environment (temperature, lighting)	Level of staffing
Off-specification	Security or safety systems	Training
Contaminated	Distractions in the environment	Competency or experience
Improper storage conditions	Particulates	Supervision
Labeling or identification	Contamination	Conflicting goals
Incorrect amount or quantity		Compliance with procedures
Improper transportation or handling		Personality issues
Expiration date exceeded or unknown		Physical ability or function
Problem with product design		Cognitive ability and function
Wrong materials		Knowledge deficit
		Communication with peers or supervisor

The Fishbone Diagram is used only to identify the potential Xs, not to prioritize them. Subsequently, more suitable tools (the *Cause & Effect Matrix, Failure Mode and Effects Analysis,* and *Multi-Vari Study*) narrow them down to the critical few. There should be

[39] From SBTI's Lean training material.

absolutely no action items coming from the Fishbone Diagram other than to transfer the Xs to the next tool.

22: HANDOFF MAP

OVERVIEW

The Handoff Map is one of the simplest tools available in a Lean Sigma project, but is powerful as both an analysis tool as well as a communication vehicle.

The approach is to start with a circle representing the world of the process. Each function is then represented on the circle as a point on its circumference. The points should be equally spaced on the circle even if the actual physical locations aren't geographically equidistant, see Figure 7.22.1.

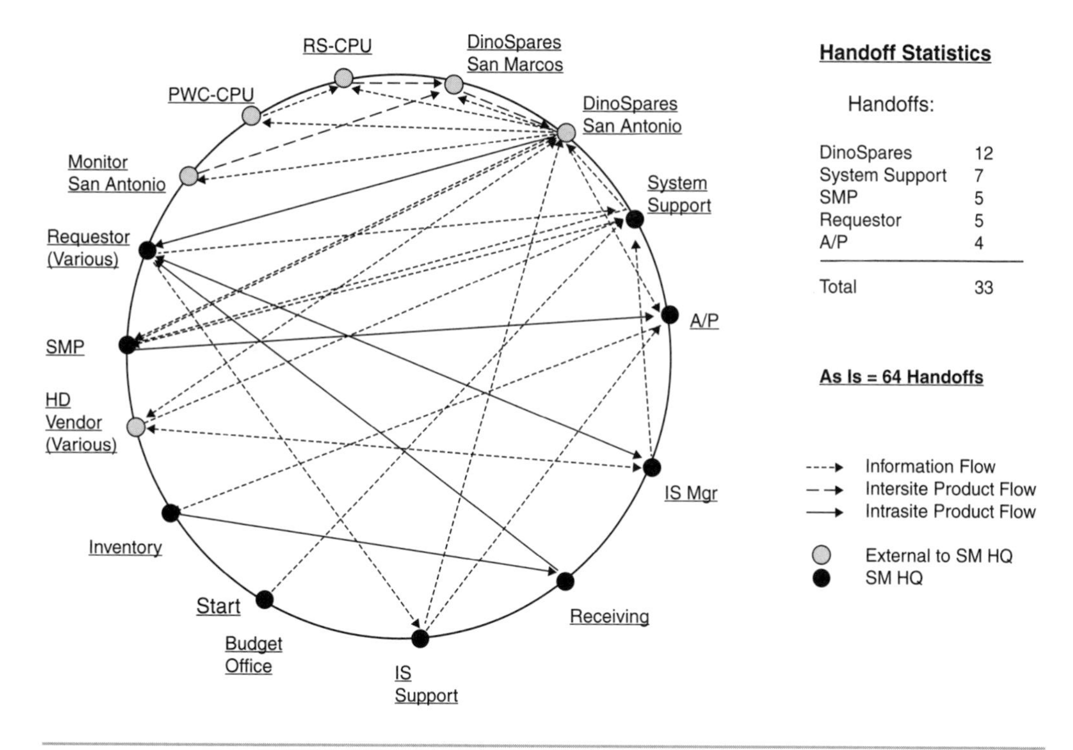

Figure 7.22.1 An example of an As-Is Handoff Map for Desktop PC Acquisition.[40]

[40] Adapted from source Sigma Breakthrough Technologies, Inc.

The process is then traced from the primary entity's perspective (often the simplest way is to trace the primary horizontal flow in the *Value Stream Map* to achieve this), and every time there is a change in functional owner, a line is placed inside the circle linking the *from* and the *to*. If there were a handoff within a function (such as a patient moved from one bed to another on a Care Unit), then a small loop would be drawn external to the circle (not shown in Figure 7.22.1) to represent this.

Often a second (or even third) path is used on the map in a different color, representing a secondary entity such as information flow.

For all the handoffs, the total number of types of handoffs and the total volume of handoffs in a given time period are calculated. Each and every handoff is an opportunity, or several opportunities, for failure; thus, the goal is to reduce the number of handoffs in the process.

LOGISTICS

Construction of the Handoff Map is done by the Team and generally takes about an hour or so to create the pictorial element. The number of types of handoffs can be read straight from the map, but to determine the total volume of handoffs might require a data collection for a week or so unless historical data is available.

ROADMAP

Two versions of the Handoff Map are created at different points in a project, the first representing the Current State or "As-Is" and the second representing the Future State or "To-Be." Sometimes a third version is created if the To-Be map isn't implemented as originally thought.

Construct As-Is

Step 1: Determine the following for the process in question:

- The scope of the process (i.e., the beginning and end points)
- The primary and any secondary entities to be mapped
- The handoff points in the process (typically functions)

Step 2: Complete the following to create the pictorial element of the Handoff Map:

- Create the circle and equally space the handoff points around its circumference.

- Take the primary entity and start at the beginning of the process; track any handoffs that occur, marking them as connections inside the circle using straight lines. If handoffs occur from a function to itself (such as the previous example of moving a patient to a different bed in the same Care Unit), then mark this as a loop on the outside of the circle, connecting a point back to itself.
- Use a different color, or if the Primary Entity map is complex, use an entirely new circle, and repeat the tracking for the secondary entity or entities, such as information or materials.

Step 3: Summarize the findings in a table on the same Map created in Step 2 listing:

- The number of handoffs by functional group per process cycle
- The volume of handoffs in a given timeframe (usually a week or a month)

Construct To-Be

The To-Be could be built directly from a To-Be *Value Stream Map* in exactly the same steps as the As-Is Map, but there is opportunity to use the tool in a slightly more elegant way, using it to achieve some handoff reduction as follows:

Step 1: Determine the following for the process in question:

- The potential new scope of the process (i.e., the beginning and end points, which are typically different from the As-Is Map)
- The primary and any secondary entities to be mapped

The handoff points are initially the same as those used in the As-Is Map.

Step 2: Complete the following to create the pictorial element of the Handoff Map as in Figure 7.22.2:

- Create the circle and equally space the handoff points around its circumference.
- Identify the categories of value-add, using information from the *Value Stream Map*.
- Work in conjunction with the To-Be process flow development (usually a *Value Stream Map*), take the primary entity and, start at the beginning of the process; map only the VA process handoffs onto the To-Be chart. As before, mark them as connections inside the circle using straight lines. A handoff within a function is questionable in terms of VA, so it is highly unlikely that there should be loops outside the circle.
- Use the same method and repeat the tracking for the secondary entity or entities, such as information or materials.

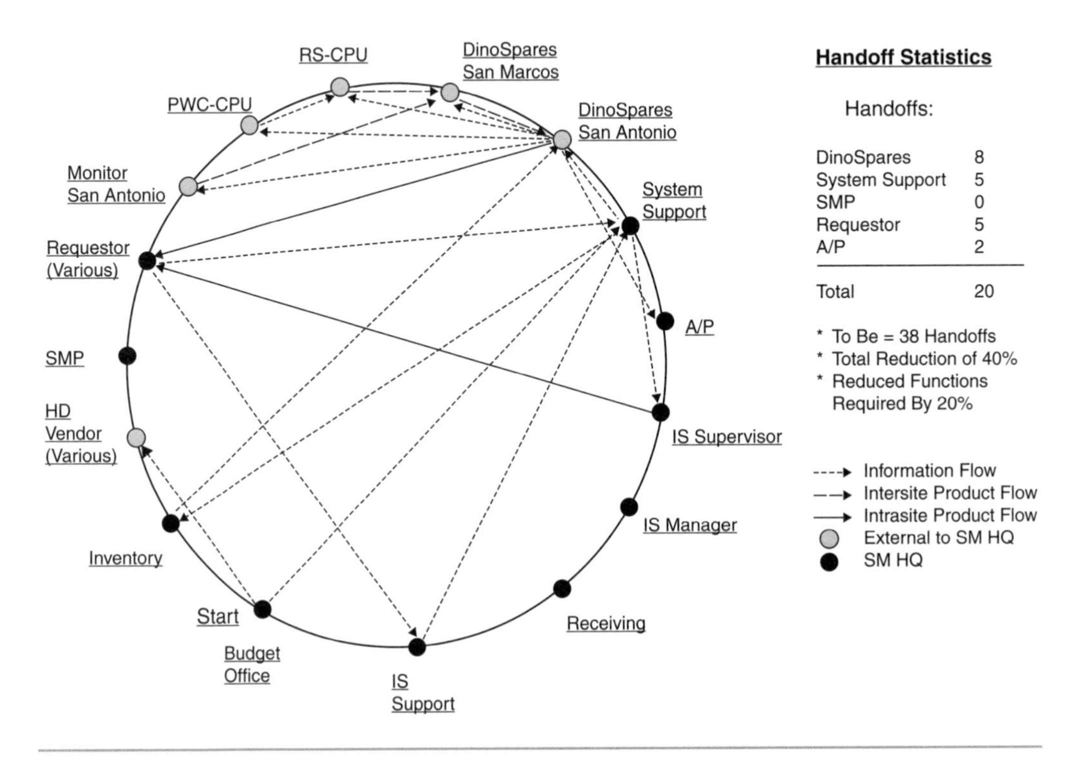

Figure 7.22.2 An example of a To-Be Handoff Map for Desktop PC Acquisition.[41]

Step 3: Summarize the findings in a table on the same Map created in Step 2 listing:

- Number of handoffs by functional group per process cycle
- Potential volume of handoffs in a given timeframe (usually a week or a month)

Interpreting the Output

The To-Be Handoff Map now graphically depicts any Global Process Cycle Time and Process Lead Time reduction opportunities. For more details see "Time—Global Process Cycle Time" and "Time—Process Lead Time" in this chapter. Using the Handoff Map in conjunction with the *Value Stream Map*, document the findings and any opportunities for streamlining.

[41] Adapted from source Sigma Breakthrough Technologies, Inc.

23: KPOVs AND DATA

OVERVIEW

Lean Sigma is different to many traditional Process Improvement initiatives in its reliance on data to make decisions. During Define the major metrics, otherwise known as the Key Process Output Variables[42] (KPOVs) or Big Ys, to measure performance of the process are identified and a ballpark historical value determined. Thenceforth, the equation $Y = f(X_1, X_2, \ldots, X_n)$ used to identify and narrow down the Xs that drive the Y is completely data driven.

Prior to undertaking that road, it is important to have a basic understanding of data and measures and how to deal with them. There are different ways to categorize data and measures:

- Continuous versus Attribute
- Results versus Predictors
- Efficient versus Effective

The type of data is important because it affects how measures are defined, how to go about collecting data, and which tools to use in analyzing the data, an example is shown in Figure 7.23.1.

Y Data \ X Data	Discrete	Continuous
Discrete	Chi-Square	Logistic Regression
Continuous	ANOVA Means / Medians Tests	Regression

Figure 7.23.1 Selecting the statistical tool based on the data type.

[42] Term originates from Dr. Steve Zinkgraf's early work at Allied Signal.

Continuous Measures

Continuous measures are those that can be measured on an infinitely divisible scale, such as weight, temperature, height, time, speed, and decibels. They are often called Variable measures.

Continuous data is greatly preferred over Attribute type data because it allows the use of more powerful statistical tests and can facilitate decisions being made with much less data (a sample size of 30 works well). Continuous data also allows measurement of *performance* (i.e., how good is it?) versus *conformance* (i.e., is it just good or bad?). If the data used in the project is Attribute type, then Belts are strongly encouraged to identify a related Continuous type measure and use that instead.

Attribute Measures

Any measures that aren't Continuous are known as Attribute or Discrete measures. Despite the lower statistical power, they are generally easier and faster to use than Continuous data and are effective for measuring intangible factors such as Customer Satisfaction or Perception.

There are several different types of Attribute data along a spectrum as follows:

- Binary has just two categories to classify the data—for example, pass/fail, win/lose, or good/bad
- Categorical or Nominal has multiple categories to classify the data, but there is no order to the categories (there is no greater than or less than involved)—for example, U.S. states (Texas, California, and so on), type of product, day of week
- Rating or Ordinal Scale has multiple categories to classify the data and there is meaningful order to the categories—for example, a Likert scale, satisfaction rating 1–5
- Count is a simple count of entities—for example, the number of defects or the number of defective items
- Percentage or Proportion is the count expressed as a percentage or proportion of the total—for example, the percentage of items scrapped or the proportion of items damaged

As the number of categories increase in Attribute data, the more like Continuous data it becomes.

The key failings of Attribute data are the need for a large number of observations to get valid information from any tests performed and also that the data can hide important discrimination. For example, giving Pass or Fail grades in course grading hides the underlying score of each student.

Results

The Lean Sigma equation $Y = f(X_1, X_2,..., X_n)$ shows the relationship between the Result (the Y) and the Predictors (the Xs). Results are measures of the process outcomes and are often called "output" factors or KPOVs. Results might be short-term (on-time delivery) or longer-term (Customer satisfaction), but they are typically lagging indicators. Also, because Results are the Ys and are driven by multiple Xs in the process, they are reactive rather than proactive variables.

Results tend to be much easier to collect than Predictors and are usually already being measured in the process. Measurement of a Result is typically done against a specification and becomes a Process Capability measure.

Predictors

Predictors are the "upstream" factors that, if measured, can forecast events "downstream" in the process; for example, an increase in raw-material order lead times might predict an increase in late deliveries. Predictors are the Xs in the equation $Y = f(X_1, X_2,..., X_n)$ and are often called "input" factors or KPIVs.

Predictors are the leading indicators of the process and generally are more difficult to identify and collect. It is unlikely that this data is available historically, so the Team has to set up a measurement system to collect it.

Due to the proactive nature of Predictors, they form the basis of the strongest process control measures.

Effectiveness

There are two categories of metrics used to describe the Big Ys in the process, namely effectiveness and efficiency. Effectiveness is an external measure of the process related to the VOC. An effective process is one that produces the desired results, from the Customer's perspective. To understand effectiveness, the questions to ask relate to

- How closely are Customers' needs and requirements met?
- What defects have they received?
- How satisfied and loyal have they become?

Effectiveness measures include

- Delivery performance, such as percent on time in full (OTIF)
- Quality performance, such as problems consumers report with their new product during the first three months of ownership, defects per unit (DPU), or customer percent defective

- Price, such as relative price index
- Customer Satisfaction, such as Press Ganey or Gallup Patient Satisfaction scores

Efficiency

Efficiency is the second category of metrics used to describe the Big Ys. Efficiency is an internal measure of the process and relates to the level of resources used in the process to achieve the desired result. To understand efficiency, questions to ask relate to how the process performs in the eyes of the business:

- What is the process cycle time?
- What is the process yield?
- How costly is the process at meeting Customer requirements?

Efficiency metrics include

- Process Lead Time
- Work content, such as assembly time per unit
- Process cost, such as labor and materials cost per unit
- Resources required
- Cost of poor quality (COPQ), such as Defects per Million Opportunities (DPMO), First Pass Yield, Scrap, Rework, Re-inspect, and Re-audit
- Percentage of time value-added

Each process and sub-process should have at least two Efficiency requirements established:

- A ratio of input-to-output value, such as cost per entity processed, process yield
- A measurement of Cycle Time, such as hours to process an entity

Operational Definitions

After a metric has been identified, the work isn't complete. To be a useful metric, a clear, precise description of the metric has to exist. This is critical so that everyone can evaluate and count the same way and there is a common understanding of and agreement on the results. In Transactional processes in particular, Operational Definitions are often the only real way to maintain control and, if poorly defined, are commonly the biggest cause of Measurement System issues.

To describe a measure fully, it is useful to create an Operational Definition for it including

- Who measures (by role, not person's name)
- What they measure (entities, measurement units)
- When they measure (timing and frequency)
- Where they measure (physical location in the process to make measure)
- How they measure (technique, steps involved)
- Why they measure (what is the measure used for)

ROADMAP

The roadmap to identifying, defining, and capturing data is as follows:

Step 1: Identify the measures that translate what is happening in the process into meaningful data. Determine if each measure is a KPOV or KPIV, if it is a measure of Effectiveness or Efficiency, and if it is Continuous or Attribute.

Step 2: Create a clear Operational Definition of the characteristic to be measured (who, what, when, where, how, and why).

Step 3: Test the Operational Definition with process Stakeholders (all those that affect and are affected by the process) to ensure consistency in understanding. Revise the Operational Definition if necessary. Any metric chosen should be both repeatable and reproducible (see "MSA—Validity," "MSA—Attribute," and "MSA—Continuous" in this chapter).

Belts sometimes believe that there is historical data readily available to complete their project. Historical data can be extremely useful when available; however, it is typically not based on the same operational definitions, hard to use (not stratified the right way, not sortable, and so on,) and generally incomplete. Active data capture is the norm in the vast majority of Lean Sigma projects, so the Team has to create a Data Collection and Sampling plan.

Step 4: Identify stratification. Before collecting the data, it is important to take time to consider how the Team wants to analyze the data after it is collected. For example, determine which Xs are to be investigated with respect to their effects on the Ys, and so on. To facilitate this, the Xs need to be built into the data collection plan and essentially are means to stratify the data. They might include things such as time, date, person, shift, operation number, Customer, SKU, buyer, machine, subassembly #, defect type, defect location, component and defect impact and criticality, and so on. Be aware though that the number of sub-categories has a big impact on the amount of data required from which to make statistical inferences.

During the Analyze Phase the data is examined and correlated based on these and other Xs. The Xs should come from the *Cause & Effect Matrix*.

Step 5: Create a sampling plan.[43] Data collected is only a sample from the process; it never represents every entity to be processed (known as the population). Even if 100% of the data points are collected for the process during a certain period of time, it is still only a sample of the whole population. In fact, this issue drives the use of statistics in the roadmap. Statistics are necessary because there is only a sample taken from the population. The Belt then uses the properties of the sample to draw inferences (predictions, guesses) as to the properties of the population. Clearly there is some guesswork (statistics) involved, but this is a faster, less costly way to gain insight into a process or large population.

The trick is to have the best sample from which to make inferences. Thus the sample needs to be a miniature version of the population—just like it, only smaller. To achieve this, there are generally two main considerations when sampling—sample quality and sample size.

Sample Quality is a measure of how well the sample represents the population and meets the needs of the sampler. Lean Sigma statistical methods generally assume random sampling in which every entity or member of the full population has an equal chance of being selected in the sample. Whichever sampling approach is taken, the Team must strive to

- Minimize bias in the sampling procedure. Bias is the difference between the nature of the data in the sample and the true nature of the entire population. A good example of a bias mistake occurred recently for a Belt in a client hospital. The Team chose to track sample medication orders using bright orange paper versus the regular white orders. Given the choice of which to process first, operators always picked orange over white; thus, the timings for the sample were lower than they should have been if the had been truly representative.
- Avoid "convenience" sampling. Difficult Customers are hard to capture data from but represent a valuable source of information.
- Minimize "non-response bias." If the likely responders hold different views to the non-responders, then there is a bias. This is common in pre-election surveys when the polls often show a large swing away from the incumbent party because dissatisfied electors tend to take time to air their views to the sampling Team.
- Minimize data errors and missing data.

It is best to use a standard Sampling Strategy, as listed in Figure 7.23.2.

[43] For more insight see *Statistics for Management and Economics* by Keller and Warrack.

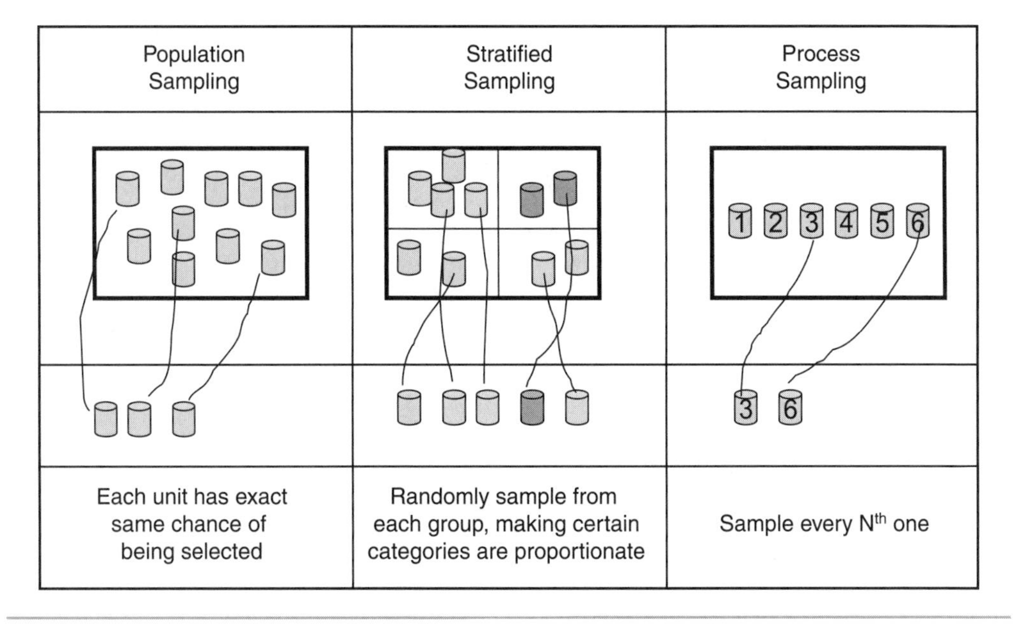

Figure 7.23.2 Standard Sampling Strategies.

Sample Size is important because, generally, precision in sampling results increases as the Sample Size *n* increases. Unfortunately, this is not a one-to-one relationship. In fact, mathematically it increases in proportion to $\sqrt{n}$ and so doubling sample size doesn't double accuracy. For example, when the sample size increases from 1 to 100, inaccuracy decreases by only 1/10th, and, therefore, it is important to keep sample sizes relatively small.

For Continuous data meaningful Sample Sizes are

- Estimating Average: >10 data points
- Estimating level of Variation: >30 data points

For Attribute Data meaningful Sample Sizes are

- Estimating Proportion, or Percent: ~100 data points

Step 6: Based on the Sampling Plan, create a data collection form and tracking system. This comprises the who, what, where, when, and how the data is captured. Some common methods include

- Check sheets—Data is collected in Attribute form as a series of check marks corresponding to types of defect or similar, see Figure 7.23.3.

- Frequency Plot Check sheet—An extension of the Check sheet in which the defects are recorded directly into a Dot Plot, see Figure 7.23.4.
- Concentration Diagram Check sheet—Used to mark defects on a physical representation of the product or process, see Figure 7.23.5. The pattern of defects builds to show where the problems occur most often.
- Data sheet—Used to record continuous data, or a mix of both Continuous and Attribute data. It is best to create the Data sheet directly in a spreadsheet or statistical software in preparation for data analysis, see Figure 7.23.6.
- Traveler Check sheet—Used to record information about individual items (applications, order forms, products, patients, and so on) as they move through the steps of the process. The easiest way to describe this is to imagine being stapled to the process entity and going along for the ride! A traveler is useful when the process is complex, and an understanding is needed for what happens to each entity as they flow through the process. However, this approach does require additional data preparation steps to bring together all the travelers at the end of data collection.
- Surveys—For more information see "Customer Surveys" in this chapter.

Whichever data collection format is chosen, it should include data source information, such as Time, Lot Number, Shift, Site, Data Collector's Name, and so on. It is also important to include a Comments section to allow for the recording of any informative conditions. The capture method should always be kept simple and easy-to-read.

Cathy's World Famous Key Lime Pies – Check Sheet for Defects									
Project: Type of defects in finished pies	Data Collected by: Dave		Dates: Apr 20–29		Location: Cathy's Kitchen		Lot Size: 240		
Defect Type:	20-Apr	21-Apr	22-Apr	23-Apr	24-Apr	27-Apr	28-Apr	29-Apr	Total
Too much lime	1	5	0	3	5	4	0	2	20
Too little lime	3	0	7	3	1	0	2	2	18
Crust too crumbly	2	1	2	2	4	0	4	2	17
Pie too large	0	3	0	3	2	3	1	4	16
Pie too small	2	1	0	0	1	1	0	0	5
Not sweet enough	1	5	0	3	6	4	0	2	21
Bite taken out of pie	1	2	1	0	1	0	1	0	6

Figure 7.23.3 An example of a Check sheet.

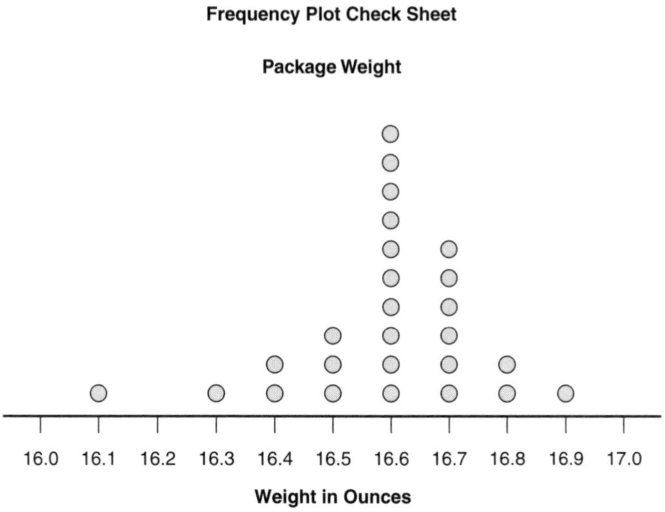

Frequency Plot Check Sheet

Package Weight

16.0 16.1 16.2 16.3 16.4 16.5 16.6 16.7 16.8 16.9 17.0

Weight in Ounces

Figure 7.23.4 An example of a Frequency Plot Check sheet.

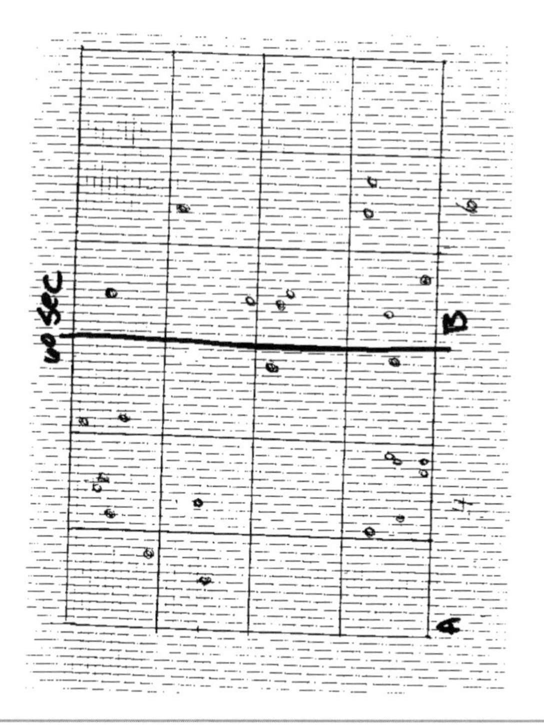

Figure 7.23.5 An example of a Concentration Diagram.

Operator	Shift	Script	Sales	Time
Bob	AM	1	109.57	24.07
Jane	PM	1	173.37	26.01
Jane	PM	1	124.00	24.57
Jane	AM	1	217.38	23.70
Walt	PM	1	154.88	27.64
Jane	AM	1	91.30	27.56
Jane	AM	1	123.99	25.16
Bob	AM	1	138.47	25.23

Figure 7.23.6 An example of a Data sheet.

A Data Collection Plan Summary sheet is a useful addition to the project, as shown in Figure 7.23.7. The Team then has a complete list of the who, what, where, when, and how for the data collection.

Step 7: Create procedures for completing the data collection forms. Any instructions should be visual and understandable by all. It is useful to include pictures of the form, describe what goes in each box and include examples of completed forms. Choose the data collectors carefully and train them on the procedures using the instructions to ensure consistent data collection.

Step 8: Test the data collection method. Do a short dry run of the data capture (a few data points) to identify problems and make adjustments. After completion, assess the accuracy, repeatability, and reproducibility of the data collection system (see "MSA—Validity," "MSA—Attribute," and "MSA—Continuous" in this chapter).

Step 9: Collect the Data as per the Data Collection Procedure. The Team must follow the Sampling Plan consistently and record any changes in operating conditions not part of the normal or initial operating conditions. Any events out of the ordinary should be immediately written into a logbook. The Belt for the project should check that data collection procedures are followed at all times. When the desired sample size is reached, stop data collection.

Step 10: Enter the collected data promptly into a database, such as a spreadsheet or statistical software. Make a backup copy of the electronic file. Keep all the paper copies to be archived as part of the project report.

What to Measure?				How to Measure?			Sampling Plan			
KPOV or KPIV	Measure	Type (Att or Cont.)	Operational Definition	Measurement Method	Additional Data Tags	Data Collection Method	Who	Type	How Many	When
Xs or Ys?	Describe the measure here.	Attribute or Continuous data?	Put a clear definition of what you are measuring here that all have agreed to.	How are you getting the measure? Inspection by counting? Physical Measurement?	Are there other ways to stratify the data, not included as a KPIV?	Check sheets, electronic,	Who will be doing the sampling?	Population, Stratified or Process?	How many per sample?	When should it be done?
KPOVs	Invoice Cycle Time	Continuous	From when the customer order is received until the customer receives the invoice	Data until send to customer will be taken out SAP. Time from electronic invoice until receipt will be measured by adding an electronic receipt to emailed invoice.	customer type, division, facility, shift, week of month	SAP to Excel	Invoice supervisor	Process	8 per day	Once an hour
KPIVs	Admin									
	Invoice Type									
	Invoice Line Items									

Figure 7.23.7 An example of a Data Collection Summary sheet.

24: LOAD CHART

OVERVIEW

A Load Chart is used to visually represent two primary project objectives:

- Understand the ability of a process to meet the pace of Customer demand
- Evaluate the distribution of labor content in the process

The Chart itself is comprised of two elements, as shown in Figure 7.24.1. The horizontal line represents the pace of Customer demand in the form of the Takt Time. For more details see "Time—Takt Time" in this chapter. In effect this is the time that the process has available to process one entity. For some processes, a differing number of entities progresses through different steps; thus, the line might be staggered and not perfectly horizontal.

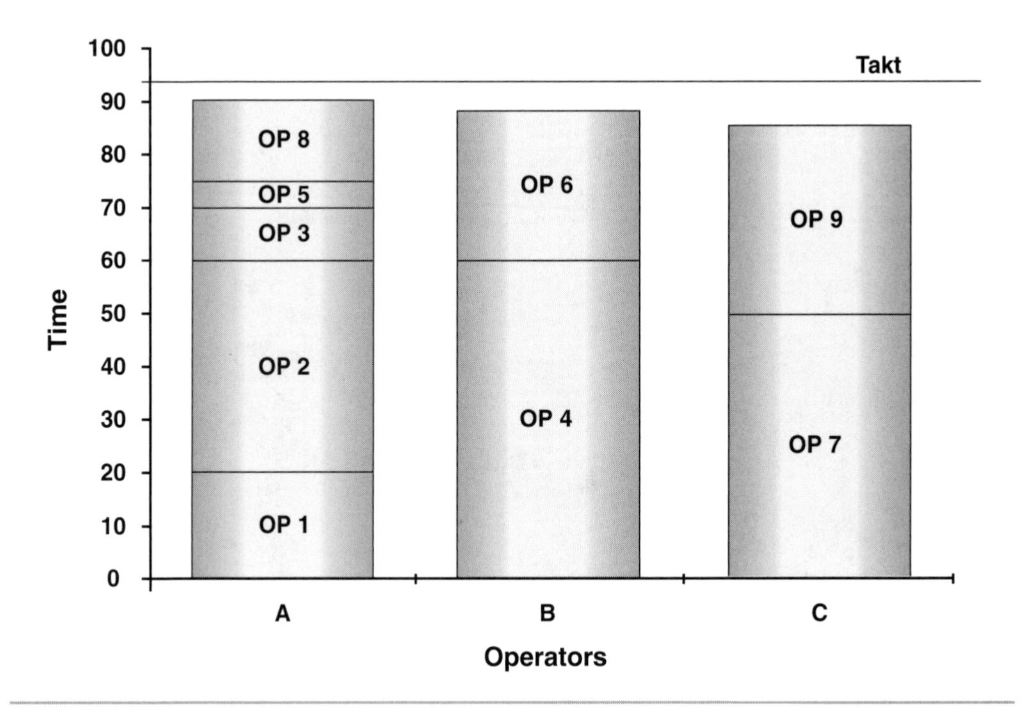

Figure 7.24.1 A Load Chart as applied to the operators in a process.

The second set of elements in the Load Chart is the vertical bars representing the Cycle Time of each operator in the process. For more details see "Time—Individual Step Cycle Time" in this chapter. If an operator bar rises above the Takt line, then the operator is cycling slower than the Takt Time; therefore, it is not cycling fast enough to keep up with Customer demand. The highest bar in the Load Chart is the bottleneck in the process. If any bar remains well below the Takt line, then that step cycles quicker than Takt and the process generates excess inventory or sits idle from time to time.

The goal clearly is to ensure that the process steps all meet Takt and all the bars sit just below the Takt line. In the goal scenario, there is no internal bottleneck and the Customer demand is the determining factor.

Logistics

Constructing a Load Chart doesn't have to be a team sport, but it does require a significant data capture, which might require Team help. For more information see "Time—Takt Time" and "Time—Individual Step Cycle Time" in this chapter.

After the data capture is complete it is a simple matter to construct the Load Chart in a spreadsheet, taking perhaps 10–15 minutes.

Roadmap

The steps to constructing the Load Chart are as follows (note that the first two steps have more than likely been completed earlier in the project roadmap):

Step 1: Identify the entity types flowing through the process and the volume of demand of each type in a given period, usually a month.

Step 2: Identify the steps in the process (usually done in a Process Map) and group the steps by Operator so that all steps done by Operator 1 are together, and so on (see Figure 7.24.1).

Step 3: Calculate the Takt Time for each Operator (see the "Time—Takt Time" in this chapter). This is done by identifying the Available Work Time for the Operator and dividing it by the demand (number of entities) for the Operator, both for the given time period. This becomes the height of the horizontal line in the Load Chart.

Step 4: Calculate the Individual Step Cycle Time for each step, that is how long it takes the Operator on average to complete the process step. Total the Individual Step Cycle Times for each operator for the steps they undertake.

For example, Operator B in Figure 7.24.1 completes operations 4 and 6. Operation 4 takes the Operator 60 seconds to complete and Operation 6 takes 28 seconds to complete. The Total Cycle Time for Operator B is 60 + 28 = 88 seconds. This becomes the bar height in the Load Chart.

Step 5: Draw the Load Chart using the numbers generated in Steps 3 and 4.

INTERPRETING THE OUTPUT

Figure 7.24.2 shows a typical Load Chart for a process prior to improvement. Each worker performs a fixed operation at each workstation and Cycle Times are not balanced. Operators 2 and 4 are cycling slower than Takt and the process cannot meet Customer demand. More than likely there is expediting or overtime involved in reducing backlog on this process.

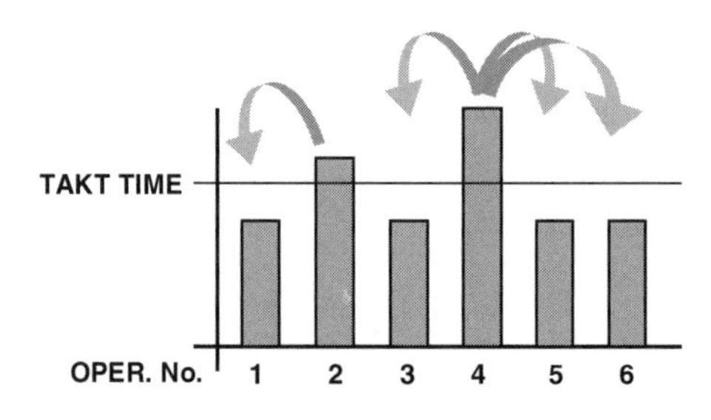

Figure 7.24.2 Load Chart for an imbalanced process (line).

Likewise, Operators 1, 3, 5, and 6 are cycling below Takt and thus are generating inventory or sitting idle or some combination of the two. It is common in a situation like this to see Operator 1 processing at full speed and a stockpile of inventory sitting in front of Operator 2. Operator 2 might receive criticism for not working hard enough and probably is a source of defects due to the hurried nature of the work. Operator 3 might appear to be the "star" of the line. Operator 3 appears completely on top of the work and can relax and still create a stockpile in front of Operator 4, the bottleneck in the process. Operator 4 is working flat out and still can't keep up and probably is another source of defects. Operators 5 and 6 are idle for a large portion of their time and grab and process everything they can from Operator 4 as soon as it is ready.

The solution lies in spreading the workload. Initially, the primary focus is initially on any bars that rise above the Takt Line. For these Operators, the work needs to be broken down and any NVA elements removed. If the bars are still above the line, then work needs to be offloaded to adjacent operators, or better still a complete redistribution of labor undertaken.

Figure 7.24.3 shows the resulting situation after the balancing has taken place. All Operators workloads are slightly below the Takt Line to balance output to demand.

It is possible to determine the number of Operators required in the process based on the Takt Time and the work content (or Cycle Time) of the steps, as follows:

$$\text{Number of Operators} = \frac{\text{Sum of Individual Step Cycle Times}}{\text{Takt time}}$$

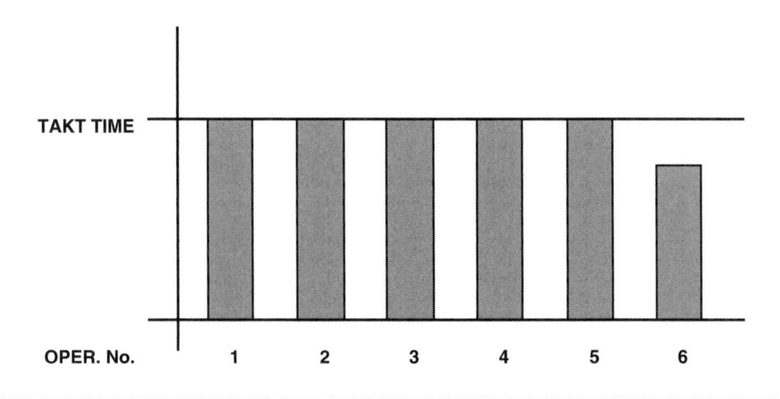

Figure 7.24.3 Load Chart after line balancing.

Of course, before doing this it is best to remove any NVA work in the process steps to minimize the total work done in the process and reduce the number of Operators required.

OTHER OPTIONS

Load Charts can be used on a larger scale by applying them to areas, rather than to individual Operators. For example, in a hospital Surgery Center, it is possible to have an individual bar for

- Pre-admission testing
- Intake and preparation
- Operating suite
- Post-anesthesia care
- Recovery

In this example, each area can be examined to determine if there is available capacity to increase throughput.

The difference in construction in this case is that the height of the bar would be the Process Cycle Time for that area, that is the speed at which a single patient can be processed. The Takt line would be calculated in the same way as before from the working hours of the area and the patient demand or the area.

Load Charts can also be applied to machines or equipment as shown in Figure 7.24.4. In this case, the bars comprise:

- Load time per cycle
- The machine Cycle Time (for a single cycle)
- Unload time per cycle
- Setup time averaged for a single cycle (i.e., if a 20-minute setup were required every 10 units, then 2 minutes would be applied as the Setup for a single cycle)

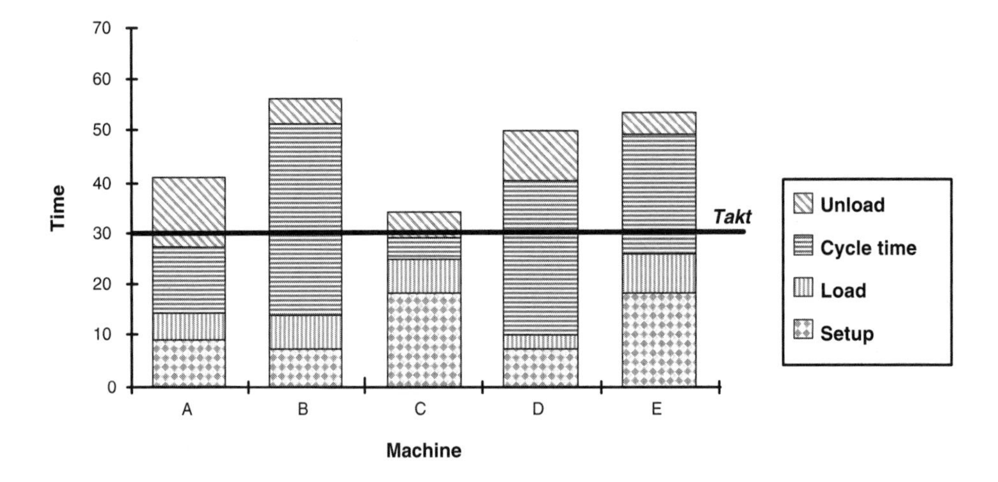

Figure 7.24.4 Application of a Load Chart to machines or equipment versus Operators.

The initial improvement targets are usually the Loads, Unloads, and Setups because these are NVA by definition.

By examining Available Machine Work Time and machine demand for a given period, the Takt Line is calculated in the same way as before.

25: MSA—Validity

Overview

Process variation affects how resulting products and services appear to Customers. However, what you (and ultimately the Customer) see as the appearance does not include only the variability in the entity itself, but also some variation from the way the entity is measured. A simple example of this is to pick up a familiar object, such as a pen. If you were to measure the diameter of the pen and perhaps judge if the lettering on the pen was "crisp" enough, and then you handed the same pen to three other people, it is likely that there would be a difference in answers among everyone. It is also highly likely that if someone handed you the same pen later (without you knowing it is the same pen) and asked you to measure it again, you would come to a different answer or conclusion. The pen itself has not changed; the difference in answers is purely due to the Measurement System and specifically errors within it. The higher the Measurement Error, the harder it is to understand the true process capability and behavior.

Therefore, it is crucial to analyze Measurement Systems *before* embarking on any Process Improvement activities.

It is always worth a little introspection here. You should ask if, for all the experiments and analyses done in the past, was the conclusion reached really what happened or was it driven purely by a noisy Measurement System?

The sole purpose of a Measurement System in Lean Sigma is to collect the *right* data to answer the questions being asked. To do this, the Team must be confident in the integrity of the data being collected. To confirm Data Integrity, the Team must know

- The type of data
- If the available data is usable
- If the data is suitable for the project
- If it is not suitable, whether it can be made usable
- How the data can be audited
- If the data is trustworthy

To answer these questions, Data Integrity is broken down into two elements:

- **Validity.** Is the "right" aspect of the process being measured? The data might be from a reliable method or source, but still not match the operational definitions established for the project.

And after Validity is confirmed (some mending of the Measurement System might be required first):

- **Reliability.** Is the valid measurement system producing good data? This considers the accuracy and consistency of the data.

Validity is covered in this section; Reliability is dependent on the data type and is covered in "MSA—Attribute" and "MSA—Continuous" in this chapter.

To confirm Validity of data, the commonest approach is to make use of a Data Integrity Audit, which has the simple aim of determining whether the data is correct and valid. Through the Audit, the Team seeks to assure themselves that the data being used is a clear and accurate record of the actual characteristics or events of interest.

LOGISTICS

Performing a Data Integrity Audit requires all the Team to participate, together with all other personnel that are part of the subsequent data collection. Participation is in the planning and structuring of the Audit, in the actual data collection itself, or both.

The Audit itself requires a short data collection to verify that the systems and processes used to capture and record data are robust. An Audit is typically not done for a single metric at a time, but applied to a complete data capture of multiple metrics. For example, an Audit is applied to the whole data capture for a *Multi-Vari Study* or *Multi-Cycle Analysis*, rather than just for a single X, otherwise the validation process would take too long.

Planning for the Audit usually takes a Team about 60 minutes, and the Audit itself typically runs for no more than 5–10 data points captured over a period of about a day, or until a major flaw is found in the data capture mechanism. If there are no problems found from the Audit, the Team should continue with the data capture as originally planned.

ROADMAP

The roadmap to confirming Validity of a planned data capture approach is as follows:

Step 1: The Data Integrity Audit is applied to a data capture for another tool, such as a *Multi-Vari Study*; therefore, it is necessary to have the details of the other tool, specifically:

- Goal
- Target metrics
- Sampling Plan

For these, it is useful to refer to "KPOVs and Data" in this chapter.

The Data Integrity Audit is a short pilot run of the Sampling Plan with the aim of confirming validity of the data.

Step 2: For each metric in the Sampling Plan, determine how the metric can be Validated by a parallel method of capture. Audits must be independent of the data collection, processing, and reporting systems that are being assessed. This can be accomplished by any method that makes sense, but the key is that there must be a second, independent source of data to compare against the "normal" data system.

For most organizations, data is kept in computer databases; so it is useful to have a representative from the IT organization involved at this point in the project. There are times when portions (if not all) of the data processes are already being checked automatically for data integrity.

If there are no automated systems involved, then a second manual, parallel data capture is required.

For each metric in the Sampling Plan, the Team must agree on acceptance requirements for the metric (i.e., how good the metric must be to be deemed valid). A complete list of criteria should be agreed upon before the Audit is conducted so that expectations are clear.

Step 3: Begin the Data Collection as per the Sampling Plan for 5–10 data points, with the Audit data being collected in parallel as per the Audit Plan in Step 2. For the Data Collection, consider the validity of the data as follows:

- Is the recorded data what the Team meant to record? It is useful to refer back to the Operational Definition of the metric at this point (see "KPOVs and Data" in this chapter).
- Does it contain the intended information?
- Does the measure discriminate between different items?
- Does it reliably predict future performance?
- Does it agree with other measures designed to find the same thing?
- Is the measure stable over time?

If points captured are clearly invalid then stop the data collection.

The temptation during such an Audit is to give it a go and see what happens and then regroup, make tweaks, and redo the Audit. It is always best to try to do the Audit right the first time.

Performing a thorough data method validation can be a tedious process, but the quality of data generated from the Sampling Plan is directly linked to the success of the project.

Step 4: Based on the Audit results in Step 3, take any actions required to mend the Sampling Plan or data capture mechanism. Sometimes Teams are tempted, after they are presented with poor Audit results (an invalid data collection system), to try to make excuses to just continue without remedying the situation. Again the Team must remember that the quality of data generated from the Sampling Plan is directly linked to the success of the project. The consequences of invalid data validation methods always vastly exceed what would have been expended initially if the validation studies had been performed properly.

Step 5: Rerun what is effectively a confirmatory Audit.

Interpreting the Output

After the Audit is complete and the Team is satisfied with the validity of the metrics in question in the Sampling Plan, then the Reliability of each metric must be determined using an MSA such as Gage Repeatability and Reproducibility Study (see "MSA—Continuous" and "MSA—Attribute" in this chapter).

26: MSA—ATTRIBUTE

OVERVIEW

Process variation affects how resulting products and services appear to Customers. However, what you (and ultimately the Customer) see as the appearance usually does not include only the variability in the entity itself, but also some variation from the way the entity is measured. A simple example of this is to pick up a familiar object, such as this book. If you were to judge if the lettering on this page was "crisp" enough and then you handed the same page to three other people, it is highly likely that there would be a difference in answers amongst everyone. It is also likely that if someone handed you the same page later (without you knowing it was the same page) and asked you to measure it again, you would come to a different answer or conclusion. The page itself has not changed; the difference in answers is purely due to the Measurement System and specifically errors within it. The higher the Measurement Error, the harder it is to understand the true process capability and behavior.

Thus, it is crucial to analyze Measurement Systems *before* embarking on any Process Improvement activities.

The sole purpose of a Measurement System in Lean Sigma is to collect the *right* data to answer the questions being asked. To do this, the Team must be confident in the integrity of the data being collected. To confirm Data Integrity the Team must know

- The type of data
- If the available data is usable
- If the data is suitable for the project
- If it is not suitable, whether it can be made usable
- How the data can be audited
- If the data is trustworthy

To answer these questions, Data Integrity is broken down into two elements:

- **Validity.** Is the "right" aspect of the process being measured? The data might be from a reliable method or source, but still not match the operational definitions established for the project.

And after Validity is confirmed (some mending of the Measurement System might be required first):

- **Reliability.** Is the valid measurement system producing good data? This considers the accuracy and consistency of the data.

Validity is covered in the section "MSA—Validity" in this chapter. Reliability is dependent on the data type. Attribute Measurement Systems are covered in this section; Continuous Measurement Systems are covered in "MSA—Continuous" in this chapter.

An Attribute MSA study is the primary tool for assessing the reliability of a qualitative measurement system. Attribute data has less information content than variables data, but often it is all that's available and it is still important to be diligent about the integrity of the measurement system.

As with any MSA, the concern is whether the Team can rely on the data coming from the measurement system. To understand this better it is necessary to understand the purpose of such a system. Attribute inspection generally does one of three things:

- Classifies an entity as either Conforming or Nonconforming
- Classifies an entity into one of multiple categories
- Counts the number of "non-conformities" per entity inspected

Thus, a "perfect" Attribute Measurement System would

- Correctly classify every entity
- Always produce a correct count of an entity's non-conformities

Some attribute inspections require little judgment because the correct answer is obvious; for example, in destructive test results, the entity either broke or remained intact. In the majority of cases (typically where no destruction occurs), however, it is extremely subjective. For such a system, if many appraisers (the generic MSA terminology for those doing the measurement) are evaluating the same thing they need to agree

- With each other
- With themselves
- With an expert opinion

In an Attribute MSA, an audit of the Measurement System is done using 2–3 appraisers and multiple entities to appraise. Each appraiser and an "expert" evaluate every entity at least twice and from the ensuing data the tool determines

- Percentage overall agreement
- Percentage agreement within appraisers (Repeatability)

- Percentage agreement between appraisers (Reproducibility)
- Percentage agreement with known standard (Accuracy)
- Kappa (how much better the measurement system is than random chance)

LOGISTICS

Conducting an Attribute *MSA* is all about careful planning and data collection. This is certainly a Team sport because at least two appraisers are required, along with an expert (if one exists), and it is unlikely that Belts apply the Measurement System in their regular job (i.e., the Belt almost certainly won't be one of the appraisers used in the *MSA*).

Planning the MSA takes about two hours, which usually includes a brief introduction to the tool made by the Belt to the rest of the Team and sometimes to the appraisers. Data collection (conducting the appraisals themselves) can take anywhere between an hour and a week, depending on the complexity of the measurement.

ROADMAP

The roadmap to planning, data collection, and analysis is as follows:

Step 1: Identify the metric and agree within the Team on its Operational Definition (see "KPOVs and Data" in this chapter). Often the exact Measurement System in question isn't immediately obvious. For example, in many transactional service processes, it could be the initial writing of the line items to an order, the charging of the order to a specific account, or the translation of the charges into a bill. Each of these might involve a separate classification step.

Step 2: Identify the defects and classifications for what makes an entity defective. These should be mutually exclusive (a defect cannot fall into two categories) and exhaustive (an entity must fall into at least one category, which typically means use of a category "Defect Free"). If done correctly every entity must fall into one and only one category and all defect categories should be treated equally (there should be no appraiser bias for one defect type over another).

Step 3: Select samples to be used in the MSA. From 30 to 50 samples are necessary, and they should span the normal extremes of the process with regards to the attribute being measured. Entities to be measured should be independent from one another. The majority of the samples should be from the "gray" areas, whereas some are clearly good and clearly bad. For example, for a sample of 30 units, five units might be clearly defective (a single, large defect or enough smaller ones to be an obvious reject), five units might be clearly acceptable (everything correct), and the remaining samples would vary in quantity and type of defects.

Step 4: Select 2–3 appraisers to conduct the MSA. These should be people who normally conduct the assessment.

Step 5: Perform the appraisal. Randomly provide the samples to one appraiser (without him knowing which sample it is or the other appraisers witnessing the appraisal) and have him rate the item. After the first appraiser has reviewed all items, repeat with the remaining appraisers. Appraisers must inspect and classify independently. After all appraisers have rated each item, repeat the whole process for one additional trial.

Step 6: Conduct an expert appraisal or complete a comparison to a Standard. In Step 5 the appraisers were compared to themselves (Repeatability) and to one another (Reproducibility). If the appraisers are not compared to a standard, the Team might gain a false sense of security in the Measurement System.

Step 7: Enter the data into a statistical software package and analyze it. Data is usually entered in columns (Appraiser, Sample, Response, and Expert). The analysis output typically includes

- Percentage overall agreement
- Percentage agreement within appraisers (Repeatability)
- Percentage agreement between appraisers (Reproducibility)
- Percentage agreement with known standard (Accuracy)
- Kappa (how much better the measurement system is than random chance)

Interpreting the Output

Figure 7.26.1 shows example graphical output from an Attribute MSA. The left side of the graph shows the agreement within the appraisers (analogous to Repeatability), and the right shows the agreement between the appraisers and the standard. The dots represent the actual agreement from the study data; the crosses represent the bounds of a 95% confidence interval prediction for the mean of agreement as the Measurement System is used moving forward.

The associated Within Appraiser statistics are shown in Figure 7.26.2. For example, Appraiser 1 agreed with himself in seven out of the ten samples across the two trials. Moving forward, agreement would likely be somewhere between 34.75% and 93.33% (with 95% confidence). To gain a narrower confidence interval, more samples or trials would be required. To be a good, reliable Measurement System, agreement needs to be 90% or better.

The associated Appraiser versus Standard statistics are shown in Figure 7.26.3. For example, Appraiser 1 agreed with the standard in five out of the ten samples. Moving forward, agreement would likely be somewhere between 18.71% and 81.29% (with a 95% degree of confidence). To be a usable Measurement System, agreement needs to be 90% or better, which is clearly not the case here.

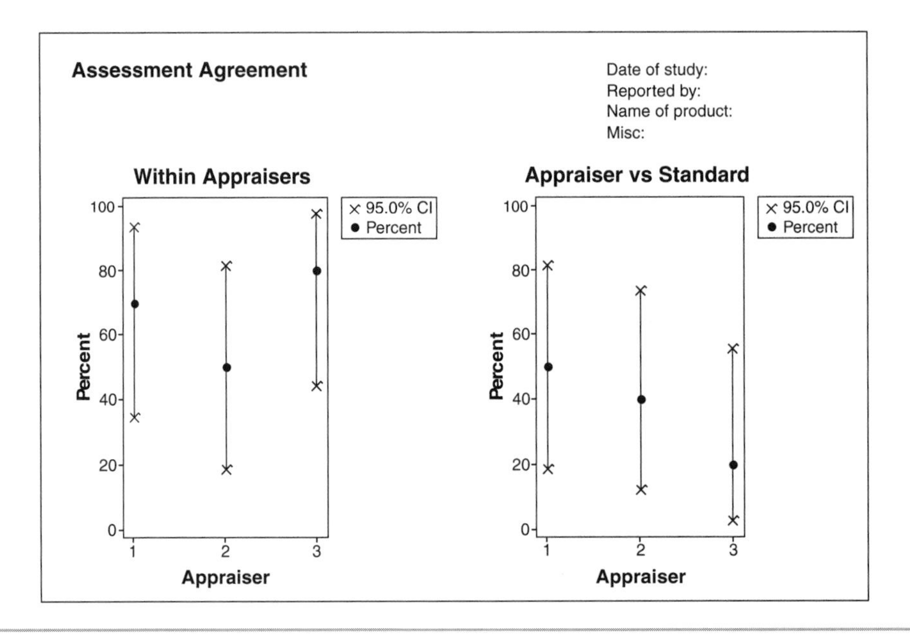

Figure 7.26.1 An example of an Attribute MSA graphical analysis (output from Minitab v14).

Within Appraisers

Assessment Agreement

Appraiser	# Inspected	# Matched	Percent	95% CI
1	10	7	70.00	(34.75, 93.33)
2	10	5	50.00	(18.71, 81.29)
3	10	8	80.00	(44.39, 97.48)

Matched: Appraiser agrees with him/herself across trials.

Figure 7.26.2 An example of a Within Appraiser Agreement (output from Minitab v14).

Each Appraiser Vs Standard

Assessment Agreement

Appraiser	# Inspected	# Matched	Percent	95% CI
1	10	5	50.00	(18.71, 81.29)
2	10	4	40.00	(12.16, 73.76)
3	10	2	20.00	(2.52, 55.61)

Matched: Appraiser's assessment across trials agrees with the known standard.

Figure 7.26.3 An example of an Appraiser Agreement versus Standard (output from Minitab v14).

The associated Between Appraiser statistics are shown in Figure 7.26.4. The Appraisers all agreed with one other in four out of the ten samples. Moving forward, agreement would likely be somewhere between 12.16% and 73.76% (with a 95% degree of confidence). To be a reliable Measurement System, agreement needs to be 90% or better, which is clearly not the case here.

Between Appraisers

Assessment Agreement

# Inspected	# Matched	Percent	95% CI
10	4	40.00	(12.16, 73.76)

Matched: All appraisers' assessments agree with each other.

Figure 7.26.4 An example of a Between Appraiser Agreement (output from Minitab v14).

Another useful statistic in Attribute MSA is Kappa, defined as the proportion of agreement between raters after agreement by chance has been removed. A Kappa value of + 1 means perfect agreement. The general rule is that if Kappa is less than 0.70, then the measurement system needs attention. Table 7.26.1 shows how the statistic should be interpreted.

Table 7.26.1 Interpreting Kappa Results

Kappa	Interpretation
–1 to 0.6	Agreement expected by chance
0.6 to 0.7	Marginal—Significant effort required
0.7 to 0.9	Good—Improvement warranted
0.9 to 1.0	Excellent

Figure 7.26.5 shows example Kappa statistics for a light bulb manufacturer's final test Measurement System with five defect categories (Color, Incomplete Coverage, Misaligned Bayonet, Scratched Surface, and Wrinkled Coating). For the defective items, appraisers are required to classify the defective by the most obvious defect. Color and Wrinkle are viable classifications (Kappa > 0.9), but for the rest of the classifications agreement cannot be differentiated from random chance. The p-values here are misleading because they are related to a test of whether the Kappa is (greater than) zero. Just look to the Kappa values for this type of analysis. The overall reliability of the Measurement System is highly questionable because the Kappa for the Overall Test is only 0.52.

Between Appraisers

Fleiss' Kappa Statistics

Response	Kappa	SE Kappa	Z	P(vs>0)
Color	1.00000	0.0816497	12.2474	0.0000
Incomplete	0.16279	0.0816497	1.9938	0.0231
Misaligned	0.28409	0.0816497	3.4794	0.0003
Scratch	0.17241	0.0816497	2.1116	0.0174
Wrinkle	0.92298	0.0816497	11.3041	0.0000
Overall	0.52072	0.0473565	10.9958	0.0000

Figure 7.26.5 An example of Kappa statistic results for multiple defect categories (output from Minitab v14).

After MSA data has been analyzed, the results usually show poor reliability for many Attribute MSAs. This is mainly due to sheer number of ways these types of Measurement Systems can fail:

- Appraiser
  - Visual acuity (or lack of it)
  - General intelligence (more specifically common sense) and comprehension of the goal of the test
  - Individual method of inspection adopted
- Appraisal
  - Defect probability. If this is very high, the appraiser tends to reduce the stringency of the test. The appraiser can become numbed or hypnotized by the sheer monotony of repetition. If this is very low, the appraiser tends to get complacent and tends to see only what he expects to see. This happens in spite of good visual acuity.
  - Fault type. Some defects are far more obvious than others.
  - Number of faults occurring simultaneously. If this is the case, the appraiser has to make the judgment into which category the defective should be classified.
  - Time allowed for inspection.
  - Frequency of rest periods for the appraiser.
  - Illumination of the work area.
  - Inspection station layout. Quite often there isn't enough space to conduct the test effectively or sometimes dedicated space is not provided at all.
  - Time of day and length of time the appraiser has been working.
  - Objectivity and clarity of conformance standards and test instructions.

- Organization and environment.
  - Appraiser training and certification.
  - Peer standards. Defectives are often deemed to reflect badly on coworkers, so the appraiser is constantly under pressure to artificially reduce the defect rates.
  - Management standards. Appraisers often report to someone who is accountable for the volume of shipped product. Reduced volumes reflect poorly on this individual, so they sometimes unconsciously apply pressure on the appraiser to ensure volumes remain high (and defects are allowed to slip by).
  - Knowledge of operator or group producing the item.
  - Proximity of inspectors.
  - Re-inspection versus immediate shipping procedures.

Given all the preceding possibilities for failure it should be apparent why Belts are strongly encouraged to move to Continuous metrics versus Attribute ones. If only Attribute measures are feasible, there are some actions that help improve reliability of the metric, but there really are no guarantees in this area:

- Set very clear Operational Definitions of the metric and defect classifications.
- Train and certify appraisers and revisit this on a regular basis.
- Break up the numbing rhythm with pauses.
  - Introduce greater variety into the job by giving a wider assortment of duties or greater responsibility
  - Arrange for frequent job rotation.
  - Introduce regular rest periods.
- Enhance faults to make them more noticeable.
  - Sense Multipliers—Optical magnifiers, sound amplifiers, and other devices to expand the ability of the unaided human to sense the defects/categories.
  - Masks—Used to block out the appraisers view of irrelevant characteristics to allow focus on key responsibilities.
  - Templates—These are a combination of a gage, a magnifier, and a mask—for example, a cardboard template placed over terminal boards. Holes in the template mate with the projecting terminals and serve as a gage for size. Any extra or misplaced terminal prevents the template from seating properly. Missing terminals become evident because the associated hole is empty.

- Overlay—These are visual aids in the form of transparent sheets on which guidelines or tolerance lines are drawn. Judging size or location of product elements is greatly simplified by such guidelines.
- Checklists—For example, pre-flight checklist on aircraft.
- Product Redesign—In some situations, the product design makes access difficult or places needless complexity or burden on the inspectors. In such cases, product redesign can help reduce inspector errors, as well as operator errors.
- Error Proofing—There are many forms of this, such as redundancy, countdowns, and fail-safe methods (see "Poka Yoke (Mistake Proofing)" in this chapter).
- Automation—Replacement of repetitive inspection with automation that makes no inadvertent errors after the setup is correct and stable. Clearly there are limitations here of cost and the current state of technology.
- Visual aids—Keep an appraiser from having to rely on memory of the standard. The appraiser is provided with a physical standard or photographs against which to make direct comparisons. For example, in the automobile industry, painted plates are prepared exhibiting several scratches of different measured widths and other visual blemishes to define defects in the paint finish.

Attribute Measurement Systems are certainly the most difficult to improve, and it is important to check continually for appraiser understanding. It is generally useful to capture data on a routine basis on the proportion of items erroneously accepted or rejected and applying *Statistical Process Control* to the Measurement System based on this.

27: MSA—Continuous

Overview

Process variation affects how resulting products and services appear to Customers. However, what you (and ultimately the Customer) see as the appearance does not usually include only the variability in the entity itself, but also some variation from the way the entity is measured. A simple example of this is to pick up a familiar object, such as a pair of glasses. If you were to measure the thickness of the middle of the left lens and then you handed the same pair of glasses to three other people, it is highly likely that there would be a difference in answers between everyone. It is also highly likely that if someone handed you the same pair of glasses later (without you knowing it was the same pair) and asked you to measure again, you would come to a different answer or conclusion. The pair of glasses itself has not changed; the difference in answers is purely due to the Measurement System and specifically errors within it. The higher the Measurement Error, the harder it is to understand the true process capability and behavior.

Thus it is crucial to analyze Measurement Systems *before* embarking on any Process Improvement activities.

The sole purpose of a Measurement System in Lean Sigma is to collect the *right* data to answer the questions being asked. To do this the Team must be confident in the integrity of the data being collected. To confirm Data Integrity the Team must know

- The type of data
- If the available data is usable
- If the data is suitable for the project
- If it is not suitable, whether it can be made usable
- How the data can be audited
- If the data is trustworthy

To answer these questions, Data Integrity is broken down into two elements:

- **Validity.** Is the "right" aspect of the process being measured? The data might be from a reliable method or source, but still not match the operational definitions established for the project.

And after Validity is confirmed (some mending of the Measurement System might be required first):

- **Reliability.** Is the valid measurement system producing good data? This considers the accuracy and consistency of the data.

Validity is covered in "MSA—Validity" in this chapter. Reliability is dependent on the data type. Continuous Measurement Systems are covered here with a tool called Gage Repeatability and Reproducibility[44] (Gage R&R); Attribute Measurement Systems are covered in the section "MSA—Attribute" in this chapter.

Gage R&R is an audit conducted on a Continuous Measurement System which is done using 2–3 people and multiple entities to measure. Each person measures every entity at least twice and from the ensuing data the tool determines

- Percentage overall agreement (% Repeatability & Reproducibility)
- Percentage agreement within individuals (% Repeatability, agreement with themselves)
- Percentage agreement between individuals (% Reproducibility, agreement with others)

[44] For more detail see "Measurement Systems Analysis—MSA" an ASQC and AIAG Publication.

Belts often think that the Gage R&R Study is just on the Gage itself, but the Study examines the *whole* Measurement System including the samples, people, techniques, and methods. This becomes important in "Interpreting the Output" in this section.

Many Belts confuse a Gage R&R Study with calibration of a piece of equipment; the two are different. Calibration considers only the *average* reading of a gage. The calibrator measures a known entity using the Measurement System 10–15 times and then takes the average of the readings. The Measurement System is then adjusted so that it is zeroed correctly (i.e., the mean is changed and any bias removed). This is generally quite straightforward for many Measurement Systems, similar to zeroing a set of bathroom scales.

Gage R&R is significantly different and often far more difficult, in that it analyzes the *variation* in the measurement system. The reason for doing this is that the variability detected in the process (entities measured) is actually comprised of the true process variation, but also the variability in the Measurement System (see Figure 7.27.1):

Total Variation = Process Variation + Measurement Variation

Or in statistical terms:

$$\sigma^2_{\text{total}} = \sigma^2_{\text{process}} + \sigma^2_{\text{Measurement}}$$

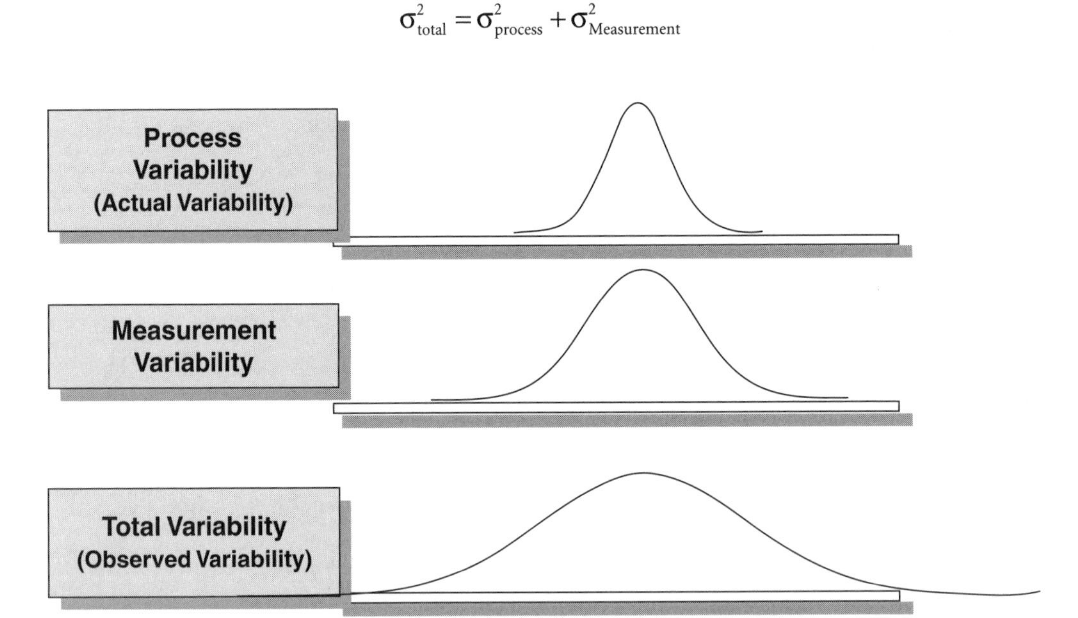

Figure 7.27.1 The effect of Measurement System Variation on the Total Variation in a process.

Thus, to effectively see the variation in the process data, the variability due to the Measurement System should be small. The purpose of a Gage R&R Study is to determine

- The size of the measurement error
- The sources of measurement error
- Whether the Measurement System is stable over time
- Whether the Measurement System is capable for the study
- Where in the Measurement System to focus improvements

As previously mentioned, the Study breaks the total observed variation in the process down into the two components of Actual variation and Measurement System variation. It also takes the Measurement System Variation and breaks it into the variation due to Repeatability plus the variation due to Reproducibility:

$$\sigma^2_{\text{MS}} = \sigma^2_{\text{Repeatability}} + \sigma^2_{\text{Reproducibility}}$$

Repeatability is the inherent variability of the Measurement System and is the variation that occurs when repeated measurements are made of the same variable under absolutely identical conditions. It is the variation between successive measurements of the same sample, of the same characteristic, by the same person using the same instrument. Figure 7.27.2 shows a graphical representation of Repeatability. Poor Repeatability causes an increase in decision error. When the same person looks at the same attribute and estimates different values then that person likely makes different decisions based on those estimates. Some of these decisions are the wrong decision!

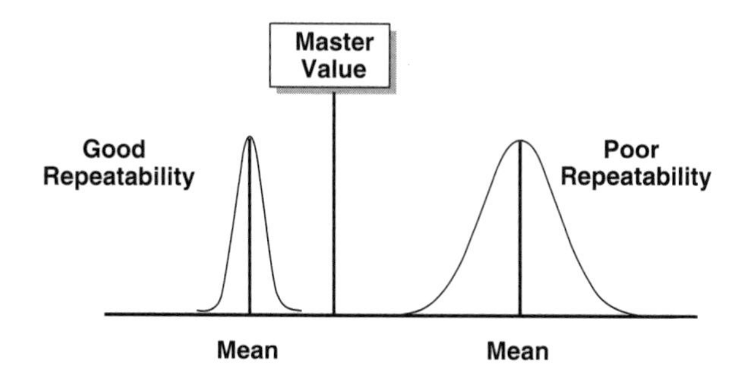

Figure 7.27.2 A graphical representation of Repeatability.[45]

[45] Source: SBTI's Lean Sigma Methodology training material.

Reproducibility is the variation that results when different people are used to make the measurements using the same instrument when measuring the identical characteristic with different conditions (time, environment, and so on). When two or more individuals return the same value for a given attribute, that measure is said to be Reproducible. A graphical representation is shown in Figure 7.27.3. When Reproducibility is not present, the value of a metric depends on who collects the measurements.

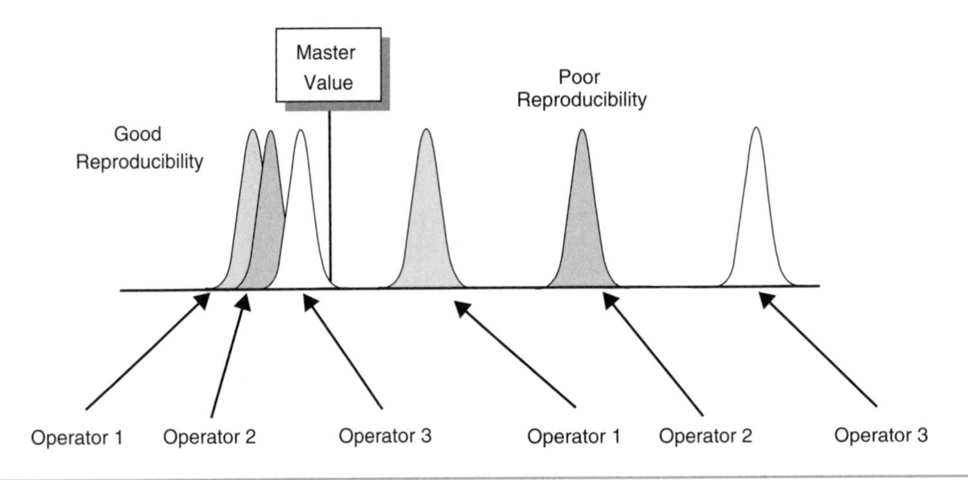

Figure 7.27.3 A graphical representation of Reproducibility.[46]

To effectively use a Gage R&R Study, it is important to understand the purpose of the Measurement System in question. Is it a "production gage" used to determine if product is in or out of specification, or is it a tool to measure a process characteristic in a project to improve process performance by reducing process variation? In the former it is important to understand the size of the Measurement System error with respect to the size of the specifications. The associated metric is known as the "Precision to Tolerance Ratio" and is defined as

$$\%\ P/T = \frac{5.15 \times \sigma_{MS}}{\text{Tolerance}} \times 100$$

The P/T Ratio represents the percent of the tolerance taken up by measurement error.[47] The metric includes both Repeatability and Reproducibility. An excellent Measurement System

[46] Source: SBTI's Lean Sigma Methodology training material.

[47] 5.15 standard deviations account for 99% of Measurement System variation. The use of 5.15 is an industry standard, but more recently some texts recommend the use of 6 standard deviations, representing 99.73% of Measurement System variation. Either value is appropriate provided that it is used consistently across the business.

has a P/T Ratio less than 10%. A value of 30% is barely acceptable. It is important to note that having the correct value for the Tolerance is crucial. In many cases, the specifications are too tight or too loose, which can be misleading.

If the Measurement System is used for process improvement, then a more appropriate metric is the %R&R, which represents the percentage of the Total Variation taken up by measurement error:

$$\% \text{ R\&R} = \frac{\sigma_{\text{MS}}}{\sigma_{\text{Total}}} \times 100$$

The metric includes both Repeatability and Reproducibility. An excellent Measurement System has a %R&R less than 10%. A value of 30% is barely acceptable.

The final metric of interest to a Lean Sigma Belt is Discrimination, which represents the number of decimal places that can be measured by the system. Increments of measure should be about one-tenth of the width of the product specification or process variation (depending on the use of the Measurement System).

LOGISTICS

Conducting a Gage R&R Study is about careful planning and data collection. This is certainly a Team sport because at least two appraisers are required, and it is unlikely that Belts apply the Measurement System in their regular job (i.e., the Belt almost certainly won't be one of the appraisers used in the MSA).

Planning the MSA takes about two hours, which usually includes a brief introduction to the tool made by the Belt to the rest of the Team and sometimes to the other appraisers. Data collection (conducting the appraisals themselves) can take anywhere between an hour and a week, depending on the complexity of the measurement.

ROADMAP

The roadmap to planning, data collection, and analysis is as follows:

Step 1: Identify the metric and agree within the Team on its Operational Definition (see "KPOVs and Data" in this chapter).

Step 2: Select samples to be used in the Study. From 6 to 12 Samples are necessary and selection of each should be independent from the others. Samples should span the normal variation of the process; for example, for a material with a mean thickness of 0.020 inches and a standard deviation of 0.001 inches samples should have thickness from 0.017 to 0.023 inches (99% of the range). Do not randomly draw samples from the process as they tend to be grouped close to the mean and not represent the full width of the process.

If the same process generates three different products with three (significantly) different thicknesses, then the Team should perform three separate studies—one for each thickness. If data for the samples were lumped together, the %R&R value would be artificially low.

Step 3: Select 2–4 appraisers to conduct the MSA. These should be people who normally conduct the assessment. If process uses only one operator or no operators at all, then perform the study without operator effects (Reproducibility effects are thus ignored).

Step 4: Select the number of trials. This needs to be at least two and the total number of data points (samples × appraisers × trials) should be greater than 30. For example, for five samples and two operators, it would be best to use three or four trials to generate 30 or 40 data points.

Step 5: Calibrate the gage, or assure that it has been calibrated.

Step 6: Perform the appraisal. Randomly provide the samples to one appraiser (without them knowing which sample it is) and have them measure the item. After the first appraiser has measured all the entities, repeat with the remaining appraisers. Appraisers must measure independently and out of sight of other appraisers to minimize potential bias. After all appraisers have measured each item, repeat the whole process for the required number of trials.

Step 7: Enter the data into a statistical software package and analyze it. Data is usually entered in columns (Appraiser, Sample, and Response). The analysis output typically includes

- Repeatability
- Reproducibility
- %R&R
- P/T Ratio

INTERPRETING THE OUTPUT

Statistical software packages generally produce both analytical and graphical analysis information. Each graph shows a different piece of the puzzle. Belts often try to read too much into each of the graphs; it is the combined story from all of the graphs that describes the Measurement System. Figure 7.27.4 shows an example Xbar-R Chart from a Gage R&R Study (for more details see "Control Charts" in this chapter).

For the Xbar-R Chart, if the averages for each operator are different, then the reproducibility is suspect. The majority of the points on the chart should fall outside the control limits consistently for all operators. If there are no points outside the control limits, it is generally because samples were not selected to cover the full range of the process (i.e., there was not enough Part-To-Part variation).

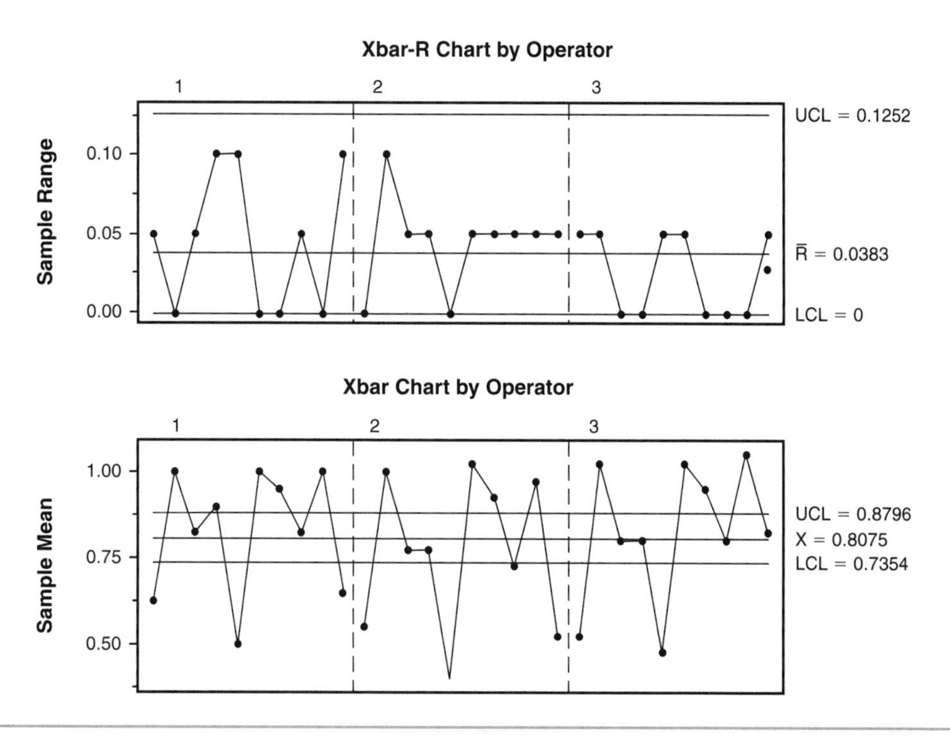

Figure 7.27.4 An example of a Gage R&R Xbar-R Chart (output from Minitab v14).

The Range Chart should show a process that is in control. The Ranges are the differences between trials and should not show any special causes of variation (i.e., remain in control). If a point is above the UCL, the operator is having a problem making consistent measurements. The Range Chart can also help identify inadequate discrimination; there should be least five distinct levels within the Control Limits. Also, if there are five or more levels for the range but more than 1/4 of the values are zero, then Discrimination is suspect. Repeatability is questionable if the Range Chart shows out-of-control conditions. If the Range Chart for an operator is out-of-control and the other Charts are not, then the method is probably suspect. If all operators have ranges out-of-control, the system is sensitive to operator technique.

Figure 7.27.5 shows an example of an Operator-Part Interaction Plot. For a reliable Measurement System, the lines should follow the same pattern and be reasonably parallel to each other. Crossing lines between operators indicates significant interactions. Also the part averages should vary enough that the differences between parts are clear.

Figure 7.27.6 shows an example of a By Operator graph, which shows the average value (Circle) and the spread of the data for each operator. The spread should be similar across all operators and there should be a flat line across the means of the operators.

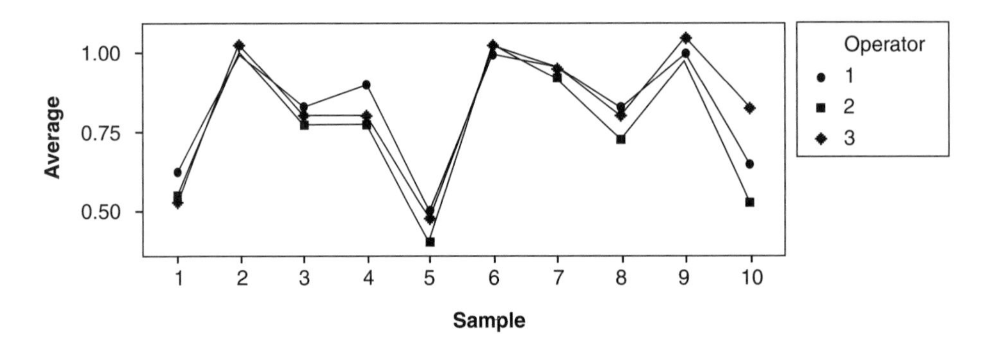

Figure 7.27.5 An example of Gage R&R Operator-Part Interaction Plot (output from Minitab v14).

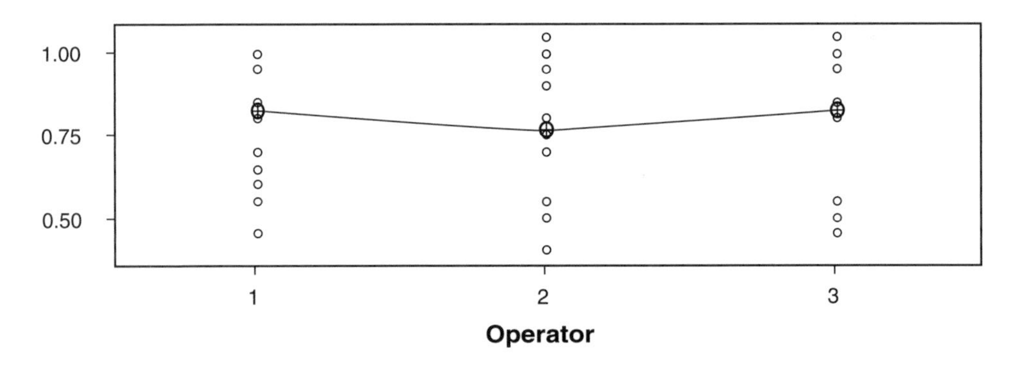

Figure 7.27.6 An example of a Gage R&R By Operator Plot (output from Minitab v14).

Figure 7.27.7 shows an example of a By Part graph. The graph shows the average (circles) and spread of the values for each sample. There should be minimal spread for each part (all the circles on top of each other), but variability between samples (different means).

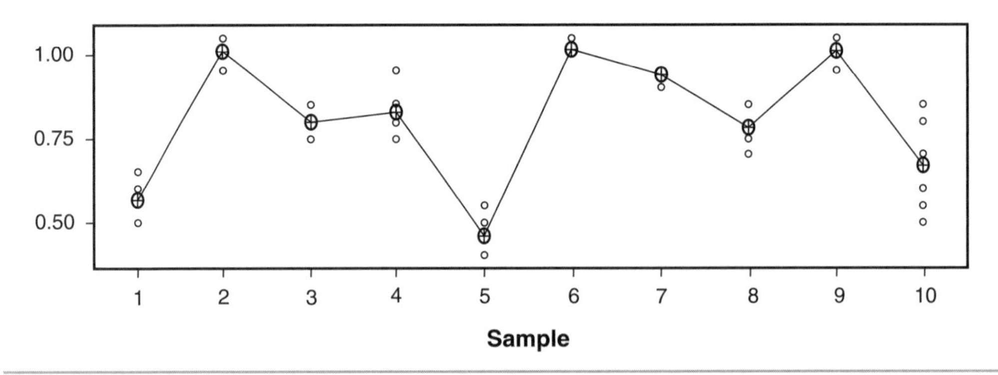

Figure 7.27.7 An example of a Gage R&R By Part Plot (output from Minitab v14).

Figure 7.27.8 shows an example of a Components of Variation graph. The Gage R&R bars should be as small as possible, driving the Part-to-Part bars to be larger. This is better understood by looking at the analytical representation of the same data, which is shown in Figure 7.27.9.

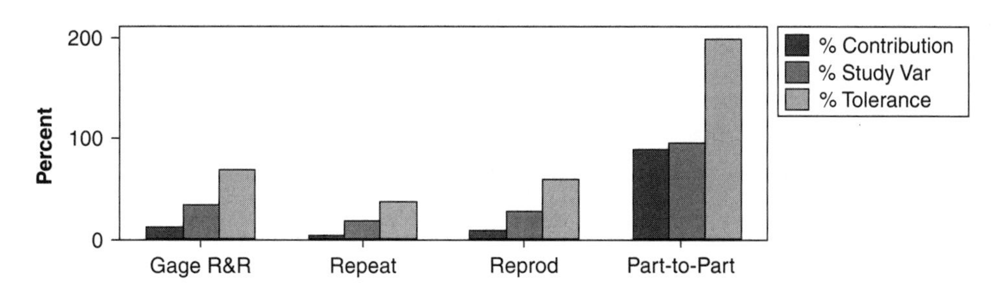

Figure 7.27.8 An example of a Gage R&R Components of Variation Plot (output from Minitab v14).

Source	Std Dev (SD)	Study Var (5.15*SD)	%Study Var (%SV)	%Tolerance (SV/Toler)
Total Gage R&R	0.066615	0.34306	32.66	68.61
Repeatability	0.035940	0.18509	17.62	37.02
Reproducibility	0.056088	0.28885	27.50	57.77
Operator	0.030200	0.15553	14.81	31.11
Operator* Sample	0.047263	0.24340	23.17	48.68
Part-to-Part	0.192781	0.99282	94.52	198.56
Total Variation	0.203965	1.05042	100.00	210.08

Number of Distinct Categories = 4

Figure 7.27.9 An example of Gage R&R analytical results (output from Minitab v14).

The key metrics to look at in Figure 7.27.9 are

- The P/T Ratio (listed as the %Tolerance) at 68.61%—this gage is clearly not suitable as a production gage (30% is acceptable).
- The %R&R (listed as %Study Variation) at 32.66%—this gage is less than acceptable to help make improvements to the process in question (30% is acceptable).
- The majority of the variation comes from Reproducibility, which can be seen from its standard deviation (0.056088) versus Repeatability (0.035940). Appraisers aren't agreeing with one another.

- The largest portion of Reproducibility comes from an Operator-Sample interaction. In some way the Operators measure different samples in a different way. This sometimes occurs when one or more Appraisers aren't good with small parts, but can adequately measure larger parts, whereas others can measure all samples equally well.
- The Number of Distinct Categories is an indication of the Discrimination of the measurement system. If the number of categories is less than two, the measurement system is of minimal value because it is difficult to distinguish one entity from another. If the number of categories is two, the measurement system can only divide the data into two groups—low and high. If the number of categories is three, the measurement system can divide the data into three groups—low, medium, and high. A measurement system that is acceptable and useful for process improvement activities must have five or more distinct categories; ten or more is ideal.

Remember, the measurement system must be mended before collecting the data! The graphical and analytical results help guide the Team in understanding where to focus the improvement. Improvement could be as simple as (re)training appraisers or it could be a project in itself. For more details see the Problem Category for Measurement System Improvement in Chapter 3, "Global Process Problems."

OTHER OPTIONS

MSA is a broad, relatively well-documented subject area.[48] The approach shown in this section is for a straightforward non-destructive Measurement System. When considering other variations, such as destructive testing or on-line measures (where there are no operators), things become trickier. Analysis of this kind is beyond the scope of this book.

28: MULTI-CYCLE ANALYSIS

OVERVIEW

Multi-Cycle Analysis (not to be confused with *Multi-Vari Studies*) is a series of data capture tools applied to the process. Its intent is to understand the flow of the Primary Entity and any secondary entities along with operator and equipment activity, basically the interaction of all of the key elements in the process. It also provides the building blocks of data to be used in other Lean Sigma tools such as Standard Work & *Load Charts.*

[48] A useful reference here is "Measurement Systems Analysis—MSA" an ASQC and AIAG Publication.

The Team observes the operation of the elements of the process and then breaks down all activities into basic elements or tasks, recording step durations. Thus, for any process the activity of the following might be analyzed:

- Primary Entity
  - Product
  - Patient
- Secondary entities
  - Material
  - Information
- Operators
- Equipment
  - Machines
  - Rooms

After the data is captured, all the work activities are categorized as VA, NVA but Required (NVAR or required waste), and NVA (or pure waste).

Understanding the activity of the Primary Entity enables flow, which makes other wastes visible. Understanding activity of secondary entities, staff, and equipment helps to identity focused opportunities for removing work content and redundancy.

Logistics

Multi-Cycle Analysis cannot be done by the Belt in isolation. It requires the whole Lean Sigma Team plus other key individuals in the process to aid in collecting the data. As with many Lean Sigma tools, the key is in the planning and preparation.

Planning of the data capture typically takes 1–3 hours of Team time and the data capture itself takes somewhere between one day and one week depending on the drumbeat of the process.

Roadmap

The roadmap to planning and conducting the Multi-Cycle Analysis is as follows:

Step 1: Identify what the elements are in the process that should be tracked in the analysis, for example:

- Product
- Patient
- Material
- Information
- Operators
- Equipment
- Machines
- Room

Step 2: For each element in Step 1, identify the possible activities that it could be undertaking. Some examples are shown in Table 7.28.1.

Table 7.28.1 Possible Activities a Process Element Might Undertake

Product	**Patient**	**Paperwork/Chart**
Processing	Processing	Processing
Transportation	Transportation	Transportation
Storage	Waiting	Waiting
Inspection/Testing	Assessment	Inspection
	Motion (self propelled)	Changing Form/Media
	Paperwork	Replicating

Room	**Machine**	**Operator**
Processing	Processing	Processing
Empty Not Ready	Empty Not Ready	Transportation (carrying the Primary Entity)
Cleaning/Set Up	Cleaning	Waiting
Empty Ready	Maintenance	Assessment/Inspection
Full Idle	Loading	Replenishing
Check/Inspection	Full Idle	Motion
	Unloading	Paperwork
	Check/Inspection	

Step 3: From the *Value Stream Map* for the element considered, breakdown all activities into basic elements or tasks and list them in the data capture template as shown in Figure 7.28.1. There is one data sheet captured per element considered; so for example, if the process involves a Product, an order, and a room, then there are three data capture sheets, one for each.

Multi-Cycle Analysis

Entity ______________________ Date: ______________________

Process ______________________ Prepared by: ______________________

Task #	Type	Keep/ Omit	Operation Description		1	2	3	4	5	6	7	Summary Statistics		
												Avg.	Std Dev	Slowest
				Start Time										
1				Finish Time								0	0	0
				Time to Complete	0	0	0	0	0	0	0			
				Start Time										
2				Finish Time								0	0	0
				Time to Complete	0	0	0	0	0	0	0			
				Start Time										
3				Finish Time								0	0	0
				Time to Complete	0	0	0	0	0	0	0			
				Start Time										
4				Finish Time								0	0	0
				Time to Complete	0	0	0	0	0	0	0			
				Start Time										
5				Finish Time								0	0	0
				Time to Complete	0	0	0	0	0	0	0			
				Start Time										
6				Finish Time								0	0	0
				Time to Complete	0	0	0	0	0	0	0			
				Start Time										
7				Finish Time								0	0	0
				Time to Complete	0	0	0	0	0	0	0			
				Start Time										
8				Finish Time								0	0	0
				Time to Complete	0	0	0	0	0	0	0			
				Start Time										
9				Finish Time								0	0	0
				Time to Complete	0	0	0	0	0	0	0			
				Start Time										
10				Finish Time								0	0	0
				Time to Complete	0	0	0	0	0	0	0			
				Start Time										
11				Finish Time								0	0	0
				Time to Complete	0	0	0	0	0	0	0			

Task Type Codes:

VA = Value Added
NVA = Pure Waste
BVA = Required Waste

Figure 7.28.1 Raw data capture template.[49]

[49] Source: SBTI's Lean Methodology training material.

Step 4: Observe the operation and time each step for seven cycles for the element concerned. This can be done directly by the recorder using a stopwatch and noted on the appropriate sheet, or the process could be recorded with a video recorder and the timings documented later. This whole timing activity can be done in parallel streams if there are enough Team members available, one per element.

Step 5: For each activity recorded in Step 4, categorize the activity into one of the following and document this in the Type column.

- Value Added
- Non-Value Added but Required (necessary waste or required waste)
- Or Non-Value Added (pure waste)

Step 6: For each activity recorded in Step 4, complete the Summary Statistics on the right side of the template, calculating the Average time (this is best calculated as the truncated mean of the seven cycles by omitting the highest and lowest and taking the mean of the middle five data points), the variation in times (usually the Range or Standard Deviation), and the slowest time.

Step 7: Transfer the activity names and summary statistics to an appropriately constructed activity template, an example of which is shown in Figure 7.28.2. Each template lists the possible element activities as per Table 7.28.1. In the example, the activity columns P, T, M, and so on, are explained in the legend on the bottom right. The activities categories come directly from the possible patient activities listed in Table 7.28.1.

Step 8: Compare actual activities to work instructions and document any inconsistencies. Document all improvement suggestions as the Team examines the data. These are not acted on directly from the data at this stage, but are pooled into potential solutions as the project progresses into the Improve Phase.

INTERPRETING THE OUTPUT

Interpretation depends on the type of element concerned, but generally is as follows:

- **Primary Entity (Product, Patient, and so on).** As a percent of total entity throughput time, identify the measure of actual value added time.
- **Operator.** As a percent of the total available work time, identify the measure of actual VA time. Activities that equate to NVA time might include:
  - Expediting
  - Status updates
  - Walking (motion)
  - Waiting for an entity
  - Looking for supplies
  - Manual operations

Patient Activity Analysis

Process: Minor Treatment **VIDEO START TIME:** ______ **Date:** 4/5/2006

Patient Type: All

Operation							Analysis									Cost		Benefit	
Process Step	Keep or Omit	Description	Code	Cumulative (Mean)	Mean	Std Dev	P	T	Distance	W	AI	INF	M	Distance	O	Lo	Hi	Lo	Hi
1		Take Initial impression	VA	2	2	0.3	2												
2		Wait in line	NVA	14	12	3				14									
3		M/G gives blue form	BVA	15	1	0.1									1				
4		Pt fills out blue form	BVA	25	10	7						25							
5		Pt gives blue form back to M/G	BVA	26	1	0.5									1				
6		M/G enters name in Medsys	BVA	29	3	4	3												
7		Pt waits in waiting room	NVA	57	28	17				28									
8																			
9																			
10																			
11																			
12																			
13																			
14																			
15																			

Code
NVA: Non-Value Added
BVA: Business Value Added
VA: Value Added

T: transport patient
M: patient move (no transport)
P: process
AI: assess / inspect
W: wait
INF: patient filling in information / paperwork
O: other

Figure 7.28.2 Example activity template for a Patient as they pass through a process.[50]

[50] Adapted from SBTI's Lean Methodology training material.

- Sorting
- Rework errors
- Reviewing/inspecting
- Approvals
- Capturing unnecessary data
- Batch processing
- Data input
- Transporting an entity
- Performing a changeover
- Loading equipment
- Positioning entities
- Waiting while the machine cycles
- Unloading equipment

- **Equipment (machines, rooms, and so on).** Study here focuses on the flexibility of the equipment, for instance:
  - Determine the time required to perform a set-up or changeover.
  - Identify the cross-functionality the equipment has and whether it be used on several entity types
  - Identify how easy is it for the equipment to change usage type

The Multi-Cycle Analysis forms the basis data for subsequent Lean Sigma tools; it is not acted on directly. Elements are analyzed separately from the templates, but they are solved together as the project progresses into the Analyze Phase.

29: MULTI-VARI STUDIES

OVERVIEW

A Multi-Vari Study[51] is an overarching study that encompasses many other Lean Sigma tools. At the highest level a study comprises

- Data capture of Xs and the corresponding level of Ys at various times from the process (see "KPOVs and Data" in this chapter)
- Examination of the data using graphs to identify potential relationships between the Xs and the Ys
- Validation of the potential relationships using the appropriate statistical tools
- Statements of practical conclusion

Data for the Multi-Vari is captured in a passive way in that the process is allowed to run in its natural state rather than actively manipulating it as in a Designed Experiment (DOE).

[51] The name "Multi-Vari" was given to this methodology by L. A. Seder in the paper titled, "Diagnosis with Diagrams," which appeared in Industrial Quality Control in January and March, 1950.

Logistics

A Multi-Vari Study is a complex series of tools and thus planning is paramount. It is absolutely a Team activity and requires significant support from the Champion and Process Owner to ensure it proceeds to plan.

A typical Multi-Vari Study can take up to four to six hours to plan and is executed over a period of one day to two months depending on availability of data and the pace of the process in question.

Roadmap

The overarching roadmap is as follows (however, each tool used within the roadmap can have its own series of sub-steps that are described separately):

Step 1: Develop a Data Collection Plan. A Multi-Vari Study is used to understand the relationship between the Xs and Y(s) in the equation $Y = f(X_1, X_2, ..., X_n)$. Data is captured as a series of snapshots in time for the value of all the Xs and the corresponding values of the Y(s) for some 30–50 data points. Data is best captured in a directly analyzable form using a Data Sheet format as shown in Figure 7.29.1. To do this, develop a data collection plan as per "KPOVs and Data" in this chapter. The Xs should not be plucked out of thin air (a common Belt mistake); the Lean Sigma roadmap should have led the Team to this point with a reduced number of Xs from use of the *SIPOC*, VOC, KPOVs, *Cause and Effect (C&E) Matrix*, and the *Failure Modes & Effects Analysis (FMEA)*.

Shift	Operator	Cement Lot	Flame Temp	Line Speed	Seal Defects
1	1	1	550	38	1
1	1	2	560	37	2
1	1	3	560	38	8
1	1	4	550	37	9
1	2	1	550	38	4
1	2	2	580	38	3
1	2	3	570	39	8
1	2	4	570	36	12
2	1	1	570	39	5
2	1	2	560	38	5
2	1	3	560	38	8
2	1	4	550	37	14
2	2	1	580	39	4
2	2	2	560	38	6
2	2	3	550	37	12
2	2	4	550	39	17

Figure 7.29.1 An example of a Data Collection Sheet for Seal Defects.

Observe the process over a short period of time as a "proof of concept" study for the data collection plan, measuring and recording values for the KPOVs (Ys), and simultaneously recording the values of the KPIVs (Xs). Have the Team carefully observe the process and take plenty of notes. Revise the data collection plan as required.

Step 2: Collect the data as per the Data Collection plan (see "KPOVs and Data" in this chapter). This is a Team activity. Data should be collected until the process has revealed a full range of variation.

Step 3: Analyze the data collected graphical to identify any possible relationships or patterns. This is typically a Belt activity, rather than tying up the whole Team's time. The Belt reports back to the Team with the full analysis completed. Graphical tools might include

- Main Effects Plots
- Box Plots
- Multi-Vari Plot
- Time series Plots
- Scatter Plots
- Dot Plots
- Interaction Plots
- Histograms

The outputs of some of these for the example data shown in Figure 7.29.1 are shown in Figure 7.29.2.

For each possible pattern or relationship identified, the Belt documents the finding similar to the Seal Defects example here:

- Seal Defects average varies by Shift (Graphs B and C)
- Seal Defects average varies by Operator (Graphs B and D)
- Seal Defects average varies by Cement Lot (Graphs B and E)
- Seal Defects variation varies by Flame Temperature (Graph F)
- There is a cyclical pattern in Seal Defects versus Time (Graph A)

Note the word *possible*. Graphical evidence is not nearly enough to demonstrate conclusively that the preceding effects are real. Each pattern or relationship identified can be real, but each needs to be followed up with the appropriate statistical test.

Essentially the Belt is conducting an investigation of the data like a detective looking for clues, with the data and statistical test forming the evidence.

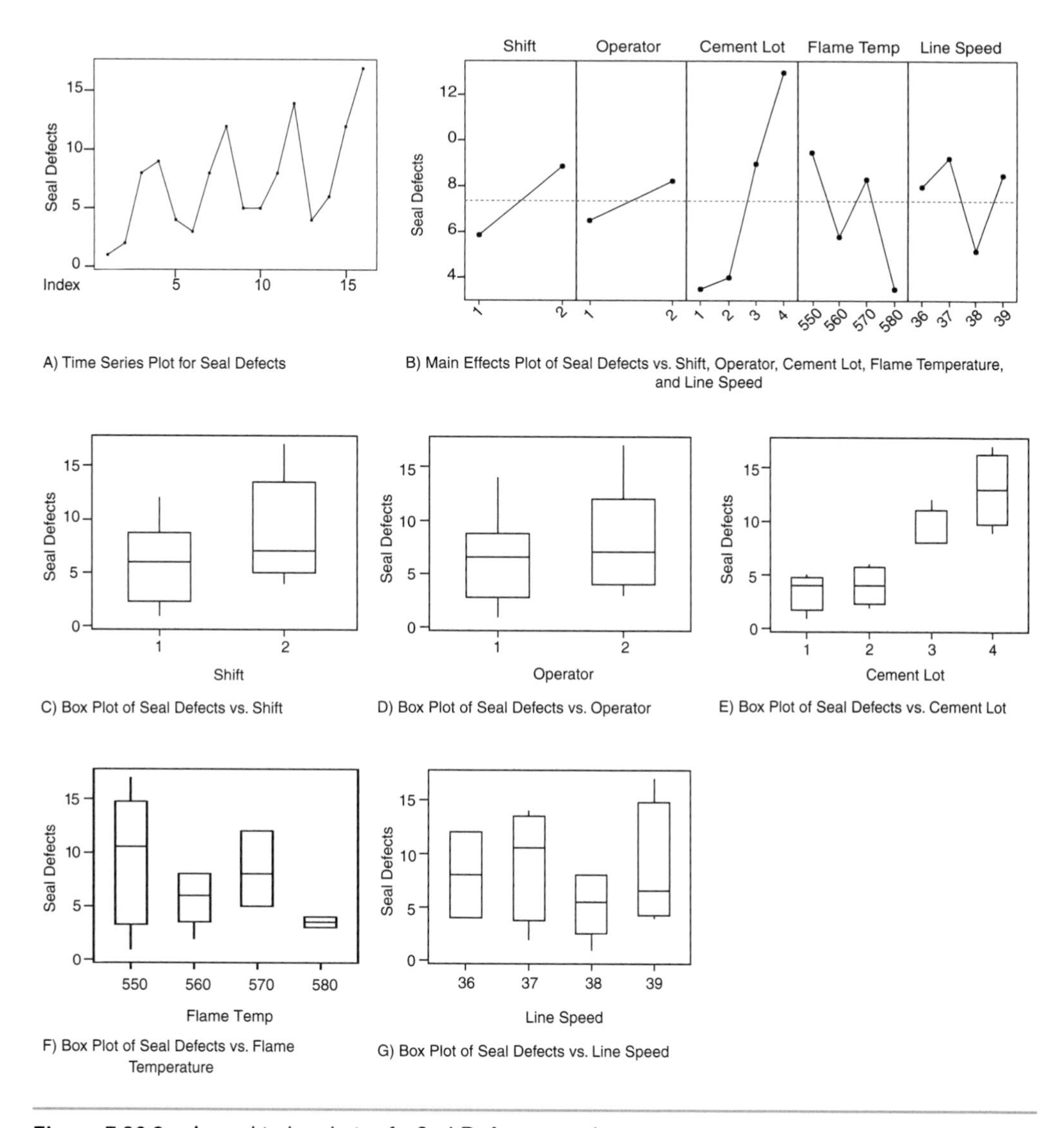

Figure 7.29.2 A graphical analysis of a Seal Defect example.

Step 4: For each and every one of the patterns and relationships identified in Step 3, a Hypothesis Test needs to be formulated in the form:

- (Null hypothesis) H_o: Nothing is going on, there is no relationship, there is just background noise, any visible effects are just due to random chance
- (Alternate hypothesis) H_a: Something is going on, the effect is real, it is not background noise

For example, for the Seal Defects example, one of the Hypotheses might be

- H_o: Cement Lot has no effect on Seal Defects
- H_a: Cement Lot has an effect on Seal Defects

Step 5: At this point a statistical test is applied to the data to determine a "p-value" or probability value. The p-value represents the likelihood of seeing a pattern or effect this strong purely by random chance if in fact nothing were going on. If for example, the p-value is 0.07, then the chance of seeing a difference in Seal Defects this big if, in this case, Cement Lot really had no effect, is 7%. The standard cutoff value is 0.05. Below this value the conclusion is that the effect is unlikely enough to be considered real ("statistically significant" in Statspeak) and it is not a good idea to accept the Null Hypothesis. If it is above this value then the conclusion should be that an effect this size is completely possible by random chance and the Null hypothesis is probably right. The p-value gives a level of confidence in the claims being made.

The catchphrase frequently used is "If p is low then H_0 must go," or in other words, if p is less than 0.05 the graph probably shows something real. Note the use of the words "probably," "could be," and "might be"—the joy of statistics!

The correct test to use is determined almost completely by the type of X and Y data, whether they are Attribute or Continuous. Figure 7.29.3 shows a simple statistical tool-selector based on data type. For example, in the example the Y Seal Defects is Attribute data (a count) and the single X Cement Lot is also Attribute data (4 lots); thus, the appropriate statistical test is a *Chi-Square Test*. Running a *Chi-Square Test on* the data would yield a p-value, which would tell you whether the effect seen in the graphs is real (statistically significant).

Step 6: Make practical conclusions. With the Team, review the analyses, determine the physical implication, and develop final conclusions. All conclusions should not be based on conjecture or intuition, should make sense from a practical standpoint, and must be supported by data.

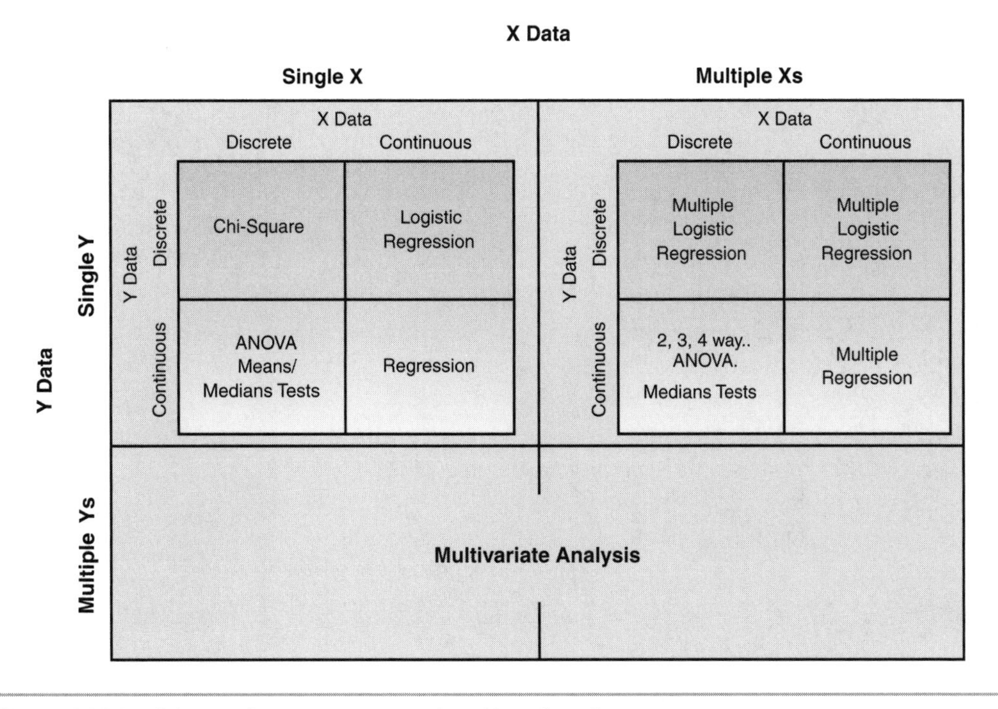

Figure 7.29.3 Selecting the correct statistical tool based on data type.

Step 7: Report the results and recommendations. For clarity, it is best to show end results in graphical format supported by the statistical test. A full Multi-Vari Report would include

- Process description
- Study objectives
- Input and output variables measured
- Sampling plan
- Process settings
- Graphical analysis results
- Statistical analysis results
- Practical conclusions
- Recommendations for further studies

30: MURPHY'S ANALYSIS

OVERVIEW

Murphy's Analysis[52] is another useful change management tool for the early stages of a project. It is applied during the early section of the Define Phase. The Team at this point needs to determine what the Customer Requirements are for the process, the difficulty being that the Team does not know what they don't know. It is difficult to interview a Customer with absolutely no understanding of the problems the process faces; so before any interviews can take place, the Team has to do enough groundwork to determine what questions to ask and how to guide the interviewing. This is where a Murphy's Analysis performs a useful function.

The Team use the Murphy's Analysis as a brainstorming tool to identify how the process can fail to function perfectly, as in Figure 7.30.1.

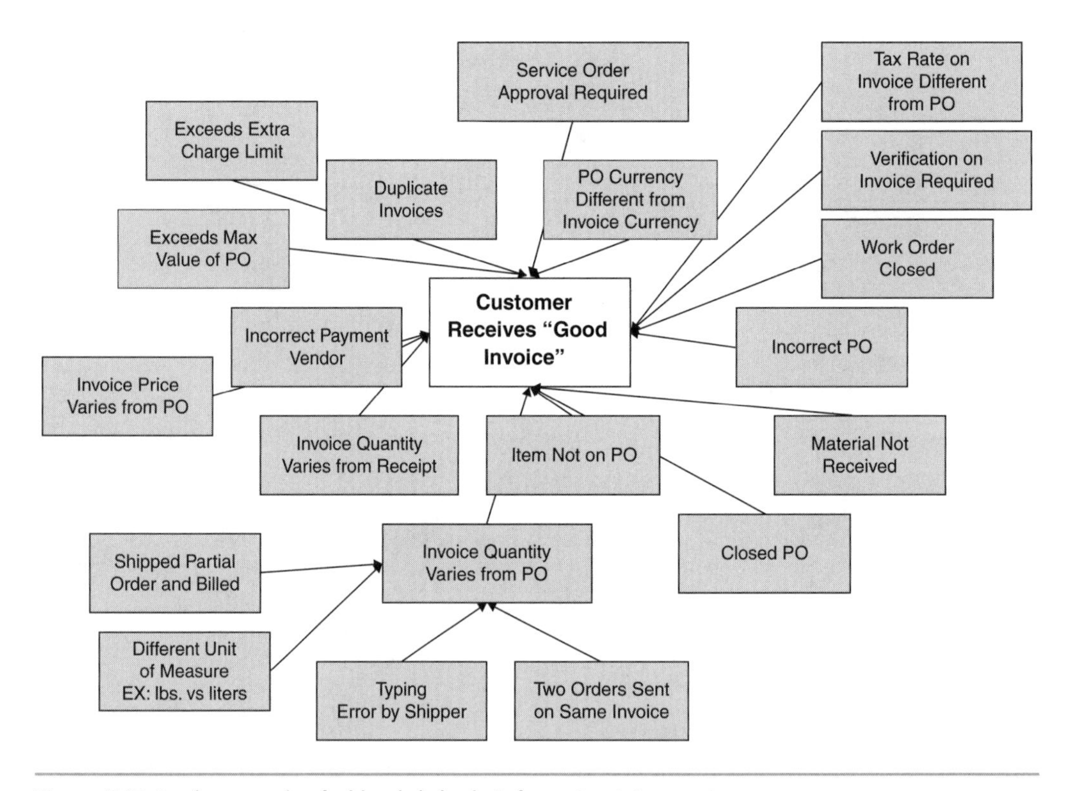

Figure 7.30.1 An example of a Murphy's Analysis for an invoicing process.

[52] Murphy's Analysis was first proposed by Dewan Simon, an SBTI Consultant, during SBTI's revamp of the Transactional Six Sigma process in 2001.

Based on the results of the Murphy's Analysis, the Team has a reasonable view of possible areas to investigate during the interviewing process.

Note that the Murphy's Analysis result is not considered in any way to be the VOC; it is only one of a series of tools used to identify the VOC.

A second useful function of Murphy's Analysis is that during the early stages of a project there is still venting that needs to be done by Team members. What better tool than one that focuses on the negatives in the process.

Logistics

Creating a Murphy's Analysis is absolutely a Team activity and should not be done by the Belt in isolation. Also, the Belt should not be tempted to create a straw man of the Murphy's Analysis; rather, it should be created from scratch with the whole Team present.

It generally takes the Team 45–60 minutes to generate a reasonably useful result. The Team needs flipchart paper, wall space, sticky notes, and pens.

Roadmap

The roadmap to creating the Murphy's Analysis is as follows:

Step 1: Start with the purpose of the process. This should have been determined during construction of the SIPOC. The purpose is typical to generate something, to bring value in some form, and is often a composite sentence for example:

- The Customer quickly receives an accurate invoice.
- Accurate lab results are available quickly.

If a purpose is not already identified, then falling back on timeliness and accuracy of the resulting product typically helps.

Place the purpose at the center of a flipchart-sized sheet of paper on a colored sticky note.

Step 2: Brainstorm how the process fails to deliver the purpose and how it might fail in the future to deliver the purpose. Write each failure mode individually on its own sticky note (use a different colored note for the purpose to keep it separate).

Step 3: For each Failure Mode, look one step back up the causal chain to identify why that failure mode occurs. For example, at the bottom of Figure 7.30.1, one potential Failure Mode is that the "Invoice Quantity Varies from PO." This is caused potentially by a number of things, including "Typing Error By Shipper."

Step 4 (optional): After the body of the Murphy's Analysis is complete as per Figure 7.30.1, capture it electronically. Then the sticky notes could be removed onto a separate sheet and affinitized into groups. This could be by subject area, but often is better stratifying the notes into High and Low Impact and High and Low Likelihood as per Figure 7.30.2.

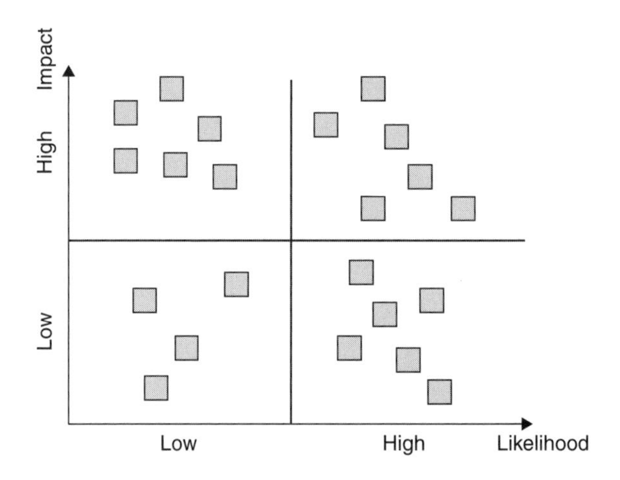

Figure 7.30.2 A Murphy's Analysis output stratified by Impact and Likelihood.

INTERPRETING THE OUTPUT

Although Murphy's Analysis is a great tool prior to *Failure Modes and Effects Analysis*, its primary purpose as per the "Overview" section is to structure the *Customer Interviewing* process.

The high impact and high likelihood failure modes should be key inputs to the questions used during the interviewing process, see "Customer Interviewing" in this chapter.

31: NORMALITY TEST

OVERVIEW

The Normality Test is used on a sample of data to determine the likelihood of the population from which the sample originates being normally distributed. The result would be a degree of confidence in the population being normally distributed (a p-value).

Normality is crucial for the majority of statistical tests examining the means and variances of samples. For example, if data becomes skewed (thus, non-normal), then the mean is probably not the best measure of center and a median-based test is probably better. The longer tail on the right of the example curve in Figure 7.31.1 drags the mean to the right; however, the median tends to remain constant.

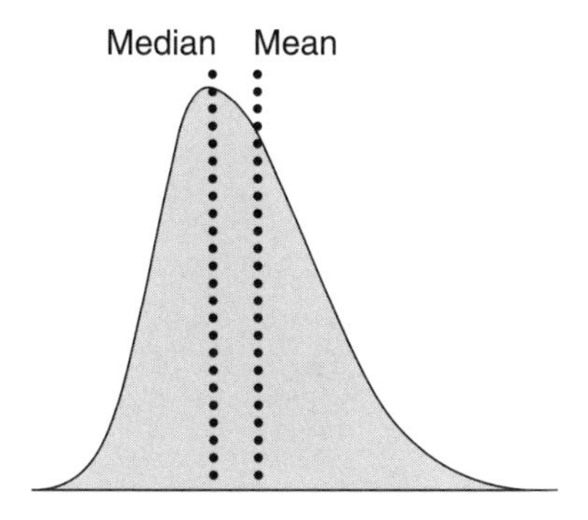

Figure 7.31.1 The effect of Normality on measures of center.

Roadmap

The hypotheses for a Normality Test are

- H_0: Population (process) data are normal
- H_a: Population (process) data are non-normal

The test is applied to a column of data (the sample) and the results obtained.

Interpreting the Output

There are a number of Normality Tests, for example those listed in Minitab include (with simple descriptions in English):

- Anderson-Darling—Examines the area between the sample data distribution and the normal distribution (smaller the better).
- Ryan-Joiner[53]—Examines the correlation between the sample data and a normal distribution (the more correlated the better). This is useful for small sample sizes.
- Kolmogorov-Smirnov—Similarly to the Anderson-Darling Test, examines the area between the sample data distribution and the normal distribution (smaller the better).

To be candid, in the world of Process Improvement there really won't be a dramatic difference in conclusion based on the tests. It is advisable to stick with one and the default in Minitab is Anderson-Darling, so I personally tend to run with that one.

[53] Similar to a Shapiro-Wilk test for those running different software.

Each test returns a test statistic, but the thing to be most interested in is the p-value, the likelihood that for the sample data a level of non-normality this large could have occurred purely by random chance even if the population were normally distributed.

Output from a Normality Test is shown in Figure 7.31.2.

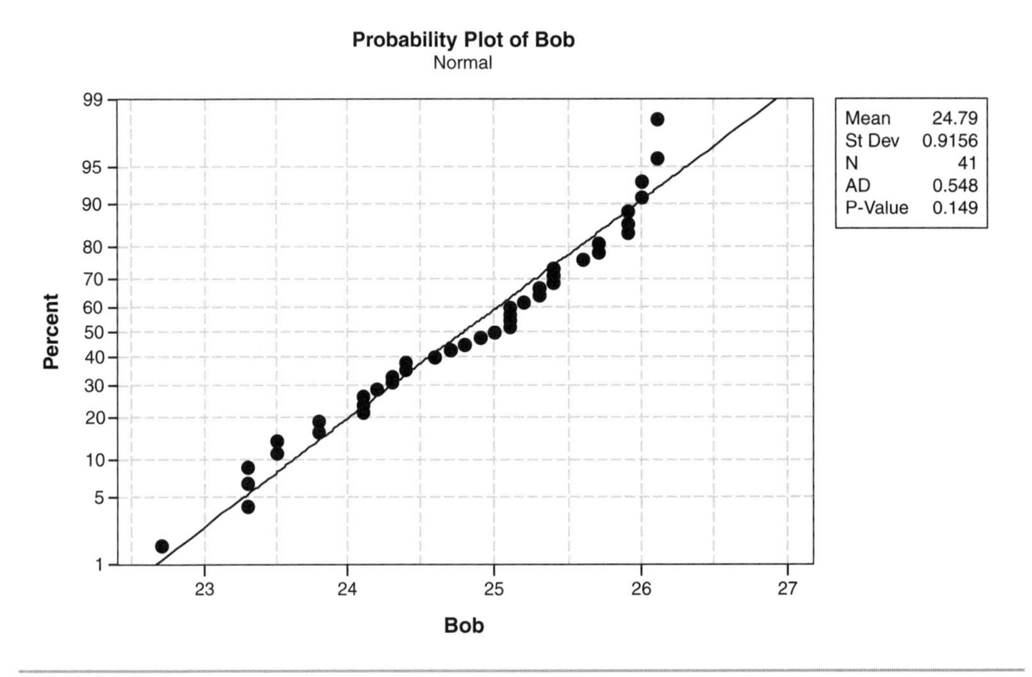

Figure 7.31.2 Normality Test (Anderson-Darling) results for a sample of Bob's time to perform a task (output from Minitab v14).

The vertical scale on the graph is non-linear and the horizontal axis is a linear scale, similar to normal probability paper. If the data were perfectly normally distributed then the points would lay exactly on the line (and the p-value, in this case, theoretically should be 1.0).

From the example results:

- The sample mean is 24.79 (and this represents the best available approximation to the population mean).
- The sample standard deviation is 0.9156 (and this represents the best available approximation to the population standard deviation).
- The sample was made up of 41 data points.

- The Anderson-Darling statistic is 0.548 (remember the lower the better) with the likelihood of seeing a statistic this large, if the parent population were normally distributed, being 14.9% (p-value).

The hypotheses for a Normality Test are

- H_0: Population (process) data are normal
- H_a: Population (process) data are non-normal

The p-value should be interpreted in the usual way:

- p less than 0.05—reject H_0 and conclude that the data are non-normal
- p greater than 0.05—accept H_0 and conclude that the data are normal

Therefore, for Bob's data, shown in Figure 7.31.2 with a p-value of 0.149, which is clearly greater than 0.05, the conclusion should be that Bob's data are normal.

32: OVERALL EQUIPMENT EFFECTIVENESS (OEE)

OVERVIEW

Overall Equipment Effectiveness (OEE) is used extensively in the maintenance and equipment reliability world to examine equipment availability. Here the tool is similar but is applied differently. If you have a maintenance background and are familiar with OEE, do *not* skip this overview; there are some key differences to consider.

Lean Sigma uses extensively the notion of entitlement, which is in effect perfection for the process. For example, a yield entitlement is 100% and for defect rates entitlement is zero. Capacity is a slightly trickier metric to apply entitlement to, but OEE helps us towards this goal.

OEE represents the percentage of capacity entitlement that the process is currently attaining or in other words:

$$\text{OEE} = \frac{\text{Actual Capacity}}{\text{Potential Capacity}} \times 100$$

OEE is broken into three elements, and in simple terms, a process achieves 100% OEE when it is up (running) doing only VA work 100% of the time, it is going as fast as it has ever gone, and it is producing 100% perfect quality entities:

$$\text{OEE} = \%\text{Uptime} \times \%\text{Pace} \times \%\text{Quality}$$

The major difference from the maintenance version of OEE is regarding the %Uptime. In maintenance terms, %Uptime is the percentage of time that the process is *available* to do *any* work, even NVA work. In Lean Sigma, %Uptime is only the percentage of time that the process *is* doing *VA* work, which typically is significantly less.

LOGISTICS

Although OEE is an analysis tool and in theory could just be done by the Belt, in practice it takes the Team to gather the data and to correctly structure the results. Data is typically captured for a minimum of 2–4 weeks to be meaningful. The time period depends on the drumbeat of the process and the frequency of things, such as changeovers and maintenance time. For some large equipment processes, data for a whole year might be more relevant.

The data is captured in three separate data captures running concurrently, one for %Uptime, one for %Pace, and one for %Quality. More often than not, the %Pace and %Quality can be calculated from historical data, so investigate them first. %Uptime is hardly ever measured in the way described previously; thus, a new (often manual) data capture is invoked. The operators themselves have to participate in the data capture, so planning and communication are key. A clear operational definition of the metric is required to explain what is being captured. Remember the data captured is just a sample and is used to look for opportunity, not measure the OEE to 3 decimal places.

After the initial analysis is complete, OEE should be captured on an ongoing basis, along with other operations metrics to understand the level of capacity performance.

% UPTIME

This is purposefully not written in the form of "downtime" which can be misleading. The tendency is only to think of downtime as when the process is broken. In reality, the process is "down" at many other times

$$\%\ \text{Uptime} = \frac{(\text{Value Added Time})}{(\text{Operating Time})} \times 100$$

Uptime is defined only as the time that the process (or process step) is up doing *VA* work. All other activities are all considered downtown, such as

- Setup
- Changeover
- Loading
- Unloading

- Idling empty
- Idling full
- Breaks for lunch, and so on
- Cleaning
- Maintenance (planned or otherwise)
- Breakdown
- Scheduled downtime
- Off-shift hours (see Variants)

Again, the metric is used to look for opportunity, not as a typical operations metric to judge supervisor performance for example.

This simple distinction can bring a whole new perspective to a process. Processes that were previously seen as nearing 100% utilization are often less than 50% utilized.

Uptime data is best captured by the operator over a period of two weeks or so, putting the activity time into predetermined buckets representing the possible activity classifications. This is usually a manual data capture with the operator writing times and activities on a table throughout the day. If the purpose is clearly explained to the operators, they typically don't try to "massage" the data because it's in their interest not to do so. For that reason, usually little Hawthorne Effect is seen.

After the data is captured, it is usually analyzed in a simple *Pareto Chart* or Pie Chart as shown in Figure 7.32.1.

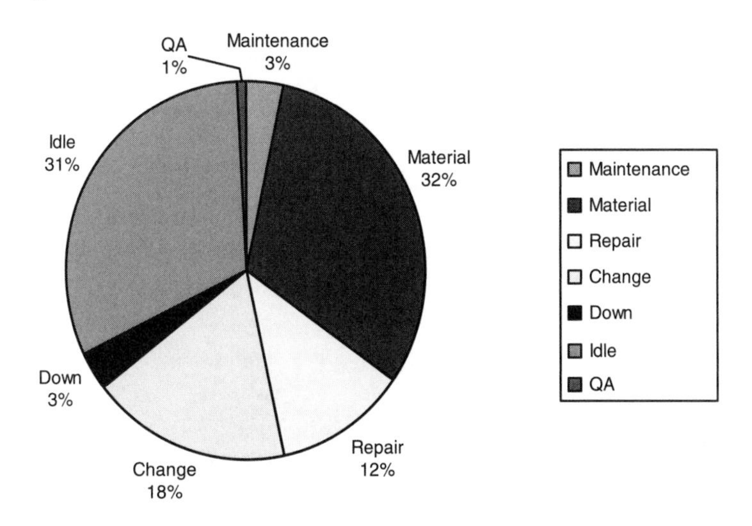

Figure 7.32.1 An example of downtime classification.

Variants

Most processes don't work 24 hours a day, 7 days a week. This often represents the simplest capacity expansion option, because no new capital is required and operating cost per unit is reduced because there are no additional warm-ups and cool-downs required.

From the point of view of the OEE metric, this falls into the %Uptime element and is usually dealt with by introducing additional metrics:

- OEE (current shift)—As before
- OEE (24 hours, weekdays only)—Versus 24 hours a day shift pattern but only for weekdays, not weekends
- OEE (24 hours, 7 days a week)—Versus a 24 hour a day operation including weekends

Obviously many variants are possible depending on the business problem in hand.

% PACE

This is usually the trickiest of the three OEE sub-metrics to measure and Belts often confuse it with %Uptime and %Quality. The simplest way to think of this is that sometimes processes slow down; they aren't fully "down," in that entities are still flowing through the process. The flow rate is reduced for some reason, maybe due to resource or material constraints, or to ensure that good product is being made (hence the confusion with %Quality).

To calculate %Pace, it is necessary to determine the Peak Rate of the process in entities per unit time. This is done based on a very short duration (usually 1–4 hours of processing). After the instantaneous Peak Rate is found it is labeled as "entitlement" for as long as it is not surpassed.

Then it becomes a matter of determining the typical processing rate. This is calculated by determining the average number of entities processed during the actual working time of the process (i.e., only during the uptime) usually over the period of a week or two.

$$\%\text{ Place} = \frac{(\text{Current Instantaneous Processing Rate})}{(\text{Ideal Processing Rate})} \times 100$$

Variants

Another, entirely equivalent, method of determining the %Pace is to judge the process by the time it takes to process a single entity. Another entitlement measure is introduced for the processing time, known as the Ideal Cycle Time, which is the time between entities exiting the process if there were literally no constraints, such as resource, and so on.

This is equivalent to the instantaneous Peak Rate expressed in time (the faster the rate, the shorter the time).

After the Ideal Cycle Time is known, it is simply a case of determining the average Cycle Time (based again on capturing data only during the uptime of the process) and working out the ratio of the two numbers:

$$\%\text{ Pace} = \frac{(\text{Ideal Cycle Time} \times \text{Processed Amount})}{(\text{Operating Time})} \times 100$$

or simply:

$$\%\text{ Pace} = \frac{(\text{Ideal Cycle Time})}{(\text{Average Processing Time})} \times 100$$

As previously, the Average Cycle Time is determined by gathering data over one to two weeks for a typical process.

% Quality

The quality rate is probably the simplest of the three sub-metrics and is given by the equation:

$$\%\text{ Quality} = \frac{(\text{ Processed Amount} - \text{Defect Amount})}{(\text{Processed Amount})} \times 100$$

In simple terms it is the percentage of all the entities produced in a given time that were produced correctly. If a defective entity is generated then effectively there is a loss of capacity of one good "A-Grade" entity.

This kind of data is usually collected as a matter of course for most quality groups and a months worth of data more than suffices.

Special Cases

The toughest part of OEE is deciding into which sub-category to place a capacity loss. Table 7.32.1 helps with the most common categorizations.

For some processes, the %Pace and %Quality are inextricably linked (by chemistry for instance) and the quality degrades as the pace is increased, because less time is given to the processing. In these cases, it is usually best to make a composite metric of Pace and Quality such as Kg of good product per hour. Uptime would still be dealt with separately. The new OEE equation is

$$\text{OEE} = \%\text{Uptime} \times \%(\text{Rate of Quality Product})$$

Table 7.32.1 Classification of Lost Capacity Elements in OEE

% Uptime	% Pace	Quality
Changeover from one entity to the next	Slow downs	Scrapped entity
Changeover from one entity type to another	Intermittent processing (stoppages aren't long enough to register as a measurable downtime)	Defective entity
Setting up the process		Downgraded entity (this is considered a capacity loss because downgraded product is typically sold at a much lower margin)
Unloading an entity		
Loading an entity		
Downtime		
Rework time		
Secondary processing not required if the entity was processed right the first time, e.g., blending poor product into good product		

An instantaneous best Rate of Quality Product would have to be identified from historical data, or calculated from basic scientific principles (mass continuity, and so on).

Interpreting the Output

OEE yields two incredibly valuable pieces of information:

- The capacity that is possible from the process
- If the capacity is low, then where the best place to look to improve it is (Uptime, Pace, or Quality)

This information guides the next steps in the project. If the OEE is running at greater than 90%, then the process really doesn't have much opportunity to get better, or any change would be tricky to find. If, however, the OEE is less than 50%, which is common (the typical OEE value for a line is around 5%–10%), then there is ample opportunity to apply a project to the process to capture some additional capacity. This assumes that there are additional sales for the capacity, because without it, the whole exercise is pretty much a waste of time.

If additional potential capacity is identified, then the tool guides the Team to the best place to capture the capacity. The rule is always to take the easiest path. Unless there is a blindingly obvious shortfall in one category, it is often best to look in all three categories, because it is easier to find a percentage here and there, rather than to find it all in the same place. If looking to all three categories, then it is useful to have separate project streams for Uptime, Pace, and Quality. Two of these could perhaps be led by additional Green Belts working under the Black Belt, while the Black Belt takes the trickier category.

Other Options

OEE can also be applied to people, although this has to be dealt with carefully from an organizational viewpoint, not many employees like to examine the percentage of VA activity that they conduct. Application to a person is useful if that person is a constrained resource (generally with a scarce technical expertise). An example of this is shown in Figure 7.32.2 where a Diagnostic Technician conducts procedures in a room with specialized equipment. The OEE for the room and equipment is required as well as the OEE for the Technician, because the combined resources room, equipment, and technician are required to conduct a procedure.

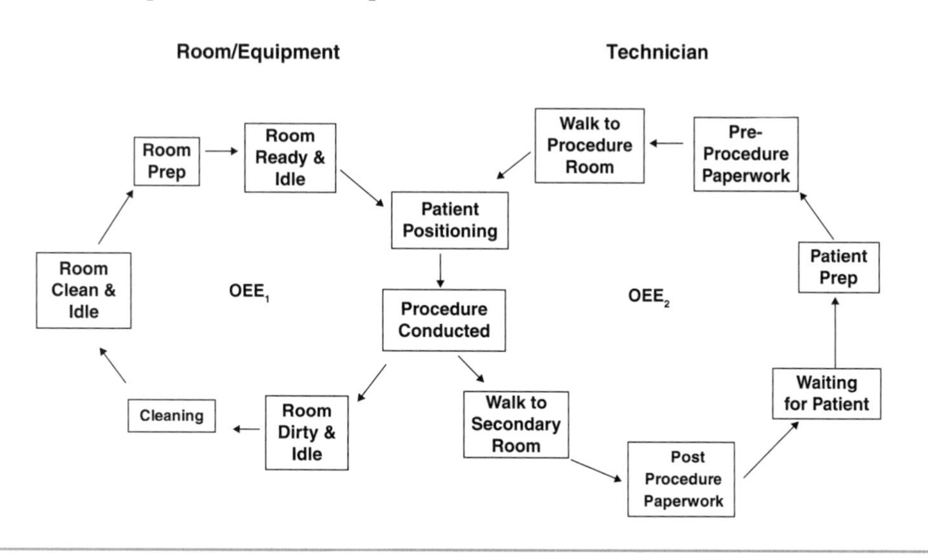

Figure 7.32.2 Example of interaction between loops.

33: PARETO CHART

OVERVIEW

The ubiquitous Pareto Chart is a simple but extremely useful Lean Sigma tool. The tool is applicable at any point in a project where a narrowing of focus is required based on volume or count data, for example, to help a Belt identify key areas to focus the project on, to help scope it better. Most projects start out broadly scoped, focusing on all problems that the process exhibits. Such projects are almost impossible to manage, and, hence, the use of a tool to identify what the biggest opportunities are and focusing there. Typically, the tool is applied to defects in the process, as shown in Figure 7.33.1, to allow a rescope by major defect, in this case perhaps focusing on Missing Screws, where more than 60% of all defects emanate from this one defect.

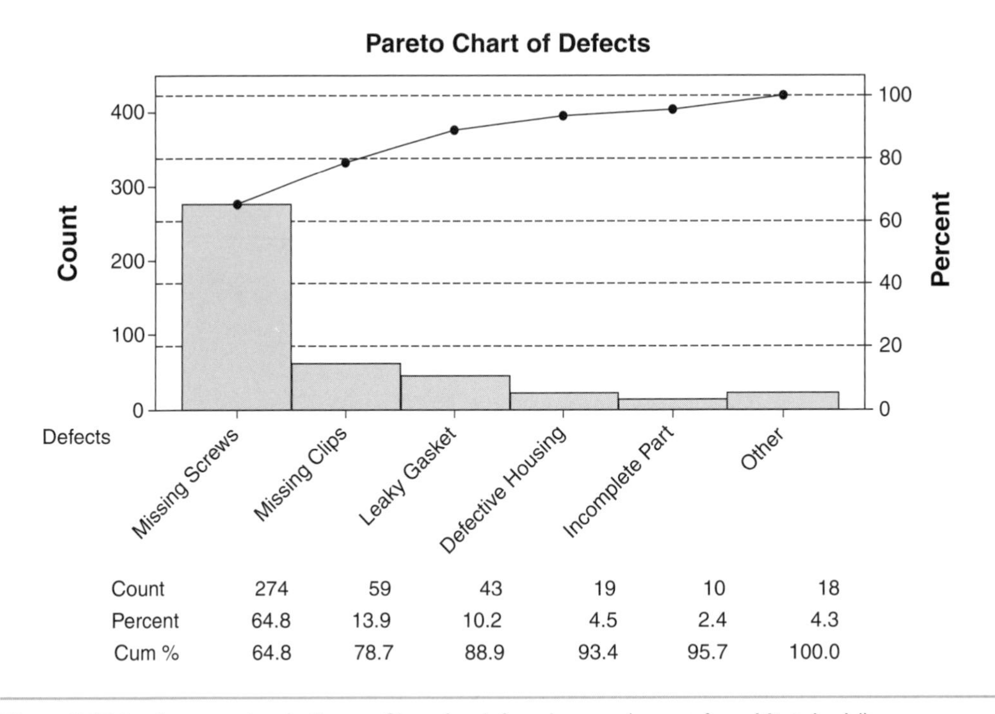

Figure 7.33.1 An example of a Pareto Chart for defects by type (output from Minitab v14).

The scoping factor does not have to be defect type though; it could be, for example:

- Patient type
- Transfer destinations

- Product type
- Document type
- Suppliers

Whatever makes a useful split of the data to separate out the key focus areas from the noise.

Roadmap

The roadmap is as follows:

Step 1: Identify the entity type to be scoped, typically in the form of counted items. This is effectively the Y in the equation $Y = f(X_1, X_2,..., X_n)$.

Step 2: Identify the scoping factor or factors that could represent a significant split of the data, such as the Xs in the same equation $Y = f(X_1, X_2,..., X_n)$. Having a solid operational definition of what constitutes the entity and similarly each scoping factor is crucial. If the Team can disagree on which box an entity falls into, then there will be problems later.

Step 3: If historical data does not exist, then determine an appropriate data collection plan. Pareto generally relies on attribute data, and at least 50–100 data points are necessary to generate a meaningful result. Data should be from a period long enough to include regular process variation.

Step 4: Follow the data collection plan to capture the Y and associated X data.

Step 5: Apply Pareto Charts to the Y data, split by the appropriate X. There is one chart per X.

Interpreting the Output

Interpretation is reasonably straightforward. The purpose was to try to split a large quantity of Y data into meaningful pieces by an X (a "scoping factor" for want of a better name), so that the project could continue in a more focused way. After Step 5 in the roadmap, the Team should have a number of Pareto Charts splitting the data by various factors. A useful Pareto is one that exhibits one or possibly two large bars explaining the majority of the counts and then the rest of the counts are split across multiple bars. A Pareto similar to Figure 7.33.1 is extremely useful to a Team, whereas a Pareto, such as Figure 7.33.2, is not, because no discrimination is found between the defect types.

Sometimes the Belt might wish to split the data one step further, as shown in Figure 7.33.3. This is valid provided that there is enough data to go around. Figure 7.33.3 has little data represented in any of the four charts; thus, no meaningful conclusions could really be drawn. Each sub-Pareto needs at least 50–100 data points to make it meaningful.

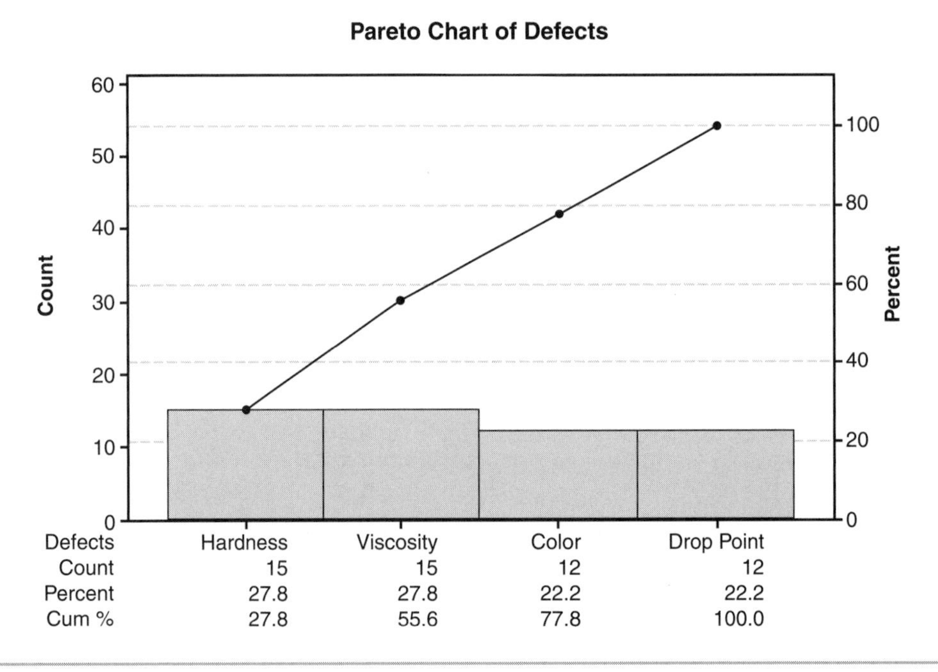

Figure 7.33.2 An example of a Pareto Chart by defect type.

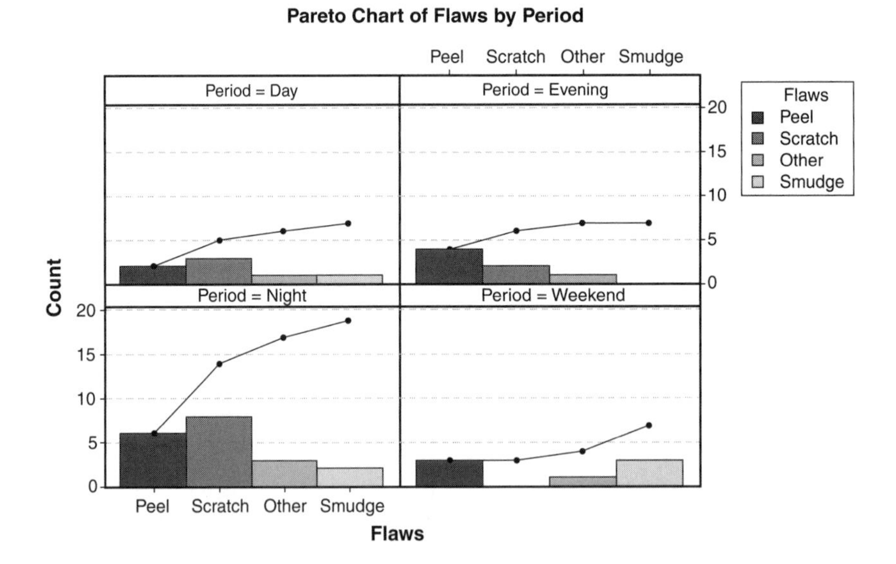

Figure 7.33.3 An example of a Pareto Chart split by two factors, Flaw and Period (output from Minitab v14).

34: Poka Yoke (Mistake Proofing)

Overview

Poka Yoke or Mistake Proofing is an important Control tool. It is also the strongest form of Control, except for designing out the process step completely. In simple terms, Poka Yoke ensures that an action cannot be done incorrectly and defects are prevented from ever occurring.

Mistake Proofing has been sometimes referred to as Fool Proofing, or Idiot Proofing. It is certainly best to steer well clear of these terms due to the negative connotations. Also, from my experience, it seems to be that the cleverest "Fools" and "Idiots" make the biggest mistakes!

Defects in processes rarely tend to "just happen." There is usually a sequence of events that lead up to the defect, and this is the focus of Poka Yoke. The Team must be able to track back through the series of causal events that can lead up to a defect and place a Control as early as possible in that chain.

Traditional defect Control systems often focus on detection as shown in Figure 7.34.1. The defect has to occur before any remedial actions are taken.

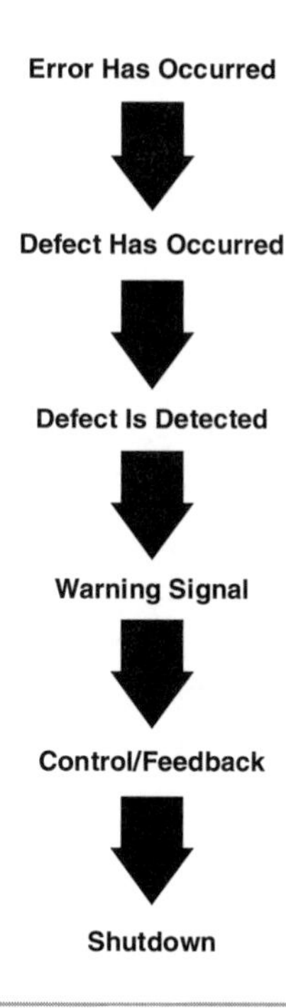

Figure 7.34.1 Traditional defect control path.

Systems of this form rely on the ability to successfully detect a defect *after* it has occurred. Only then are any actions taken to control the situation.

Poka Yoke takes the more proactive route by using the philosophical approach of $Y = f(X_1,X_2,\ldots, X_n)$. If it is possible to identify Xs that drive most of the behavior in the Y and can cause the Y the exhibit defects, then by focusing on the Xs it should be possible to even prevent the defect occurring at all.

Thus, if there is an error in an X and it is detected, then it is possible to predict the impending defect and react ahead of time, as shown in Figure 7.34.2. Work in Poka Yoke, therefore, is focused heavily on error detection as early as possible in the chain of events, to give enough of an early warning to an operator prior to a defect ever happening.

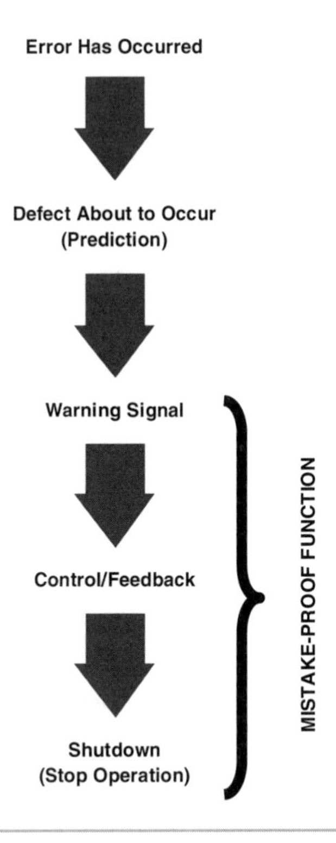

7.34.2 Poka Yoke defect control path.

Logistics

Poka Yoke is one of a series of Control tools, which are done as a team sport and best performed by the people closest to the process, running the process on a day-to-day basis.

It is not typically done as a separate event, but is usually built into the Team meetings during the Control Phase.

Roadmap

The following roadmap assumes a reactive effort in that a problem already exists, such as during a Process Improvement project. Ideally, however, mistake proofing is best used in a proactive mode during Process Design to design out the defect from the beginning. The roadmap is iterative at times and involves some trial and error; such is the nature of preventing humans from making mistakes.

Step 1: Describe the defect or potential defect. Refer back to the VOC work done in the Define Phase, along with the FMEA. Mistake Proofing takes time and resource, so focusing on the highest Risk Priority Numbers (RPNs) in the FMEA pays dividends. If this is a defect that currently occurs in the process then determine the approximate defect rate.

Step 2: Identify the operation where the defect is or could be discovered and also the operation where the defect is made. These are often different operations.

Step 3: Detail the sequence of process steps and actions as documented in the standard operating procedures or work instructions:

- From the operation where the defect is made
- To where the defect is discovered

Step 4: Watch the actual process steps and actions being done and identify any steps that differ from the standard.

Step 5: Identify the error conditions that could be contributing to the defect—for example, tools, environment, gauging, incorrect information, and so on. Referring to the observations made in Step 4 (using the *5 Whys* approach) ask why the error happens repeatedly, to track back in the causal chain of events until the root cause (or source error) is identified. Do not try to develop a device until you determine the root cause.

Step 6: Identify feasible mistake-proofing devices required to prevent the error or defect. This is the most creative element of the tool and can be the trickiest. In training, Belts are often asked to identify as many mistake-proofing devices as possible, for example, in a car, then consider how each achieves its goal to get their creative juices flowing. There is no easy guide to generating ideas, but some hints include

- Keep the device as simple as possible. Complex devices tend to fail more regularly, completely defeating their purpose.
- Keep the device as tangible and physical as possible. Don't rely on a device that requires intelligence to operate.
- The most inexpensive solutions tend to work the best. In fact the most elegant solutions often cost nothing or are even cheaper than the existing approach.
- The device must give immediate or prompt feedback. Any delay can cause the operator to miss the sign.
- Devices that give a prompt *reaction* work best to prevent the mistake.
- Create devices with focused application. Don't try to solve all the potential pitfalls with a single complex device. Create small, focused devices.
- Always get input from the right people. This is generally the actual process operators, but it is often worthwhile to bring in an equipment vendor or someone similar.

Belts in transactional businesses or service industries sometimes struggle to see the analogies to their processes. In fact, Poka Yoke tends to give greater return in these kinds of processes because it often eliminates layer upon layer of "patches" that have been put on the process over time. Some example devices might include

- Use templates to provide a box for entry of each character, or demonstrating what should appear in a field, for example, DD-MM-YY.
- Use masks to make non-required fields invisible to the user so incorrect entries are not placed in them. For example, placing a mask over the copier, so that unneeded buttons are hidden. Ideally the mask is more than just visual and the buttons themselves are switched off.
- Replace freeform text fields in databases with pull-down menus (combo boxes) to force selection from a designated list. To demonstrate this, it is an interesting exercise to pick a common client or product in a company database and count the number of variants.
- Use protected fields in databases so that accidental entries and overwrites are prevented.
- Use color-coding of documents to emphasize distinction; for example, overnight parcel shipment documents are typically a different color than those for ground shipments.
- Use bar coding to identify items. Note that this method relies heavily on getting the correct bar code on the entity in the first place, which is often overlooked.
- Create dedicated tools, for example, a single computer dedicated to shipping manifests.
- Create security permissions for the ability of a user to edit, delete, and read to prevent accidental deletion and overwrite.

- Create decision criteria aids, such as decision trees and expert systems. These are used heavily by product support groups for IT; for example, the person on the other end of the phone asking you a series of intelligent questions to pinpoint the exact reason why your printer has failed is, in fact, just following a standard series of questions.

For a complex problem, more than one solution is evolved with the notion of trying each.

Step 7: Give it a go. The Poka Yoke term here is to "try-storm" the device, quite an apt description. If it doesn't work, try again with a subtle variant or with another idea that came from Step 6.

Step 8: Mistake proof the mistake-proofing device. After something workable is in place, try to see ways to make it simpler and impossible to override or remove.

After the roadmap is complete for one device, it has to be repeated until you account for all high RPNs in the FMEA. It is fortunately an acquirable skill and Teams quickly become adept at identifying quick, simple mistake-proofing devices for the Xs in the process.

35: Process FMEA

Overview

Failure Mode & Effects Analysis (FMEA) is a type of risk management. There are at least four types of FMEAs used in the Lean Sigma world, each considering the failure of different project related elements:

- Market FMEA. Considers the risk of failure of a product offering entering a particular market segment during the development process through things like Customer acceptance, technology immaturity, and competitive landscape.
- Design FMEA. Considers the risk of failure of a product design or system design characteristic (a measure of performance of the product) during the development process due to supplied raw material capability, process capability, and the robustness of the design itself.
- Project FMEA. Considers the risk of failure of meeting a project deliverable in any type of project whether it be Lean, Six Sigma, or a major Engineering project.
- Process FMEA. Considers the risk of failure of an input to a process step.

This Lean Sigma guide is aimed at process improvement and the focus here is the Process FMEA.

LOGISTICS

This is a Team activity and should include representation for all the key areas affecting the process. Choose those individuals who live and breathe the process every day, not just process owners who might not be close enough to it to know the details.

It typically takes a morning to complete the preliminary FMEA, but as a tool, FMEA should be revisited at every project meeting for at least ten minutes to identify new risks and to follow through on the action items.

It is tempting for the Belt to create a straw man of the FMEA prior to working with the Team. This is typically a mistake because the Team doesn't feel buy-in to the FMEA and the overall result is usually not as rigorous as it needs to be.

For the initial construction it is usually best to use sticky notes and then transfer the creation into a spreadsheet. The scoring can be done by projecting the computer image onto a screen.

ROADMAP

The stages to the construction are as follows (see Figure 7.35.1):

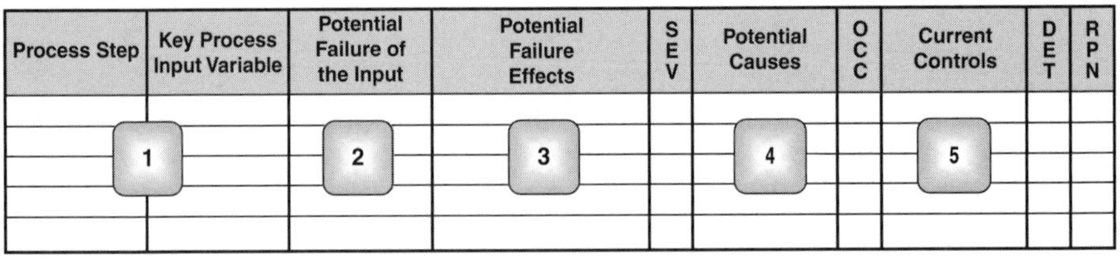

Process Step	Key Process Input Variable	Potential Failure of the Input	Potential Failure Effects	SEV	Potential Causes	OCC	Current Controls	DET	RPN
1		2	3		4		5		

Figure 7.35.1 Steps 1–5 for construction of the Process FMEA.

Step 1: List the process steps and their associated Xs in Column 1 and 2. These come directly from the *Cause & Effect Matrix* and represent an already slimmed down and ranked list. This gives immediate focus to the more important Xs.

Working systematically one X at a time, complete Step 2 through Step 10.

Step 2: For the X in Column 2, list all the ways that the X can fail, known as the Failure Modes. The most common mistake here is to list *why* it would fail as opposed to *how*. Example Failure Modes for any X are

- X is too high (light too bright)
- X is too low (light is too dim)
- X is intermittent/variable (light flickers)
- X is missing (no light at all)

Any X can have multiple Failure Modes and each must be dealt with separately by adding additional rows in the FMEA. A good place to look for potential Failure Modes is the *Murphy's Analysis.*

Belts with Transactional projects often struggle at this point, specifically when the X they are considering is a resource (person) who does a particular process step. The pitfall here is that a person is not really a single X, but a group of Xs, which should have been expanded when creating the *Process Variables Map*. A better X related to the person might be their skill level (Failure Modes: too highly skilled, too low skilled, patchy skills, no skill) or their availability perhaps (Failure Modes: intermittent availability, constant low level of availability, not available at all).

Step 3: For each Failure Mode, list the Effects that the Failure Mode would have on the downstream Customer(s). These can remain grouped in one cell in the spreadsheet as you consider them as a whole and focus on the worst case of them as you proceed (i.e., don't create additional rows in the worksheet at this stage).

Step 4: For each Failure Mode list the Cause of the Failure Mode. Each Failure Mode can have multiple Causes. Unlike the Effects column, the Causes are all dealt with separately, and an individual row should be created for each Cause (see Figure 7.35.2). The Cause is the "why" the Failure Mode occurs.

At this point we have a three step causal chain: the Cause, which causes the Failure Mode, which in turn causes the Effect.

Step 5: For each individual Cause list the *current* set of Controls for the causal chain Cause → Failure Mode → Effect. Be specific about the Controls. Sometimes Belts mistakenly list that no Controls are in place when in reality the operator might, for example, see a visible impact, thus forming the basis of a detection mechanism albeit informal.

Step 6: After a process step has been examined, transfer the Failure Modes, Effects, and Causes from the sticky notes to an Excel spreadsheet ready for scoring. The Severity, Occurrence, and Detection rates are scored based on a scoring matrix. There are many of these available (see Figure 7.35.2). If at all possible, a standard should be set for certain types of projects across the organization, or better still one standard for all projects.

Step 7: For each Failure Mode score the severity of the *worst* case of the list of effects associated with the Failure Mode as per the scoring table (see Figure 7.35.2 and Figure 7.35.3). Severity ratings are from 1 (best case) to 10 (worst case).

Step 8: For each Cause (usually multiple per Failure Mode), rate the likelihood of occurrence as per the scoring table. Occurrence ratings are from 1 (best case) to 10 (worst case).

Step 9: For each Control group (one group per Cause), list the detection rate of the combined group of Controls. Detection ratings are from 1 (best case) to 10 (worst case). The Controls are considered on their ability to impact the causal chain Cause → Failure Mode → Effect. The further back up the chain (in the direction of the Cause), you move towards prevention; thus a lower score is given. The further down the chain (in the direction of the Effects), you move towards the Customer detecting the chain of events for you; thus, a higher score is given. If absolutely no controls are in place at all (not even operator vigilance), then the score is a 10.

Rating	Severity of Effect	Likelihood of Occurrence	Ability to Detect
10	Lose Customer	Very high: Failure is almost inevitable	Cannot defect
9	Serious impact on customer's business or process		Very remote chance of detection
8	Major inconvenience to customer	High: Repeated failures	Remote chance of detection
7	Major defect noticed by most customers		Very low chance of detection
6	Major defect noticed by some customers	Moderate: Occasional failures	Low chance of detection
5	Major defect noticed by discriminating customers		Moderate chance of detection
4	Minor defect noticed by most customers		Moderately high chance of detection
3	Minor defect noticed by some customers	Low: Relatively few failures	High chance of detection
2	Minor defect noticed by discriminating customers		Very high chance of detection
1	No effect	Remote: Failure is unlikely	Almost certain detection

Figure 7.35.2 An example of a Standard Scoring Table.

Process Step	Key Process Input Variable	Potential Failure Mode of the Input	Potential Failure Effects	SEV	Potential Causes	OCC	Current Controls	DET	RPN
				7		8		9	10

Figure 7.35.3 Steps 7–10 for construction of a Process FMEA.

Step 10: The Risk Priority Number (RPN) is calculated as RPN = Severity × Occurrence × Detection

Step 11: Sort the whole FMEA on the RPN Column so that the highest RPNs come to the top.

Starting with the highest RPNs and working downwards, row by row, complete Step 12 for each row until a point is reached down the FMEA that is considered to have a negligible effect on the process (i.e., down in the noise).

RPN	Actions Recommended	Owner	Due Date	Cost	Benefit	Actions Taken	SEV	OCC	DET	RPN

Figure 7.35.4 Step 12 for construction of a Process FMEA.

Step 12: Identify what action is required to reduce the RPN (see Figure 7.35.4). Focus is usually in the order Detection (Controls) first, then Occurrence, then Severity. Each action should have

- An owner
- A due date
- An associated cost
- An associated benefit

At this point the preliminary FMEA is complete; however, this is an evolving tool that should be revisited at every Team meeting to sign off on completed actions and to add new Failure Modes, and so on, if appropriate. After an action has been completed, the Team recalculates the Severity, Occurrence, Detection, and subsequent RPN (which should have lowered or otherwise the action was worthless). The lower RPN should be enough to drop it out of the danger zone and down into the noise. If not, then this particular Cause → Failure Mode →Effect chain should be revisited after the higher RPNs have been actioned.

INTERPRETING THE OUTPUT

The primary output of any FMEA is the set of actions and the implementation of those actions—anything else is just theory. If the FMEA is not to be acted upon then its construction was a complete waste of time.

An FMEA might span multiple pages with many, many related action items. It is not required that the Team complete all of the actions. At some point down the FMEA (based on the RPN), actions "get down in the noise." Only the meaningful (higher leverage) actions should be completed, or the Team goes on ad-infinitum.

36: PROCESS VARIABLES (INPUT/OUTPUT) MAP

OVERVIEW

A Process Variables Map or otherwise known as an Input/Output Map is a graphical representation of the process. It is typically used as the primary tool to identify all the Xs in a process.

It is crucial to identify all the Xs using this tool, because no other tools appearing later are designed for the same purpose. If the Team misses an X now, it is not identified to be considered later.

LOGISTICS

This is a Team activity and should include representation for all the key areas affecting the process. Choose those individuals who live and breathe the process every day, not just Process Owners who might not be close enough to it to know the details.

The activity typically takes a total of a morning to complete. It is best to create the map physically on the wall using flipchart paper and sticky notes, rather than directly into a piece of software, such as Visio. Using software makes it tricky for all the Team to participate.

A common mistake is for the Belt to want to create a straw man of the map first and bring that to the Team meeting to finalize construction. This approach creates problems with Team involvement and buy-in to the result. The recommended approach is to construct the map from scratch during the Team meeting.

ROADMAP

The roadmap to constructing a Process Variables Map is as follows:

Step 1: The Team should revisit the *SIPOC* first to ensure that they fully agree on

- The scope of the process in hand—The Team must agree on the beginning and end points of the process. Be specific and don't assume that this is obvious. Try to use unambiguous triggers when the process starts and ends. Often it's useful at this stage to clarify the entity that runs through the process too.
- The key performance measures, sometimes known as KPOVs or Ys—Often it is useful for the Team to imagine that they were outsourcing this process, and, if so, how would they measure the performance of competing suppliers. These should have been more fully defined from the *Customer Requirements Tree*. Sometimes a complex process is in place to deliver little or simple output. Clearly a good question to ask at this stage is "Do we really need this process at all?"

- The major inputs known as KPIVs or Xs to the process—These are typically raw materials, labor, information, and energy in some form.

Step 2: List the primary process steps (one per sticky note) to achieve the process as defined in the *SIPOC* and place them in a vertical line down the middle of the flipchart paper (as per Figure 7.36.1). The steps should exhaustively describe the full scope of the process (i.e., don't miss any steps, particularly at the beginning and the end of the process).

The level of detail here is usually the biggest stumbling block in the construction. Typically processes have 5–10 major steps in their Process Variables Map. It would have to be a large and complex process to warrant any more than that. Also note that every step should contain at least some VA or Business VA, or in other words don't create a process step called "Delay."

Step 3: Starting at the first step in the process, list all the outputs and measures of performance (KPOVs) for each step on one or more sticky note. Some of these should align with the major outputs of the whole process as defined in the *SIPOC*. If all major Ys from the *SIPOC* do not appear at least somewhere at the detailed level in Step 3, then a major process flaw has just been discovered (i.e., the process is required to deliver something that doesn't appear anywhere along the way through it). This happens more than we care to admit.

Continue step by step through the process until all process steps are complete.

Step 4: Starting again at the first step of the process, list all the inputs (Xs) to each step (step by step), both physical entities *and* factors that affect the output of the process step. For physical entities, don't just list the name of the entity, list the Xs that describe what it is about them that affect the output. For example, if water is the entity, then Xs could be water feed rate, water temperature, and so on. One key example of this to ensure is done correctly is listing Operator as an X. What is it about the operator? Is it technical competency, experience, handedness (left or right), age, height, whether they wear glasses, and so on? Xs need to be specific enough to be actionable. In Figure 7.36.1, for example, the Order Processor's skill level and availability are the factors that affect their ability to enter the order. In this case, the Team might have considered others such as eyesight, ability to concentrate, distractions, and so on.

Continue step by step through the process identifying all the Xs until all process steps have been completed.

Items to note are that outputs (Ys) of one step can be the inputs (Xs) to the next and the same Xs can also appear in multiple steps.

Although Figure 7.36.1 shows only part of a Map and, therefore, a limited number of Xs, a completed Process Variables Map has somewhere from 100 to 400 Xs in it and takes up at least two flipchart pages.

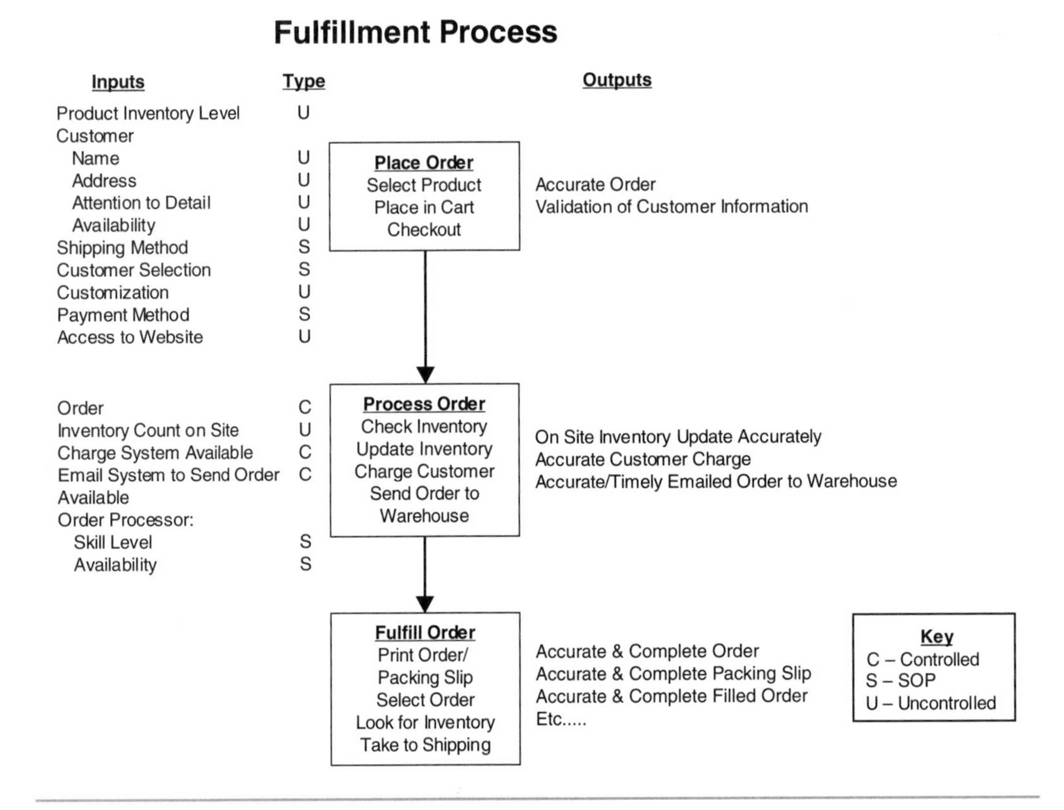

Figure 7.36.1 Process Variables Map under construction.

One major pitfall here is for the Team to discount Xs (filter them out) at this stage. The Map is a tool to identify *all* the Xs. No Xs should be filtered at this stage; that comes later with the *Cause & Effect Matrix* and so on.

Step 5: For each X, identify whether the X is currently controlled C or uncontrolled U. The key word here is *currently*. All Xs are theoretically controllable in some way, even gravity, provided you had enough money to do so. The important question is whether or not you successfully control or even choose to control the X at the present time.

It is often useful to include the third, intermediate category of Standard Operating Procedure SOP. This is a weaker version of "Controlled" because, in effect, you have set the expectation that the X is maintained at some level, but it is at the operator's discretion to do so.

Interpreting the Output

There is little interpretation required for the output of the Process Variables Map because the tool is used only to identify all the Xs and no decisions are made at this point.

Some texts recommend taking action directly from the Map relating to the lack of controls on an X. Clearly this is feasible, but it is questionable based on whether or not the X has been identified as being important (at this stage in any roadmap, this is not usually the case).

37: Project Charter

Overview

The purpose of Define is to ensure that

- The project scope is clear
- The Customer/market value is understood
- The business value is understood
- The measures of process performance are agreed and baselined
- Clear breakthrough goals are set for the measures
- There is business support to do the project in the form of a Project Leader and Team (and all resources are freed up appropriately to do the project)
- There are agreed Champions and Process Owners for the project to ensure barriers are removed
- This is the right project to be working on

The Project Charter is a 1–2 page document to ensure all these elements are considered. An example Charter is shown in Figure 7.37.1 and enlarged in Figures 7.37.3–5. The Charter, in essence, represents a project contract between the Belt, Team, and Champion, defining the problem, scope, value, goals, and support required.

The Charter is not just a tool for the current Team conducting the project, it is a tool used to help

- Others in the business understand what the project is about
- Future Teams examining the same process, but perhaps focusing on different elements
- Other project Teams working on different processes looking for pointers to improve their own projects

PROJECT CHARTER

PROJECT NAME: Emergency Department (ED) Throughput

<table>
<tr><td>Belt Name:</td><td>Laura Belt</td><td>Champion:</td><td colspan="4">Bob Champion</td></tr>
<tr><td>Process Owner:</td><td>Kim Owner</td><td>Revision:</td><td colspan="4">5/20/05</td></tr>
<tr><td>Element</td><td>Description</td><td colspan="5">Team Charter</td></tr>
<tr><td>1. Process:</td><td>The process in which opportunity exists</td><td colspan="5">South City Emergency Department (all acuity levels)</td></tr>
<tr><td>2. Project Description:</td><td>What will the project achieve?</td><td colspan="5">Increase throughput and decrease length of stay in the ED without reduction in the quality of care (per clinical quality scores and patient satisfaction scores)</td></tr>
<tr><td>3. Organizational priority:</td><td>Which strategic drivers(s) will the project affect and why?</td><td colspan="5"><u>Productivity</u>: more patients seen using same staffing levels
<u>Growth</u>: increased patient throughput
<u>Service</u>: shorter wait times for patients
<u>Quality</u>: clinical impact due to shorter wait times for patients</td></tr>
<tr><td>4. Project Scope:</td><td>Which part of the process will be investigated? Where does the process begin and end? List any area that are not in scope.</td><td colspan="5">Initial scope will be for the entire ED process from the patient walking through the door to when they exit the ED to home or the next point of care. This will subsequently be narrowed to focus on the areas of maximum opportunity once initial measurement is conducted. The project will initially consider all acuity levels.</td></tr>
<tr><td rowspan="5">5. Metrics:</td><td></td><td>Metric</td><td>Baseline</td><td>Goal</td><td>Entitlement</td><td></td></tr>
<tr><td rowspan="2">What will be the impact on the key process metrics?</td><td>Patient Length of Stay</td><td>185</td><td>100</td><td>?</td><td>Minutes</td></tr>
<tr><td>ED Volume</td><td>31,500</td><td>40,000</td><td>?</td><td>Patients / yr</td></tr>
<tr><td rowspan="2">What will be the impact on any secondary process metrics?</td><td>Left Without Being Seen</td><td>4.5</td><td>1</td><td>0</td><td>%</td></tr>
<tr><td>Wait Time</td><td>47</td><td>10</td><td>0</td><td>Minutes</td></tr>
<tr><td>6. Expected Benefits:</td><td>Who are the affected stakeholders and what benefits will they see? (hospital, patient, physician, etc.)</td><td colspan="5"><u>Patient</u>: Reduced length of stay, increased quality of care (quicker response to being seen by physician, diagnosis, timeliness of treatment).
<u>Staff</u>: Increased productivity and efficiency.
<u>Physicians</u>: Increased patient volume (revenue).
<u>Hospital</u>: Increased revenue, decreased case cost.</td></tr>
<tr><td>7. Team Members:</td><td>Who are the full-time team members and any expert consultants?</td><td colspan="5"><u>Full Time</u>:
Jane Smith – ED RN
Norman Conquest – ED RN
Alex Dean – ED Tech
Jesse Jones – ED-Unit Support Partner
<u>Ad Hoc</u>: E.D. physicians/staff, Paramedics, Clinical Quality
<u>External</u>: Physicians, Patient Financial Services, Security, Radiology Transport</td></tr>
<tr><td>8. Departments Affected:</td><td>Which departments or functions does the process cross and thus will be affected if it changes?</td><td colspan="5">ED, Lab, Radiology, Security, Patient Financial Services, Patient Registration, Environmental Services, Inpatient Nursing Units</td></tr>
<tr><td>9. Schedule:</td><td>Project Start Date</td><td>June 20, 2005</td><td colspan="2">Project Completion Date</td><td colspan="2">December 31, 2005</td></tr>
<tr><td>10. Support Required:</td><td>Will you need any special capabilities, hardware, etc.?</td><td colspan="5">$1.6m construction costs.</td></tr>
</table>

Figure 7.37.1 An example of Project Charter.

With these possibilities in mind, it is important that the Charter is clear and readily understandable by anyone in the business who reads it.

Logistics

At a Program level, there are typically a multitude of potential projects for a business to undertake. Once a potential project is identified, it needs to enter the project selection process to be judged in its merits versus all other projects.[54] To have enough information to make this judgment, an owner of the project (who typically would later become the Project Champion) would construct the draft Project Charter.

After the project is selected, the Champion reviews the draft Charter with the nominated Project Leader (Black or Green Belt) to build in more detail and specifically to agree on the appropriate Team members. After this is complete, the Belt and Champion review the Charter with the selected Team during the first Team meeting to make the final tweaks prior to commencement into Define.

The Project Charter is a living document through the Define and Measure Phases of the project, although generally the only elements to be updated during this period would be the metrics and scope (as more understanding of the process is gained), along with minor Team changes if there is a significant scope change.

After Measure, the Charter typically becomes "locked-in" for the remainder of the project.

Roadmap

The Project Charter is generally completed in linear order from top to bottom as follows:

Header Information

This includes basic information about the project. It is critical that each Charter has a designated Belt *and* Champion identified. All the Process Owner(s) must be identified, and after the draft Charter is complete, they should be bought into the objectives of the project. The Revision Date reduces confusion over the latest version of the Charter.

1. Process

This should identify the process in which the opportunity exists. It needs to be specific enough to identify location (site) if appropriate.

[54] For more detail on project identification and selection see Steve Zinkgraf's book *Six Sigma—The First 90 Days*. For problematic processes in which the target project is not obvious, see the Discovery Roadmap in Chapter 6, "Discovery—Tools applied to Identify Projects," in this book.

2. Project Description

This should describe what the project is intended to achieve. It should be based around key performance goals and generally takes the form of "Improve the metric from A to B," for example, "Improve the RTY for the process in unit ABC from 80% to 95%" or "Improve the number of lines on-time and complete from 92% to 98% for resupply." The description should only be a statement of the problem or goal and should not infer or include a solution.

3. Organizational Priority

This should identify which strategic drivers (the highest level business metrics) the project affects and why. Movement of the key performance metric (listed in the Project Description) should relate back to the high-level organizational goals. This can be done with the aid of a Goal Tree, as shown in Figure 7.37.2.[55] If there is no clear linkage, then the viability of the project should be in question. Typically each organizational goal is listed along with the impact of the project on that particular goal, as shown in Figure 7.37.3.

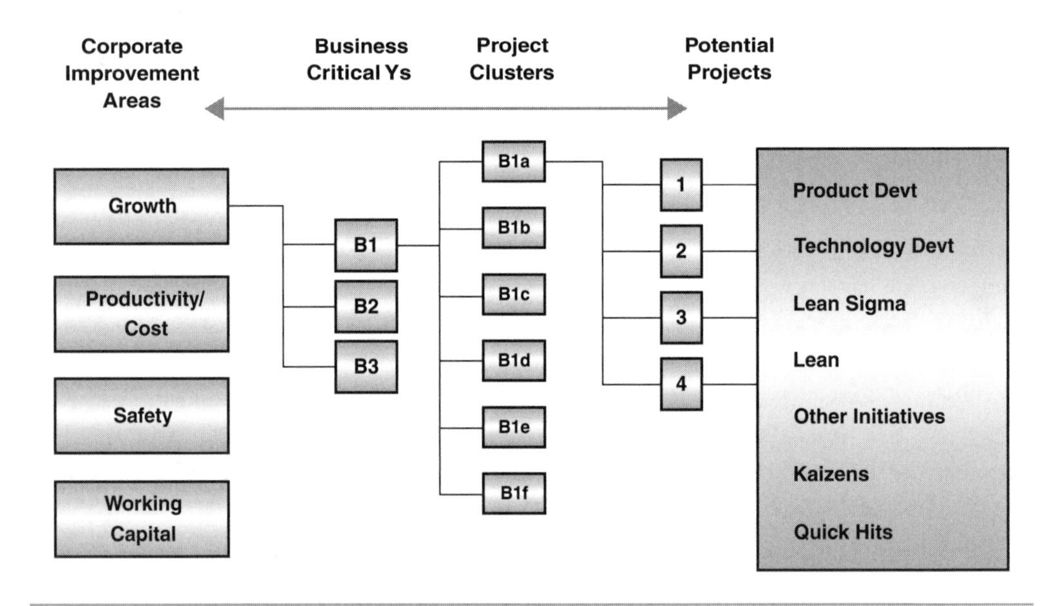

Figure 7.37.2 Linking the project goals to the organizational goals.[56]

[55] For more details, see Steve Zinkgraf's book *Six Sigma—The First 90 Days.*

[56] Actual construction of a Goal Tree is part of the project identification and selection process and is beyond the scope of the book. It is described in detail in Steve Zinkgraf's book, *Six Sigma – The First 90 Days.*

PROJECT CHARTER
PROJECT NAME: Emergency Department (ED) Throughput

Belt Name:	Laura Belt	Champion:	Bob Champion
Process Owner:	Kim Owner	Revision:	5/20/05
Element	**Description**	**Team Charter**	
1. Process:	The process in which opportunity exists	South City Emergency Department (all acuity levels)	
2. Project Description:	What will the project achieve?	Increase throughput and decrease length of stay in the ED without reduction in the quality of care (per clinical quality scores and patient satisfaction scores)	
3. Organizational priority:	Which strategic drivers(s) will the project affect and why?	***Productivity***: more patients seen using same staffing levels ***Growth***: increased patient throughput ***Service***: shorter wait times for patients ***Quality***: clinical impact due to shorter wait times for patients	
4. Project Scope:	Which part of the process will be investigated? Where does the process begin and end? List any area that are not in scope.	Initial scope will be for the entire ED process from the patient walking through the door to when they exit the ED to home or the next point of care. This will subsequently be narrowed to focus on the areas of maximum opportunity once initial measurement is conducted. The project will initially consider all acuity levels.	

Figure 7.37.3 An example of a Project Charter elements 1–4.

4. Project Scope

This identifies the part of the process to be investigated and where this process section begins and ends. It is imperative that the Team lists any areas that are not in scope. The scope is one of the most important sections of the Charter to ensure the Team stays focused on the task in hand, but also to truly gain consensus of the size of the undertaking. One of the most demoralizing things a Belt or Team can experience is to think that they have completed a project, only to be told they have missed a key part of the process later by the Project Champion or Process Owner.

To avoid this it is important to set additional boundaries for the Team to remain focused. This is done by identifying "in frame" and "out of frame" criteria, for example:

- Process steps included
- Product lines
- Customer segment
- Technology
- Project budget

A common mistake made early in Lean Sigma Programs is to make the scope of a project too large (sometimes known as "boiling the ocean"). Knowing what is too large is a skill that is tricky to teach and generally in the early stages of a Program it is best to rely on an external consultant or skilled Master Black Belt to act as the guide.

5. Metrics

Lean Sigma as a methodology that is heavily data-based. This begins with the Charter. The Metrics section captures the desired impact on the key measures of process performance. Successful projects tend to focus on improving a limited set of key metrics (typically one to three). Other key metrics might be listed as secondary "balance" metrics and might be listed with the intent of not adversely affecting them. For each metric there should be well-defined Baseline, Goal, and Entitlement values.

In the early stages of the project (before the Measure Phase), the baseline is taken from whatever data is available historically or from a short run study of the key metrics. The reason to understand the current performance level at this point is to judge whether the project is worth doing. This can typically be justified with a high-level understanding of the key metrics, but the baseline is later replaced with more rigorous data from the Measure Phase.

The use of Entitlement as a Goal-setting concept is one of the strongest components of Six Sigma (and subsequently Lean Sigma) and underpins the breakthrough mentality. Entitlement, for a metric, is the best value that the metric could ever achieve; the ultimate in performance, for example:

- Scrap: 0%
- Yield: 100%
- Capacity: running the fastest the process has ever run, for 100% of the time at perfect quality (known as 100% *OEE*)

When Goals were set in traditional process improvement projects, it was generally done by looking at current performance and demanding a small improvement, 1% or 5% for example. Another alternative might have been to conduct benchmarking of competitors and setting the goal as the best of their performance levels. There are two things wrong with these approaches, namely:

- Most operators in a process can "hold their breath" for a 1% improvement, meaning that the result is gained by working a little harder. After the project is complete, however, the process reverts back to its initial performance.
- If a process is improved to the level of the competition, then what (or even so what)? Will Customers value the improvements made enough to switch? Is it even enough?

Use of Entitlement really drives a different mentality in Goal setting. It might be impossible to achieve Entitlement itself (often not), but a goal of a change of halfway from current performance to Entitlement[57] or even three-quarters of the way is absolutely possible in most processes and really drives a different mindset. This kind of change cannot be achieved by "holding our breath" and genuinely requires a fundamentally different approach. It also sends a strong message to the Team, Operators, and the rest of the organization that this is a serious endeavor and requires significant effort and resource.

The values of Baseline and Goal (and sometimes Entitlement) can change throughout the project based on gaining a deeper understanding by use of the tools. If values change, then ensure that the Charter is updated and the Revision date is amended.

6. Expected Benefits

This should list all the affected Stakeholders and the benefits they will see from the project. Benefits generally fall into three categories:

- Financial—Monetary measures, for example, an increase in revenue or profitability or a decrease in working capital
- Quantifiable—Measurable on a continuous scale, but not readily translated into Financial returns, for example, a reduction in Lead Time or Length of Stay or perhaps an increase in document accuracy
- Observable—Not related to a continuous scale, but form a step-change in the process, for example, gaining the ability to bid in a market or a process certification (the organization either has it or not)

Any Financial benefits need to be clearly defined with a mathematical description of how to calculate the impact in dollars. The calculation formula must be accepted and supported by the Program Finance representative. If the method of calculation is not well defined in the beginning of the project, there is a struggle at the end to determine the financial impact.

Each project must have a measurable, tangible benefit to the Customer or it isn't the right project to undertake. When the project is completed, the Team should be able to go to a Customer and confirm that the project impact was "felt."

7. Team Members

This should list the full-time Team members and any expert consultants (these can be listed as "ad-hoc"). Team members are typically part-time assistants to the effort; the Black Belt is the only full-time member. The pitfall here is to make the Team too large.

[57] For example, for a current scrap level of 88%, the goal would be halfway to the 100% Entitlement and hence 94%. Likewise, three-quarters of the way would be 97%.

The optimum Team size is four to six key members, but recognize that not all Team members need to participate in all of the activities.

5. Metrics:		Metric	Baseline	Goal	Entitlement	
	What will be the impact on the key process metrics?	Patient Length of Stay	185	100	?	**Minutes**
		ED Volume	31,500	40,000	?	**Patients / yr**
	What will be the impact on any secondary process metrics?	Left Without Being Seen	4.5	1	0	%
		Wait Time	47	10	0	**Minutes**
6. Expected Benefits:	Who are the affected stakeholders and what benefits will they see? (hospital, patient, physician, etc.)	***Patient***: Reduced length of stay, increased quality of care (quicker response to being seen by physician, diagnosis, timeliness of treatment). ***Staff***: Increased productivity and efficiency. ***Physicians***: Increased patient volume (revenue). ***Hospital***: Increased revenue, decreased case cost.				

Figure 7.37.4 An example of Project Charter elements 5–6.

The importance of including this section on the Charter is so that

- Proper recognition can be given to the Team members throughout the project and after completion.
- The Team members feel ownership and accountability for the project outcome.
- After the project is complete, anyone wanting to gain more insight into the project can contact any of the listed Team members.

With these points in mind it is important not to just identify Team members by title only, but to specify both name and title.

The required level of Team member time must be released to the project by their management, and the Team members themselves must be informed of their participation in the project as promptly as possible.

8. Departments Affected

All departments affected by the project need to be consulted. To this end, this section should list all departments or functions that the process crosses and will be affected if it changes. Process Owners in the affected departments or functions should have sight of and ability to make inputs to the Charter.

9. Schedule

Although a detailed project schedule needs to be maintained separately by the Belt, it is important to agree to the time boundaries for the project (i.e., the Project Start Date and

End Date). Projects early in the life of the Lean Sigma Program take somewhere from four to six months (including the training period) to complete. Later when Belts have had project exposure, the duration should be 3–4 months. It is possible to accelerate this further using tactically placed Kaizen[58] events, an approach referred to as "K-Sigma" which is beyond the scope of this book.[59]

10. Support Required

The Team should list needs for any special capabilities, hardware, and so on. In fact, this is somewhat of an "acid-test" for the Project Team. Any capital requirements listed early in the project are generally indicative of a solution in mind. This is not a good thing, because the Charter should reflect only the problem and not any preconceived notions of solution. Typically, this box remains empty, unless the Team deems a particular skill set or additional tools necessary to conduct the project.

7. Team Members:	Who are the full-time team members and any expert consultants?	***Full Time***: Jane Smith – ED RN Norman Conquest – ED RN Alex Dean – ED Tech Jesse Jones – ED-Unit Support Partner ***Ad Hoc***: E.D. physicians/staff, Paramedics, Clinical Quality ***External***: Physicians, Patient Financial Services, Security, Radiology Transport		
8. Departments Affected:	Which departments or functions does the process cross and thus will be affected if it changes?	ED, Lab, Radiology, Security, Patient Financial Services, Patient Registration, Environmental Services, Inpatient Nursing Units		
9. Schedule:	Project Start Date	June 20, 2005	Project Completion Date	December 31, 2005
10. Support Required:	Will you need any special capabilities, hardware, etc.?	$1.6m construction costs.		

Figure 7.37.5 An example of Project Charter elements 7–10.

After the Charter is complete, it should be revisited by the Team to give it a "sanity check" as follows:

- In measurable terms, what is the project trying to accomplish?
- Is this project worth doing?

---

[58] A Kaizen is a rapid change event conducted by getting the appropriate process stakeholders together as a team for a period of 2–5 days and using simple Lean tools to streamline the process and element waste. The changes are implemented during the event itself.

[59] At the time of writing there are no texts written on this subject. For more detail on K-Sigma, refer to www.sbtionline.com.

- What happens if this project fails?
- Does it fit with company objectives?
- Is this a Customer-oriented project?
- Is the scope reasonable?
- What are the specific goals and stretch targets?
- Who owns the process and how are they involved?
- What is the probability of success?
- Is benchmark information available and if so, where?
- What resources are available to the Team?

Also, the Charter should be quickly revisited each time new understanding of the process is gained during Define and Measure to make subtle changes if necessary. The Champion and Process Owner should agree upon any and all changes.

The Charter typically evolves during Define and Measure, but after the sign-off of the Measure Phase it is unlikely to undergo any further changes for the remainder of the project.

As mentioned previously, completion of a strong Project Charter is a crucial part of the Define Phase of the Project and its value should not be underestimated. Weak Project Charters are almost certainly the biggest cause of any Project failures.

38: PULL SYSTEMS AND KANBAN

OVERVIEW

Pull Systems and Kanban is a large, complicated subject and many texts have substantial sections dedicated to them.[60] The intent here is to give a practical overview of Pull Systems and give the reader a roadmap to implement them.

Pull in simplest terms means that no one at an upstream process step should (process and) provide an entity until the downstream Customer asks for it—then they should provide it quickly and correctly. A Pull *System*, therefore, is a cascading series of processing and triggers (requests) from downstream to upstream activities in which the upstream supplier processes nothing until the downstream Customer signals a need.

An example of this is shown in Figure 7.38.1.

[60] For more see *Lean Thinking* by Womack and Jones or *Lean Transformation* by Henderson and Larco.

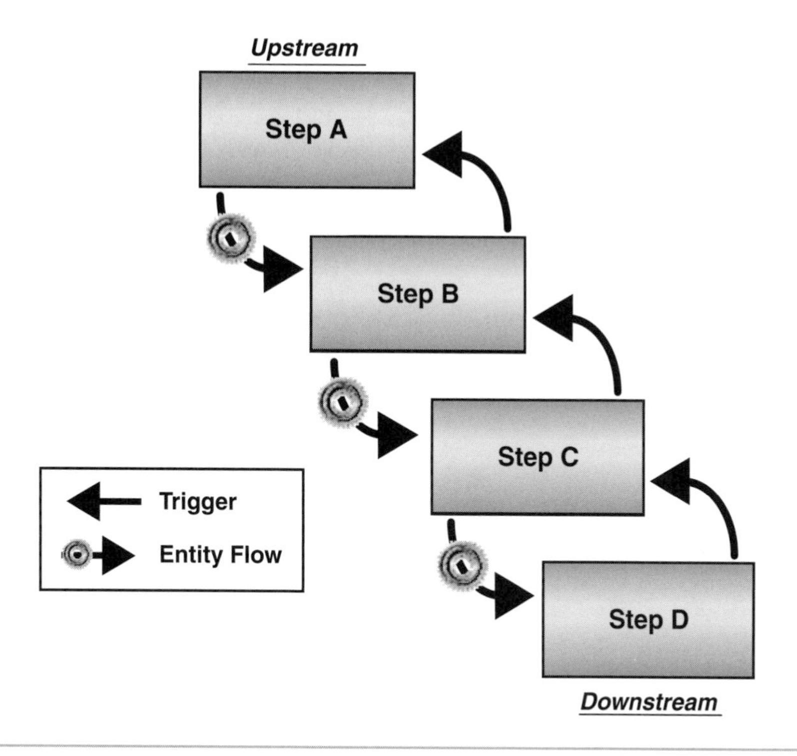

Figure 7.38.1 Graphical representation of a Pull System.[61]

The system in Figure 7.38.1 would flow as follows:

- The Customer uses an entity from Step D
- Step D is now empty and thus a trigger is sent to Step C to replenish Step D
- Step C sends an entity to Step D (replenishing it), Step D sits and waits for the next request
- Step C is now empty and thus a trigger is sent to Step B to replenish Step C
- Step B sends an entity to Step C (replenishing it), Step C sits and waits for the next request
- Step B is now empty, and so on

From the perspective of each location in the process, there is a simple set of rules that create the Pull System, as seen in Figure 7.38.2. A location should work only if their Outbox is

[61] Adapted from SBTI's Lean Methodology training material.

empty and their Inbox is full and there is a trigger to work. In all other circumstances the location would remain idle. This can be an extremely alien concept to most traditional process thinkers. Management typically worries about the poor utilization of resources, which "must certainly drive up costs." Operators feel extremely uncomfortable with being perceived as "not working efficiently."

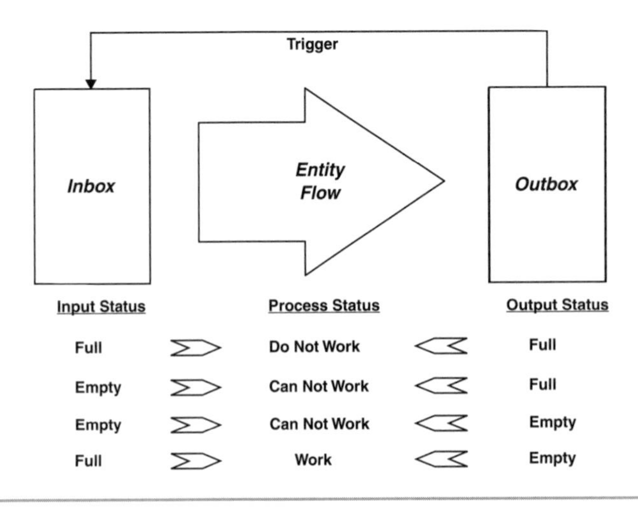

Figure 7.38.2 Location processing rules in a Pull System.[62]

Despite these concerns, a Pull System has many advantages over a traditional Push System, as can be seen in Table 7.38.1. In summary a Pull System

- Increases the speed to meet customer demand
- Reduces the amount of inventory without creating part shortages
- Decreases the amount of floor space required
- Improves the product and service quality

A Pull System, despite its many advantages, is a difficult transition to make. It requires an advanced and stable work environment to be successful; because of this, it should be one of the last pieces of the implementation puzzle to be put in place. Before installing a Pull System, the process itself must be

- **Reliable.** It should have dependable equipment with high-quality rates (generally through Mistake Proofing) and a flexible, multi-skilled workforce.

[62] Adapted from SBTI's Lean Methodology training material.

- **Organized.** The process should be well structured with minimal travel required and no delays due to materials movement.
- **Repeatable.** All work content should be done consistently, with clearly defined and understood Standard Work.
- **Balanced.** The process should be level loaded within Takt Time to generate a stable schedule. Lot sizes should be small or ideally the process should use one-piece flow. See also "Time—Takt Time" in this chapter.

Table 7.38.1 A Contrast between Push and Pull System Characteristics[63]

Push	Pull
Demand is forecasted (a prediction or guess), orders are launched in anticipation, sometimes weeks or months ahead of the delivery date	Cross-functional team designs a material plan, which accommodates today's demand (no anticipation)
Unpredictable process activity	Predictable process activity
Inflexible schedule	Flexible schedule
Shop Order launched (then wait)	No Order (ready now)
Low Inventory Turns 3–10	High Inventory Turns 20–100
Manage by schedule, hot sheets, material shortage meetings, expeditors, air freight, expedited shipments	Manage by Kanban planning
High scrap and rework costs. Quality meetings and decisions made at a senior level.	Low scrap and rework costs. Problems resolved by operators themselves.
Often delays	On-time shipments
Complex	Simple

The path to implementing a Pull System is actually in the implementation of the triggers themselves. The triggers are known as Kanbans. A Kanban is a simple and visual system, typically a small card or other visual cue, which regulates the Pull in the process by signaling upstream operation and delivery. Given that Kanbans are primarily used in Pull Systems they are also referred to as Pull Signals.

There are two basic types of Kanban signals:

- In-Process Kanban—A visual signal used to pace the movement of entities in a flow. These can be in the form of

[63] Adapted from SBTI's Lean Methodology training material.

- A prescribed location on the workstation (or floor), as shown in Figure 7.38.3. Operators are allowed to have a maximum of one entity in each of their Kanbans plus one entity at their workstation (where necessary).
- A container for the component parts of an entity. As the container arrives to a workstation it is the visual cue for that operation to be performed. Each container typically contains the work instructions themselves or a barcode to scan and pull up the appropriate work instructions from a terminal at the workstation.

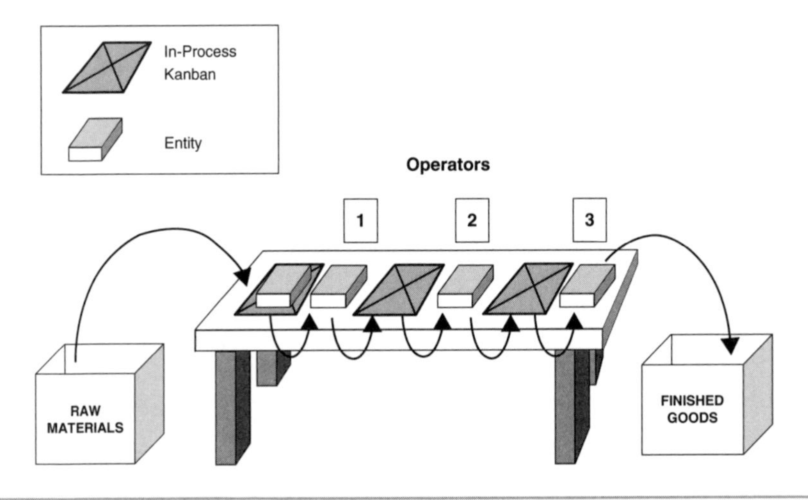

Figure 7.38.3 A location-based In-Process Kanban.[64]

- Material Kanban—A visual signal to replenish materials consumed in a process. These commonly are in the form of a card, as shown in Figure 7.38.4. The card lists amongst other things the origin, destination, entity, quantity, and any specific delivery instructions. Each entity type has its own Kanban and replenishment process; thus, there is not one material plan but literally hundreds of individual material plans. The basic flow of the Supplier Kanban is shown in Figure 7.38.5. Materials travel from the Supplier and are accompanied with the Kanban card. As the materials are used, the Kanban card passes to the Materials Planner who records their use and sends the card back to the Supplier to trigger the next materials delivery and the cycle repeats. At least two Kanban cards (usually more, see the following) need to be in the system to ensure that no stock-outs occur.

  Materials Kanbans fall into three forms depending *from* where the materials are replenished:

[64] Adapted from SBTI's Lean Methodology training materials.

- Supplier Kanbans (usually Yellow) are for materials originating at a Supplier and delivered to a storage location, a fabrication point or to the point of use.
- Operation Kanbans (usually Red) are for materials originating from a fabrication point and delivered to a storage location or to a point of use.
- Withdrawal Kanbans (usually Blue) are for materials originating from a storage location and delivered to a fabrication point or to a point of use.

CARD ID #	001	PART #	CG-001
REV #	001	KANBAN QTY	4
SUPPLY POINT	Cup Supplier	PART NAME	Plastic Cup
USAGE POINT	Final Assembly		
NOTES	Deliver with open end upwards		

Figure 7.38.4 An example of a Material Kanban card regulating flow of plastic cups in a production simulation.

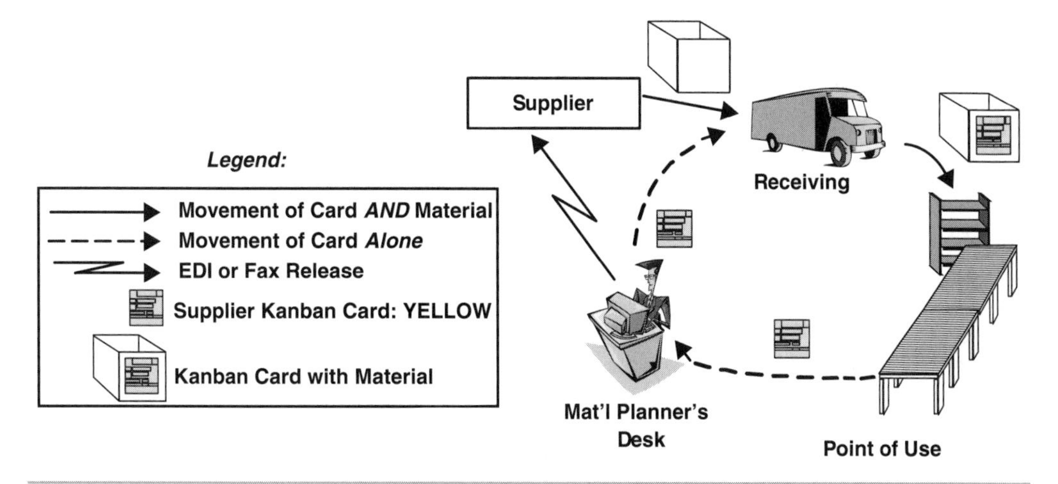

Figure 7.38.5 A Material Kanban at work.[65]

[65] Adapted from SBTI's Lean Methodology training material.

LOGISTICS

Setting up a Pull System requires considerable effort and planning and cannot be achieved just by the Team alone. The Process Owner needs to play a key role in setting up the Kanbans and ensuring personnel are fully aware of the need for the System and their accountability to it.

Planning the implementation takes the Team perhaps 2–4 hours. The implementation itself could take anything from 1 day to a month to complete depending on the complexity.[66] Use of a facilitating Consultant is probably a good idea for the first implementation.

ROADMAP

Pull Systems are successfully implemented only in processes involved in repetitive activity; they are not applicable in one-of-a-kind processes based on infrequent and unpredictable orders. The roadmap to implementation of a Pull System varies depending on whether the implementation is an external one (typically involving a Supplier or Customer), or an internal one within the Business itself. The latter of these, being the most straightforward are considered first.

Internal Implementation

Step 1: Determine the process and entity types on which to implement Pull. As mentioned previously, the process must be repeatable, reliable, balanced, and responsive. The Team should have already spent considerable time improving the process; the Pull System should be one of the last elements of the implementation.

Step 2: Establish a cross-functional implementation Team. This is made up of the existing Lean Sigma Team, but should be augmented with the following if not already represented:

- Quality
- Process Engineering
- Area Supervisor
- Materials Manager or Planner
- Maintenance
- Operator
- Union Representation (Pull requires a fundamental change in the way Operators work and thus early understanding by appropriate Union parties can make or break the implementation)

---

[66] For more information, see *Lean Transformation* by Henderson and Larco.

- Information Systems (triggers used are often computer-based)
- Lean Consultant (as mentioned previously, if this is one of the first implementations of such a System, it is highly advisable to seek consulting help to facilitate the venture)

Step 3: Train the Team in Kanban techniques as per Table 7.38.2. It is useful to run through a production simulation such as ProdSIM™[67] at this point. It is often easier to understand by being part of a simulated Pull Process prior to a real implementation.

Table 7.38.2 The Rules of Kanban and Pull Systems

The Rules of Kanban	
Flow	The consuming step should "Pull" from the providing step through use of the Kanban.
Providing	The providing step provides entities in the quantity and the order dictated by the Kanban. Nothing is made without a Kanban and only the Kanban quantity is produced (no more, no less). The providing process must be standardized and stabilized.
Quality	Every entity produced at the providing step must be of acceptable quality. Quality must be ensured *prior* to movement, so only good entities are passed.
Transferring	Nothing is transported without a Kanban. Kanban cards must always accompany the entities themselves.
Consuming	The consuming step should never request replenishment until the Kanban is empty. There should be only one partial Kanban at a time. The consuming process should be smoothly sequenced and leveled to the Takt Time.
Continuous Improvement	The number of Kanbans should decrease over time. The Replenishment Time from the provider should be a focus for a reduction effort.

Step 4: Confirm the Goals and Objectives within the Team, so all know exactly what is required of the implementation. The Belt should briefly explain how the solution was determined (a potted history of the Lean Sigma project) and then walk the Team through the "future state" process map. Take plenty of time for discussion and questions—this pays dividends later.

Step 5: Gather information for each entity type (some of this might have been done earlier to construct the *Demand Segmentation*):

- Daily Rate at maximum level of demand
- Average Daily Demand

[67] ProdSIM™ is the registered trademark of The Change Works, providers of change simulations and consulting services.

- Replenishment Time (see "Time—Replenishment Time" in this chapter)
- Kanban Sizes (see Step 6)
- Bills of Materials (BOMs)

For non-dedicated processes (more than one entity type flows through the process), identify

- Set-up times
- Changeover times
- Volume and mix for each entity type (how much of what is demanded from the process)

The data collected is revisited on a regular basis as Kanban sizes are adjusted or demand changes, so it is useful at this point to construct a single, central database for all the Pull data, if it doesn't already exist.

Step 6: Calculate the Kanban requirements, specifically

- Lot size (Standard Container Quantity)—This is predominantly driven by space availability adjacent to the working area and consideration of the Replenishment Time. If the Replenishment Time is long, then it might be necessary to have a larger Lot Size or to maintain an inventory in a warehouse or storage area close to the consuming process. The lineside Lot Size would never usually go beyond 1–3 days of stock.
- Number of Kanbans needed—There are multiple versions of formula to calculate the number of Kanbans, but the generic formula is as follows:

$$\#\text{Kanbans} = \frac{\text{ADD} * \text{ Replenishment Time } * (1 + \text{Safety Factor})}{\text{Standard Container Quantity}}$$

  where ADD is the Average Daily Demand and the Safety Factor represents the inventory buffer to ensure the line never starves. The Safety Factor is usually set at 1 or more in the early stages, but as the process settles and becomes more robust it is reduced. There must always be a minimum of 2 material Kanbans for each entity type.

After the Lot Sizes and Number of Kanbans are determined, it is useful to test the set up by physically simulating the process, or with a small simulation model. It is clearly better to learn of mistakes prior to going live with the Pull System.

Step 7: Make the lineside preparations. Develop a Kanban floor space layout using 5S principles. Order any supplies required, such as storage equipment. Allocate space on the floor ready for the Kanban.

Step 8: Train all stakeholders (anyone who affects or is affected by the change) in the Kanban process. This includes anyone who touches a Kanban, as well as their management.

Step 9: Go live. There is inevitably a need to handhold operators through the first few hours. Have someone from the Team at the lineside at least for the first day and through the first changeover of entity type.

External Implementation

External implementation of a Pull System shares many of the preceding steps, with the added complexity of dealing with Suppliers, Purchasing, and Accounts Payable.

Step 1: Determine the process and entity types on which to implement Pull. As mentioned previously, the process must be repeatable, reliable, balanced, and responsive. The Team should have already spent considerable time improving the process; the Pull System should be one of the last elements of the implementation.

Step 2: Establish a cross-functional implementation Team. This is made up of the existing Lean Sigma Team, but should be augmented with the following if not already represented:

- Quality
- Process Engineering
- Area Supervisor
- Materials Manager or Planner
- Maintenance
- Operator
- Union Representation—Pull requires a fundamental change in the way Operators work and thus early understanding by appropriate Union parties can make or break the implementation
- Information Systems—Triggers used are often computer based
- Lean Consultant—As mentioned previously, if this is one of the first implementations of such a System, it is highly advisable to seek consulting help to facilitate the venture
- Purchasing—Pull requires a significant change to the way product is purchased, typically a move to making "blanket orders" and then "calling off" from the order as materials are used
- Accounts Payable—Payments are typically made to the Supplier on use, as opposed to delivery
- Supplier—As an integral part of the Pull System, the Supplier should be involved as early as possible

Step 3: Train the Team in Kanban techniques as per Table 7.38.2. It is useful to run through a production simulation such as ProdSIM™,[68] at this point. It is often easier to understand by being part of a simulated Pull Process prior to a real implementation.

Step 4: Confirm the Goals and Objectives within the Team, so all know exactly what is required of the implementation. The Belt should briefly explain how the solution was determined (a potted history of the Lean Sigma project) and then walk the Team through the "future state" process map. Take plenty of time for discussion and questions—this pays dividends later.

Step 5: Start implementing "Blanket Purchase Orders" if these are not already implemented. For the Supplier to deliver materials on time and in full to the line, it is important that they have as much visibility as possible into the demand schedule and a commitment from the consumer. This commitment is in the form of, for example, a Blanket Order of materials for the next six months. The materials are not billed for immediately or shipped immediately; rather they are billed on use and delivered as per the Kanban mechanism.

Step 6: Create the procedure for broadcasting the trigger to the supplier. This can be done with Kanban cards (the card triggers the build) or there are Kanban Software packages available to create an electronic Kanban as it is sometimes called to trigger the build at the Supplier in anticipation of the card arrival. The broadcast could simply be by Electronic Document Interchange (EDI), fax, or e-mail. Determine also the timing of the broadcast relative to the consuming process and establish how the trigger is created and accepted without fail each cycle.

Step 7: Establish Supplier profiles. Not all Suppliers are created equally—some provide materials that are crucial to the performance of the product and must meet stringent criteria; others provide more commodity-type components. The delivery and quality performance of Suppliers of "critical" components should be investigated and those Suppliers are typically then "Certified" if their performance meets need. Purchasing is then instructed that they cannot buy from non-certified Suppliers on critical components.

Step 8: Gather information for each entity type (some of this might have been done earlier to construct the *Demand Segmentation*):

- Daily Rate at maximum level of demand
- Average Daily Demand
- Replenishment Time
  - Kanban Collection Time
  - Supplier Lead Time (from actual data, not from Supplier promises!)

[68] ProdSIM™ is the registered trademark of The Change Works, providers of change simulations and consulting services.

  - Transit Time
  - Receiving
  - Quality Time
- Kanban Sizes (see Step 9)—Supplier involvement in this is important.
- Bills of Materials (BOMs)

For non-dedicated processes (more than one entity type flows through the process) identify:

- Set-up times
- Changeover times
- Volume and mix for each entity type (how much of what is demanded from the process)

The data collected is revisited on a regular basis as Kanban sizes are adjusted or demand changes, so it is useful at this point to construct a single central database for all the Pull data, if it doesn't already exist.

Step 9: Calculate the Kanban requirements, specifically the

- Lot size (Standard Container Quantity)—This is predominantly driven by space availability adjacent to the working area and consideration of the Replenishment Time. If the Replenishment Time is long then it might be necessary to have a larger Lot Size or to maintain an inventory in a warehouse or storage area. The Point Of Use (POU) Lot Size would never usually go beyond 1–3 days of stock.
- Number of Kanbans needed—There are multiple versions of formula to calculate the number of Kanbans, but the generic formula is as follows:

$$\#\text{Kanbans} = \frac{\text{ADD} * \text{ Replenishment Time } * (1+\text{Safety Factor})}{\text{Standard Container Quantity}}$$

where ADD is the Average Daily Demand and the Safety Factor represents the inventory buffer to ensure the line never starves. The Safety Factor is usually set at 1 or more in the early stages, but as the process settles and becomes more robust it is reduced. There must always be a minimum of 2 material Kanbans for each entity type.

After the Lot Sizes and Number of Kanbans are determined, it is useful to test the set up by physically simulating the process, or with a small simulation model. It is clearly better to learn of mistakes prior to going live with the Pull System.

Step 10: Make the lineside preparations. Develop a Kanban floor space layout using *5S* principles. Order any supplies required, such as storage equipment. Allocate space on the floor ready for the Kanban.

Step 11: Train all stakeholders (anyone who affects or is affected by the change) in the Kanban process, including the Supplier's personnel. This includes anyone who touches a Kanban, as well as their management.

Step 12: Go live. There is inevitably a need to handhold operators through the first few hours. Have someone from the Team at the lineside at least for the first day and through the first changeover of entity type.

Step 13: Set up a Continuous Improvement Team for Kanban (see "Other Options" in this section). The complexities of an external Kanban make it almost impossible to execute perfectly from the Go Live. Improvements are needed as the process settles.

Other Options

Continuous Improvement is a big part of any Pull System implementation. The focus of such activity should be on

- Kanban reduction (numbers of cards in the system)
- Improvement of Lead Times in both the supplying and consuming processes
- Supplier Replenishment Time reduction
- Transportation Time reduction
- Receiving Time reduction
- Quality Inspection Time reduction
- Setup and Changeover Time reduction in both the supplying and consuming processes
- Travel distance reduction

39: Rapid Changeover (SMED)

Overview

Rapid changeover or rapid setup is also known as Single Minute Exchange of Dies (SMED) and was devised by Shigeo Shingo[69] during the period 1950 and 1969. SMED

[69] For more information, see *A Revolution in Manufacturing: The SMED System* by Shigeo Shingo published by Productivity Press.

was developed primarily with three Japanese companies, Mazda, Matsuzo, and Toyota and since then has revolutionized the entire Automotive and Metal Fabrication industries. It is obviously at home in all discrete-manufacturing industries, but is equally applicable to most service industries, for instance, in healthcare for room cleaning and turnover and more obviously in hotels where the room turnover occurs between the noon checkout and the 3 p.m. check-in time.

By reducing changeover times, businesses gain significant benefits from reduced costs per unit, lot sizes, Process Lead Time, inventory costs, and setup errors. Businesses also gain increased capacity (and therefore reduction in the need for excess capital equipment), yield, quality, and safety. More consistent, effective changeovers and setups also simplify scheduling by creating a more predictable process.

Changeover time is the actual *clock* time, not labor time or work content, so it is possible to start a stopwatch at the commencement of the changeover and stop it at the completion. The key to reducing the changeover time is to have a more encompassing definition of the time itself as shown in Figure 7.39.1. Traditionally, changeover is often considered just to be the tool change time when in fact the lost time is much greater and is better defined as

The amount of time between the end of the last VA step of the previous run to the beginning of the first value-added step of the next run.

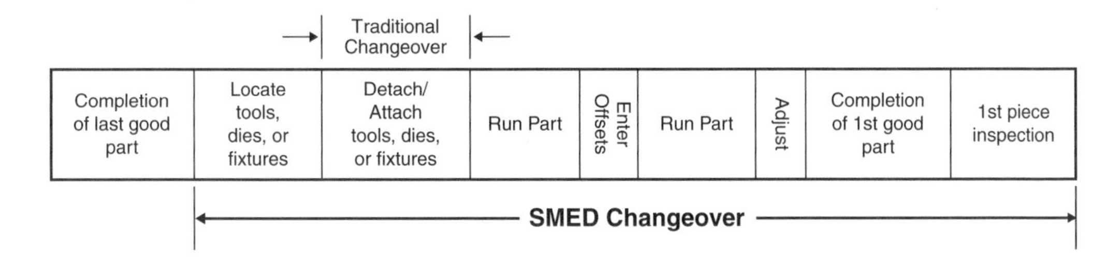

Figure 7.39.1 An example machine changeover in discrete manufacturing industry.

An example changeover is shown for a hospital treatment room in Figure 7.39.2. Traditionally the changeover was considered just to be the room cleaning, whereas the actual lost time is much greater.

The reasons for long changeovers stem primarily from inconsistency and lack of ownership. Individual operators often do changeovers as they see fit, based on their own experiences and no two operators changeover the same way. Often operators on one shift are not comfortable with a changeover or setup done on a previous shift. Therefore, changeovers are repeated and some operators even think that quality is better with a longer changeover.

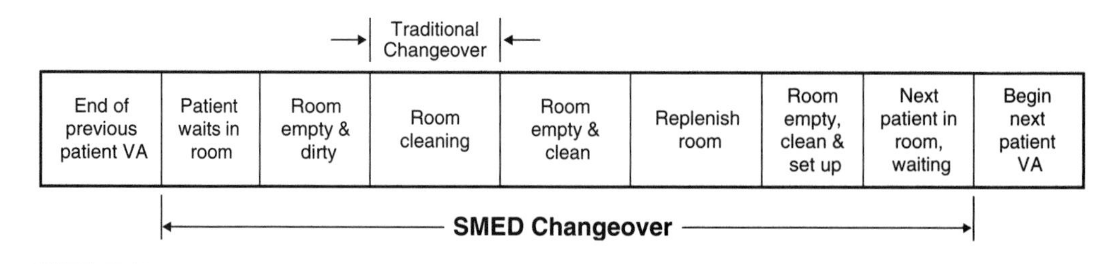

Figure 7.39.2 An example of changeover for a hospital Treatment or Operating Room.

LOGISTICS

Although changeover reduction should be used on an ongoing basis, the first work in an area is typically done as an "Event." The event duration is anywhere from one day for a small piece of equipment to perhaps a 4–5 day event in a larger more complex process. Typical event duration is three days.

The key elements to have in place prior to an event are

- **Support.** A SMED event is not a small affair and significant change is likely made. This requires solid leadership support to ensure success. The event should have a clear Owner/Champion that kicks off the event and serve as the Customer for the event.
- **Scope.** When conducting an event, the biggest failure is usually due to scope creep, so set a clear, focused, bounded process area to tackle during the event period and don't be tempted to move beyond this area.
- **People.** Having the right people involved is crucial. SMED is an active Team sport and requires people that actually do the work in the target area. Key functions to involve are
  - Operators and Supervisors—The most important members of the Team are those who actually operate the equipment and perform changeover tasks.
  - Process Engineering—Act more in a support role, adding technical expertise and implementing and building on operators' ideas.
  - Maintenance—Are often involved in the changeover itself and are also good resources for mechanical ideas and fixture design.
  - Quality Assurance—Quality checks often add significant time to changeovers and so QA involvement can help reduce this time.
- **Communication.** All parties involved in and affected by the event need to know of its existence and implications to them. The other functions listed in the previous "People" bullet also need to be informed that work needs to be done "there and

then" rather than be put on a To-Do List for future change. SMED follows the Kaizen mentality of doing it today, instead of planning it to be done over a period of weeks; hence the need for strong Leadership support. The rule in all Process Improvement is always to communicate ten times more than you would expect.

- **Equipment, Supplies, and Ancillary Items.**
  - Most of the event is done in the process area itself, but another quiet area is required to train the Team briefly on the SMED process and to serve as a work area at times through the event.
  - Facilitation equipment and supplies are necessary, such as flipcharts, pens, and perhaps, an LCD projector and screen to show the training materials and to work on the *Critical Path Analysis.*
  - Videoing the changeover before and throughout the event is crucial to analyze the activity. Ideally one video camera per changeover operator is required.
  - Work is likely done on laying out the area. An enlarged blueprint or CAD drawing of the process plan is needed.

ROADMAP

After the Team is formed and trained in SMED methods, the roadmap is as follows:

Step 1: Baseline current performance. It is important to establish an accurate baseline for changeover downtime with enough detail to make a distinction between setups, changeovers, and other downtime. This data should be reported at least monthly in the form of hours or as a percentage of available machine time.

If this data is not available historically, then it can be calculated approximately from Step 2 along with the number of changeovers made in a period.

Step 2: Video a *current* changeover. A video camera is the best tool for gathering changeover data; without it the all-important details are missed. The whole changeover needs to be recorded in one sitting, so a separate camera is needed to record each operator in the changeover in parallel.

Make sure that the cameras have on-screen timers, or if not then put a large digital clock in a strategic location in the area to track the time.

Be sure to inform the operators about the video recording ahead of time, in fact it is preferable to use the operators in the Team to be the guinea pigs for the recording if possible to alleviate some Hawthorne effect. Perhaps even video multiple changeovers and use the last one where the Hawthorne Effect is minimized. Be sure to capture the entire changeover from the end of the last VA step to the beginning of the next. If any pre- or post-changeover work is done, then expand the scope of the recording accordingly.

Focus on the details of the changeover activities as much as possible and don't be tempted only to set up a tripod and come back when the changeover is complete, too much detail is lost.

Step 3: Analyze the video recordings and document the changeover with the operators involved to complete the first four columns of an SMED Time Analysis Chart for the changeover, as shown in Figure 7.39.3.

Product: ____________ Machine/Equipment: ____________________ Date: ____
Product P/N: ____________ Load Center: ____________________

Step No.	Description	Start	Step Time	Before		After		Method of Reduction	Cost
				Int. Time	Ext. Time	Int. Time	Ext. Time		
	Total Time:								
		%Reduction							

Internal: Machine Stopped
External: Machine Running

Figure 7.39.3 SMED Time Analysis Chart.

Step 4: Classify the changeover steps as they are currently into internal and external time:

- **Internal Operations.** Those operations that are performed only when the machine is not running. Examples include changing a fixture on a machine or changing a bed ready for the next patient. Place the step time for internal operations in Column 5 in the SMED Time Analysis Chart.
- **External Operations.** Those operations that might be performed while the machine is running. Examples include organizing tools, preparing new fixtures, getting material, and so on. External operations can be performed either before or after the changeover. Place the step time for external operations in Column 6 in the SMED Time Analysis Chart.

Identifying the tasks that are currently being performed as internal operations, but could potentially be completed externally, is a key to reducing changeover time.

Step 5: Generate ideas to reduce the changeover time using the SMED Time Analysis Chart. There is no prescriptive approach here, but the primary places to focus include the following:

- Moving internal items to external by doing the following:
    - Using pre-setup activities
        - Tools cleaned and sharpened
        - Gages preset and supplies ready
        - Materials transported and sequenced
        - Tools, fixtures, and supplies positioned
        - Paperwork organized and prepared
    - Using post-setup activities
        - Clean, inspect, return, and repair tools, materials, and fixtures
        - Store all items in assigned locations
- Dividing the changeover activities by resource and apply *Critical Path Analysis* to each resource. On the Critical Path strive to do the following:
    - Eliminating all NVA activities
    - Eliminating or reducing other activities
    - Moving activities off the Critical Path
- Improving organization by the following. Note that 75% of the reduction can come from organization and *5S*.

- The area is sorted and cleaned
- All necessary tools are nearby with the use of shadow boards for hand tools
- Paperwork and checklists are posted in the area
- Visual flow of work through area with incoming and outgoing lots clearly are identified near the area
- Dedicated tool or die carts are used for each machine
- People, tools, and parts are organized in the work area before a changeover:
  - After a changeover begins, the participants should never have to leave the area
  - Everyone and everything involved with the setup is waiting for the "machine" to shut down
  - All operators follow a routine
- Reducing or removing elements in the changeover by doing the following:
  - Using one-touch equipment exchange (no adjustments, only touch the equipment once)
  - Using clamps or quick-release devices to attach tools, fixtures, air and water leads, and so on
- Eliminating adjustments and trial runs. Adjustments and test runs can account for as much as 50% of changeover time. Strive to eliminate them completely, not just reduce the time taken to perform adjustment.
  - Breaking the practice of running equipment, testing it, adjusting, running more trial product, and so on. No product should be spoiled during setup.
  - Replacing infinite adjustments with mechanical stops. Designing fixtures, tools, and associated equipment so that they are self-positioning.
  - Using digital readouts instead of dials and manual measuring tools when adjustment is absolutely necessary.
- Standardizing the area with the following:
  - Determining which entities are going to run on the machine, then evaluate the equipment to see if it can be modified to serve each without changeover.
  - Determining what tools, tool holders, and supplies are currently being used, then evaluate each to see if one can replace several. Size standardization often allows the use of common fixtures, dies, and tools.
  - Standardizing all fasteners (size and type) on all fixtures.
  - Determining if gauges used can be standardized or simplified.

- Simplifying the area by the following.
  - Using fewer parts during setup by combining functional pieces.
  - Making it obvious when the setup is complete and correct (stops, gauges, and so on).
  - Creating pre-kits of all non-standard tools and storing in staging area.
  - Ensuring tool kits are clearly marked with part numbers.
  - Ensuring information on supplies sheets are updated.
  - Making someone accountable for preparing changeover kits prior to changeover.
  - Simplifying mounting and removal of dies, fixtures & chucks.
  - Arranging all tools and supplies at point of use, by frequency of use. Operators should not have to move or turn frequently for tools.
- Mistake proofing
  - Color coding or marking all standardized tools, fixtures, jigs for identification and where they are used.
  - Using pins, blocks, or stops for quick alignment if parts, dies, or fixtures can be oriented in more than one way.
  - Using preset stops, limit switches, light switches, proximity switches and standardized settings
  - Using *Total Productive Maintenance* principles to ensure that all tools, dies, materials, and supplies are defect free and working properly.
- Problem solving. Things not happening or working the way they were designed can cause as much as half the time spent on a changeover. The goal is not to look for ways to make the setup easier in spite of these problems, but to eliminate them altogether. Anything that stands in the way of a perfect, trouble-free setup is viewed as a problem:
  - Missing or bent fixtures
  - Unsharpened, worn, or misplaced tools
  - Out-of-specification supplies
  - Mechanical failures

Step 6: Create the Improvement Plan. Document the Team's ideas and form two project lists:

- Short-term action items that can be completed immediately with little or no capital requirements. All action items should have clear owners and due dates.
- Long-term projects that are more involved and might require capital and equipment modifications. Project owners should be allocated.

Step 7: Execute the Improvement Plan to install the new changeover process.
Step 8: Control:

- Write new changeover procedures and best practices.
- Provide training across all shifts to standardize the new setup procedures.
- Document and publish the results of the new changeover process.
- Continue tracking changeover time to monitor results
- Strive for continuous improvement
- Recognize and reward outstanding improvements

Step 9: Proliferate to other machines or work-centers, so that best practices are being implemented across all machines.

None of the steps in the roadmap involve alchemy or rocket science. Most improvements involve just common sense, organization, and changes in basic attitudes.

40: REGRESSION

OVERVIEW

Y Data \ X Data	Attribute	Continuous
Attribute	Chi-Square	Logistic Regression
Continuous	ANOVA Means/ Medians Tests	Regression

Regression is one of the statistical tools used in the *Multi-Vari* approach and is probably the most powerful. A Regression analysis determines

- The statistical significance of a relationship between a Continuous X and a Continuous Y in $Y = f(X_1, X_2, \ldots, X_n)$.
- The nature of the relationship itself (i.e., the equation).

There are two basic forms of Regression:

- Simple Linear Regression, which relates one Continuous Y with one Continuous X
- Multiple Linear Regression, which relates one Continuous Y with more than one Continuous X

Regression analysis is the statistical analysis technique used to investigate and model the relationship between the variables. For both the Simple and Multiple techniques, the model parameters are linear in nature, not quadratic or any other power. Given the sheer size of the subject and the application of the tool in Lean Sigma, here the focus is primarily on Simple Linear Regression. Multiple Linear Regression is covered briefly in "Other Options" in this section.[70]

As with all statistical tests, a sample of reality is required. Generally 30 or more data points are required for the X and the corresponding value of Y at that point. Regression is a passive analysis tool and so the process is not actively manipulated during the data capture. After the requisite number of data points have been collected, they are entered as two columns into a statistical software package and analyzed.

Analyzing the data graphically using a Fitted Line Plot shows a result similar to the example shown in Figure 7.40.1. Here the X is "Age Of Propellant" in a rocket motor and the Y is "Shear Strength" of the propellant at that age. The data points are plotted on a Scatter Plot and then a straight line is fitted through them to give the best statistical fit. This is the Regression Line. There are many ways of doing this mathematically; in Regression the approach is to use Least Squares, which minimizes the total squares of all the distances from the line.

The equation of the straight line (the Regression model) is given above the graph in Figure 7.40.1 and is

$$\text{Shear Strength (psi)} = 2628 - 37.15 \times \text{Age of Propellant (weeks)}$$

Thus, in the future, for any Age of Propellant from 0 to 25 weeks,[72] it is possible to predict the physical property Shear Strength for that propellant. Also, if the Shear Strength had to be maintained above a certain level to perform correctly, then it is also possible to calculate a would-be shelf life for the propellant based on the model.

In the top right of Figure 7.40.1 are three statistics. These are in fact only three of many which are available from the full analysis results, which are shown in Figure 7.40.2. The analysis shows the same equation (model) representing the relationship between Y and X.

For the constant term and for each X in the model there is a p-value indicating whether that term is significantly non-zero. Both have a p-value of zero, which indicates

[70] For a deeper theoretical understanding of Regression, see *Introduction to Linear Regression Analysis* by Douglas Montgomery and Elizabeth Peck.

that there would be a small (almost zero) chance of getting coefficients this large (far from zero) purely by random chance. Specifically

- The p-value for the constant indicates that the Y-intercept is not equal to zero
- The p-value for the Age of Propellant indicates that the slope of the Regression line is not equal to zero

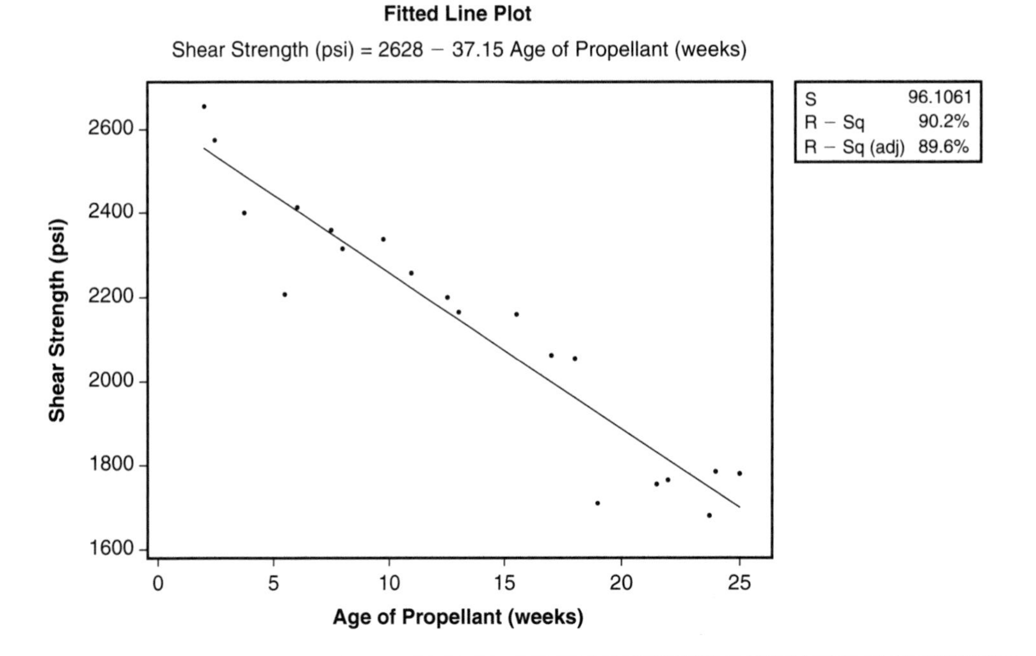

Figure 7.40.1 Example Fitted Line Plot[71] (output from Minitab v14).

The statistics on the center row are the same as those listed on the Fitted Line Plot in Figure 7.40.1:

- S is the standard deviation of the variation not explained by the model, known as the Residuals. It is the spread of the data around the Regression line.
- R-Sq (R^2) is the amount of variation in the data that is explained by the model. It is calculated from the ANOVA table at the bottom of Figure 7.40.2 by the equation

[71] Source: SBTI's Lean Sigma Methodology training material.

[72] There is no data outside of this timeframe and so no predictions should be made beyond 25 weeks.

SS(Regression) / SS(Total). Here 90.2% (calculated as 1527483 ÷ 1693738) of all the variability in the sample data is explained by the model.

- R-Sq(adj) is an indicator of whether any redundant (non-contributing) terms have been included in the model.[73] If the R-Sq(adj) falls well below the R-Sq value then there are redundant terms. Here the two are close and thus the conclusion should be that all the terms used in the model actually contribute something.

Regression Analysis

The regression equation is
Shear Strength (psi) = 2628 − 37.2 Age of Propellant (weeks)

Predictor	Coef	St Dev	T	P
Constant	2627.82	44.18	59.47	0.000
Age of P	−37.154	2.889	−12.86	0.000

S = 96.11 R − Sq = 90.2% R − Sq (adj) = 89.6%

Analysis of Variance

Source	DF	SS	MS	F	P
Regression	1	1527483	1527483	165.38	0.000
Residual Error	18	166255	9236		
Total	19	1693738			

Figure 7.40.2 Analysis results for the Rocket Propellant example.[74]

The bottom table is an ANOVA (Analysis Of Variance) table. For more details see "ANOVA" in this chapter. The ANOVA table breaks the variation into two main pieces:

- The variation explained by the model (known as the Regression)
- The variation not explained by the model (known as the Residual Error)

The calculation of these is shown graphically in Figure 7.40.3:

- The Mean of the Y data is calculated and represented by the dashed horizontal straight line in the figure.

[73] This is more applicable in Multiple Linear Regression versus Simple Regression.

[74] Source: SBTI's Lean Sigma Methodology training material.

- The Total Variation (Source Total in the ANOVA Table) is calculated by taking the square of the distance for each data point from the mean and then summing all the squares. SS(Total) = 1693738 is calculated this way.
- The Residual Error is calculated by taking the square of the distance of each data point from the Regression Line and then summing all the squares. It is the bit left over after the line has been fitted. SS(Residual Error) = 166255 was calculated this way.
- The variation explained by the model, known as the Regression, is calculated by taking the square of the distance from where the line predicts a point *should be* from the mean for every data point and then summing all the squares.

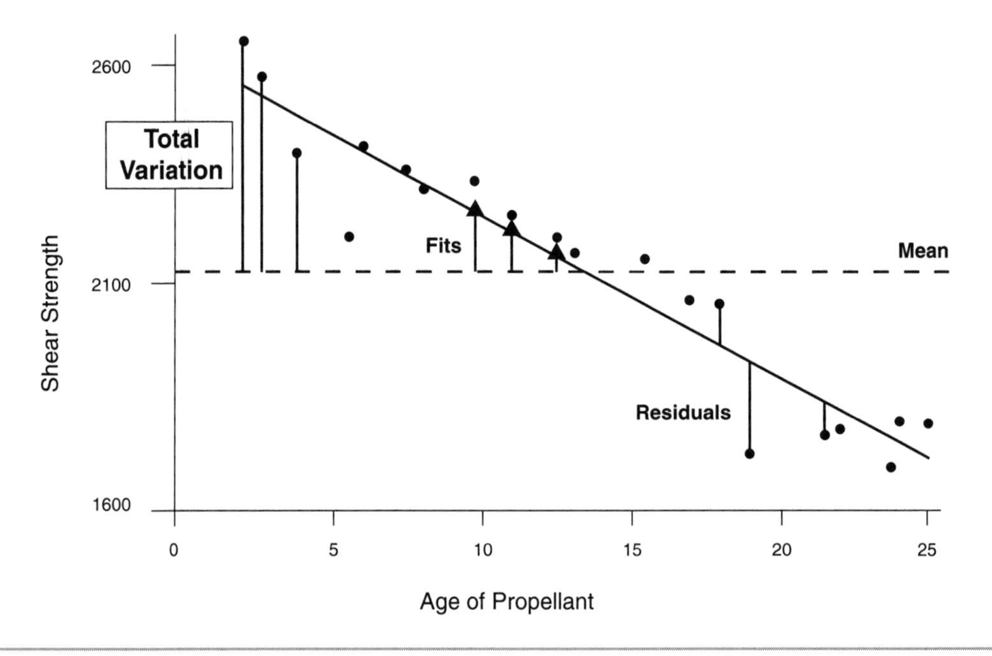

Figure 7.40.3 Graphical representation of ANOVA calculation.

From the preceding calculations it is possible to calculate a signal-to-noise ratio based on the size of the Regression (the signal) versus the background noise (Residual Error). This is the F-test in the table. Here the value of F is 165.38, which means the size of the signal due to the X is 165.38 times greater than the background noise.

The software then looks up the F value in a statistical table to discover the likelihood of seeing a difference of this magnitude.[75] The likelihood is the p-value, in this case 0.000.

[75] For more detail on exactly how this is calculated see *Introduction to Linear Regression Analysis* by Douglas Montgomery and Elizabeth Peck.

The p-value indicates the likelihood of seeing a relationship this strong in the data sample purely by random chance; this means that there is no relationship at the population level, it happened by coincidence in selecting the sample from the population. As in most statistical tests, if the p-value is associated with a pair of hypotheses, for Regression:

- H_o: Y is independent of X
- H_a: Y is dependent on X

If the p-value is less than 0.05 (as in this example) then the null hypothesis H_o should be rejected and the conclusion is that the Y is dependent on the X. Belts sometimes are misled at this point into assuming that there is a direct causal relationship between the X and the Y. There might be, but a change in X does not necessarily directly *cause* Y to move. The statistically correct explanation here is that when X moves 1 unit, Y moves by some consistent associated amount.

The analysis is not complete until the model adequacy is validated, which is done by reviewing the quality of the fit and an investigation into the variation that has not been explained, the Residuals (the bit left over). Residual evaluation gives a warning sign that the generated model might not be adequate or appropriate.

Looking at Figure 7.40.3, you know the residual is the actual value minus the fitted value, and it can be negative or positive depending on whether the data point is above or below the line. There are several measures of model adequacy with respect to the Residuals:

- The sum of the Residuals = 0
- The Residuals have a constant variance
- The Residuals are normally distributed
- The Residuals are in control

To validate model adequacy it is useful to examine the residuals graphically. To determine if the Residuals are Normal a few options are available:

- A Probability Plot can be applied as shown in Figure 7.40.4a. Residuals on the Normal Plot should form a straight line.
- A Histogram can be applied as shown in Figure 7.40.4b. The Histogram should appear to be forming a normal curve. This can be hit-and-miss and should be used in conjunction with the Normal Probability Plot.
- A *Normality Test* can be applied on the Residuals to gain a p-value. See "Normality Test" in this chapter. This is by far the best approach.

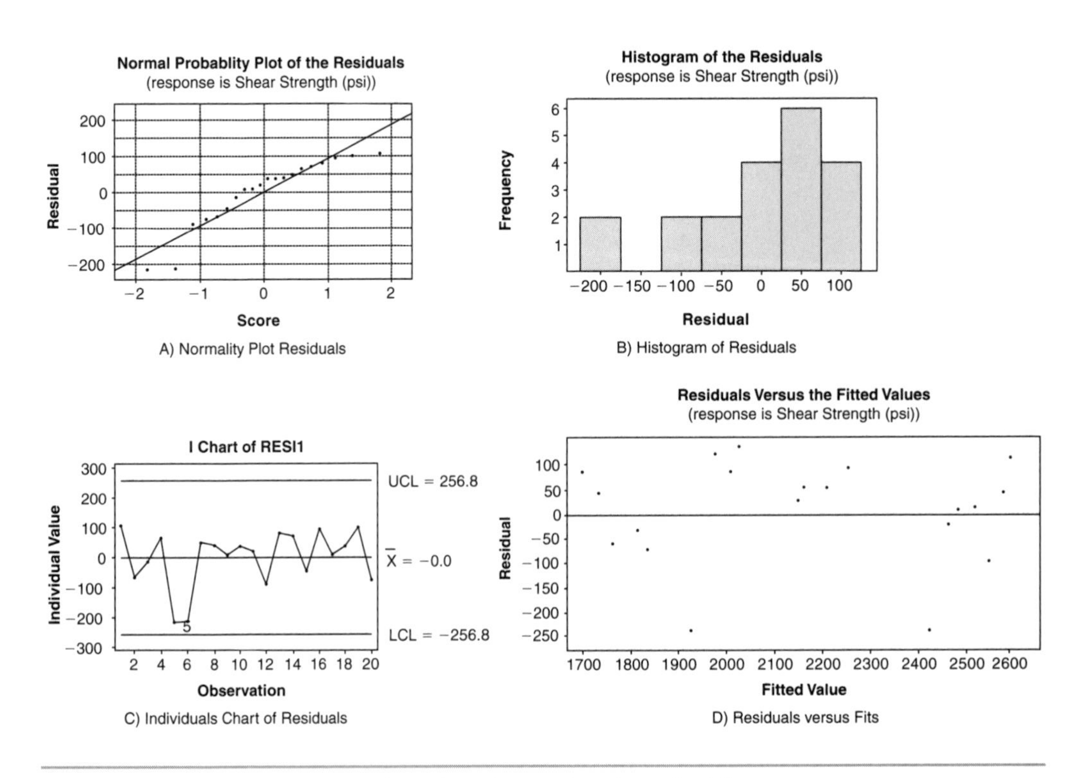

Figure 7.40.4 Graphical Analysis of Residuals (output from Minitab v14).

To determine if the Residuals are in Control, an Individuals Chart can be applied to the Residuals as shown in Graph C. Residuals that appear out of control should be studied further. Possible out-of-control issues might include Measurement Systems error, incorrect data entry, or an unusual process event. In the case of the latter, the Team should consult any notes taken during the data collection to evaluate the impact of the process event.

To determine if the Residuals have constant variance and to show that they are random (just background noise), a Residuals versus Fits Plot can be applied, as shown in Graph D. The Residuals should be distributed randomly across the Plot; any obvious patterns could indicate model inadequacy as described in Table 7.40.1.

If there are patterns in the Residuals and the R^2 value is very high, it probably presents no problem; however, if, for example, R^2 is less than 80% then there might be opportunity to create a better model based on the paths recommended in the table.

After the model is deemed to be adequate, the Team should collectively draw practical conclusions from it and present them back to the Process Owner and the Champion.

Table 7.40.1 Interpretation of the Residuals versus Fits Plot

Pattern	Residuals versus Fits	Interpretation
Residuals are contained in a straight band with no obvious pattern in the graph.	Residuals / 0 / Fits	The model is adequate.
Residuals show a funnel pattern. The variance of the errors is not constant and increases as Y increases.	Residuals / 0 / Fits	The model is inadequate. This might be resolved by transforming the Y.[76]
Residuals show a parabolic or quadratic pattern.	Residuals / 0 / Fits	The model is inadequate. This might be resolved with a higher order model (quadratic, for example).
Residuals show a bow pattern. The variance of the errors is not constant.	Residuals / 0 / Fits	The model is inadequate. This might be resolved by transforming the Y.[77]

[76] Beyond the scope of this book.
[77] Beyond the scope of this book.

Roadmap

The roadmap to conducting a Regression analysis is as follows:

Step 1: Plan the study. Identify the Ys and Xs to be considered. For each Y (preferably both the Xs and Ys) verify the Measurement System using a Gage R&R Study (see "MSA—Continuous" in this chapter). Agree on the data collection approach and assign responsibilities to the Team members (for more details see "KPOVs and Data" in this chapter).

Step 2: Pilot data collection. Validate the data collection approach as created in Step 1. Modify and retest if necessary.

Step 3: Collect the data, carefully following the agreed data collection approach. Take copious notes of process conditions and record any unusual process events. Transfer the data promptly into electronic form and make backup copies.

Step 4: Analyze the data, as per the details in "Overview" in this section:

- Create the Fitted Line Plot
- Evaluate significance of R^2 and the p-values
- Check the Residuals to validate model adequacy

Step 5: Formulate practical conclusions from the analysis, including potential follow-on studies.

Interpreting the Output

Regression in its Simple Linear form is quite straightforward to apply. There are, however, as with all tools, several pitfalls that can cause Belts problems:

- The purpose of a model is to create a prediction model for behavior of the response Y based on the predictor X. However, if the X itself cannot be predicted, then the model is useless. An example of this might be a desire to predict the maximum daily load on an electric power generation system from a maximum daily temperature model. The accuracy and usefulness of the Regression model for electric load prediction is conditional to the forecast of the temperature; the accuracy of which is patchy at best.
- Regression is an interpolation technique, not an extrapolation technique. Predictions from Regression models are made only with confidence within the confines of the data. If no data has been taken in an operating region, the model is hit-and-miss at best. To remedy this, data points should be taken over the breadth of the region in which predictions are made.
- Single points can heavily affect Regression models. In Graph A of Figure 7.40.5, the single outlier dramatically reduces the R^2 value of the model. If the outlier is a bad

value, then the model estimates are wrong and the error is inflated. However, if the outlier is a real process value, it should not be removed. It is a useful piece of data for the process. Refer to notes taken during data collection to understand the point and if possible try to recreate it.

- In Graph B in Figure 7.40.5, the single outlier increases the R^2 and regression coefficient. In this case, evaluate the model with and without the point to determine its effect. If the R^2 value greatly changes during this analysis, then that value is too influential. Conduct other data runs near that point to lower its leverage and confirm its validity.
- Regression models should represent meaningful relationships. Take for example the relationship shown in Figure 7.40.6. Data about a city showed that as population density of storks increased, so did the town's population. As much as I'd like to believe this relationship, it could equally be the reverse, mundane scenario. As the town's population increases then there are more chimneys (nesting grounds for storks); thus the stork population can increase accordingly.

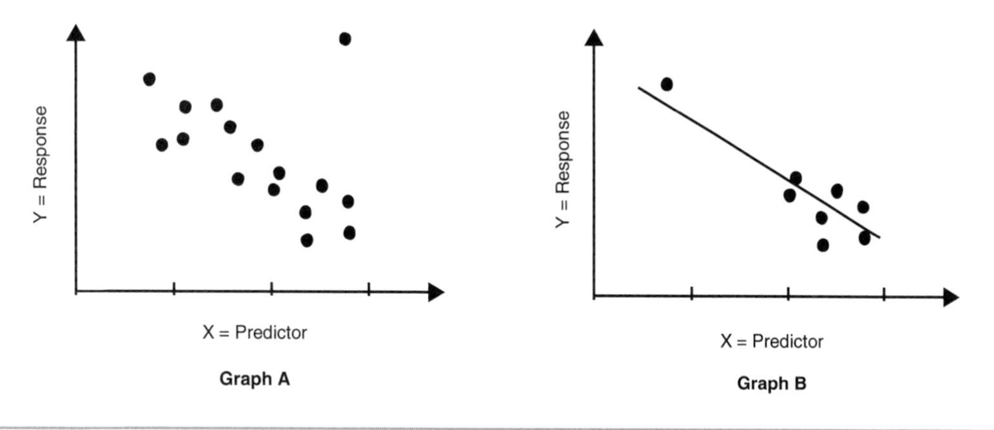

Figure 7.40.5 The effect of a single data point on the regression model.

Other Options

"Overview" and "Roadmap" in this section describe Simple Linear Regression, the investigation of the relationship between one Continuous X and one Continuous Y. Multiple Linear Regression, on the other hand, investigates the relationship between multiple Continuous Xs and one Continuous Y. The principles used are similar; however, the Fitted Line Plot no longer helps in this case. The multiple Xs are added into the Regression analysis and the same pointers are used, namely the R-Sq and p-values. The R-Sq(adj) becomes even more important in Multiple Linear Regression because there are more terms (more Xs) added into the model and certainly not all of them give any contribution.[78]

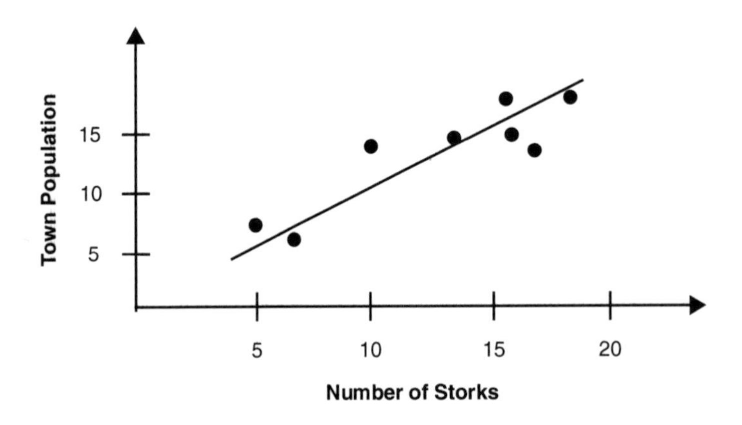

Figure 7.40.6 Incorrect causal relationships.[79]

Linear Regression is just that, linear. Sometimes the behavior of the relationship between the X and the Y is non-linear. Most software packages allow the user to select higher order models, specifically quadratic (including X^2 terms) and cubic (including X^3 terms). Belts tend to get carried away adding in higher-order terms when really the key tenet here is "the simpler the model, the better." Unless there is compelling reason to add a higher-order term, the linear model is usually preferable. Look to the R-Sq(adj) value as the model advances up an order. If R-Sq(adj) decreases for a higher-order model then the higher-order terms do not bring any additional value.

41: SIPOC

OVERVIEW

The SIPOC is a simple map of the process and, despite that simplicity, is a powerful tool in the Lean Sigma toolkit. SIPOC stands for

- Suppliers—Who provides the Inputs for the Process?
- Inputs—What goes into the Process?

[78] For much more detail on Multiple Linear Regression, see *Introduction to Linear Regression Analysis* by Douglas Montgomery and Elizabeth Peck.

[79] Source SBTI's Lean Sigma Methodology training material.

- Process—How is the Process performed?
- Outputs—What comes out of the Process (and, based on the Outputs, how it could be measured)?
- Customers—Who receives the Outputs of the Process?

SIPOC appeared in many earlier incarnations of Process Improvement, but its use in Lean Sigma is subtly different, or at least the map is recognized for what it truly achieves. SIPOC is also known under other anagrams, most commonly COPIS and POCIS. These variants are the same tool but involve filling in the columns in a different order (see "Roadmap" in this section).

When Teams are first formed there is typically a great deal of passion around objectives, potential solutions, and sometimes who is perceived to be to blame for the poor process. Sometimes it's even difficult for the Team to agree on what the process *is* and what it achieves. The SIPOC helps the Team reach consensus on the simple scope and purpose of the process and the project. Also, because the Team is actively focused on a reasonably objective tool, the early project problems of storming and venting are somewhat alleviated. It has been described as "the bright shiny object used to distract the Team while the project progresses past step zero." To that end it is a potent change management tool.

The useful outputs of the tool are

- A more unified Team
- An agreed process scope
- An agreed process purpose
- The beginnings of a list of Customers to feed into VOC work (which is expanded in the Customer Matrix)
- The beginnings of a list of Xs to feed the $Y = f(X_1, X_2,..., X_n)$ approach (which is expanded in the *Process Variables Map*)

An example SIPOC is shown in Figure 7.41.1.

Logistics

Creating an SIPOC is absolutely a Team activity and cannot be done by the Belt alone. The Belt should not be tempted to create a straw man of the map prior to the Team session; it should be created from scratch in the Team environment.

It takes the Team about 60–90 minutes to construct the map; remember, it is the process of creating the map and not so much the map itself that is important. It is typically best to create the SIPOC on a flipchart-sized page in landscape orientation using sticky notes rather than writing directly on the page.

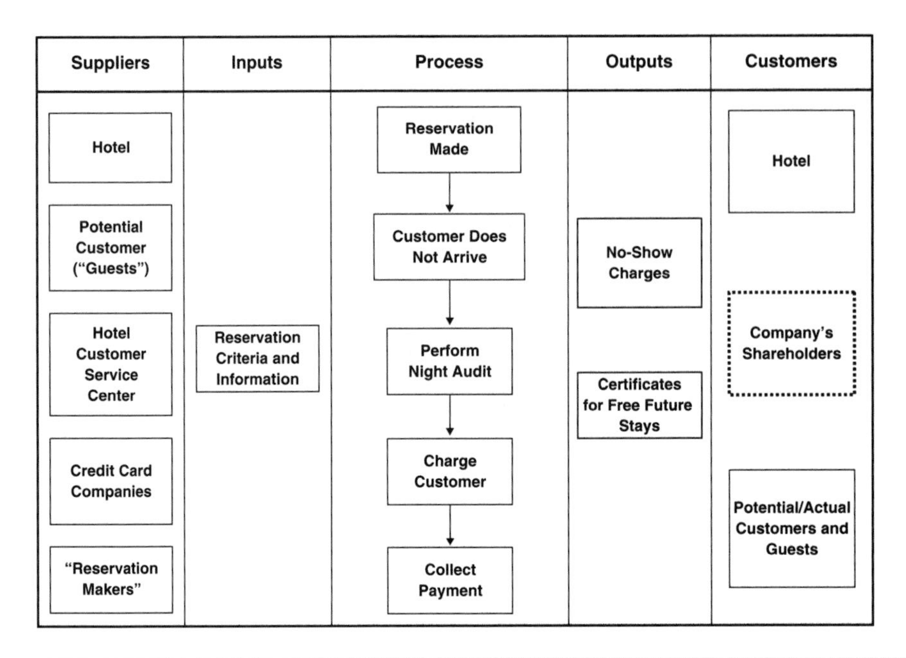

Figure 7.41.1 An example of a SIPOC for a hotel reservations process.[80]

ROADMAP

There are two accepted ways to construct the SIPOC:

- Start with the *Customers* and work to the left, always keeping the Customer as the focus (the sequence is thus COPIS). This can be confusing if the process isn't achieving its desired intent, because there is a significant mismatch between what the process does look like and what it should look like when the Team gets back to the *Process* column. Often this method is useful in Process Design, as opposed to Process Improvement.
- Start with the Process and work outward (effectively POCIS). This is generally more intuitive for Team members and is the more accepted approach in Process Improvement. This approach is used here.

[80] Source: SBTI Transactional Process Improvement methodology training material.

The steps to construction based on the second approach are as follows:

Step 1: Begin with the Process column. Brainstorm the high-level process steps and place on sticky notes on the page in the P column. A maximum of 5–7 steps are used because the Team is going to dig down later. List the steps sequentially using the structure "Verb Noun Modifier," for example "Enter Order In System."

Step 2: Identify the Outputs of the *whole* Process, not each Process Step. A common novice Belt mistake here is to try to create horizontal logic for each Process Step. Outputs, written as nouns, are typically one or more of

- Products or Services
- Information
- Decisions
- Documents

And should be tangible, in that it should be possible at any time to identify whether the process has generated an output or not.

If applicable, consider the Outputs for both internal and external Customers. For example, for an invoicing process the external Customer output might be the invoice itself, whereas the internal Customer output might be an Accounts Receivable transaction.

It is also useful at this point to begin to consider measurable outputs for the process, the potential Big Ys for the process (see "KPOVs and Data" in this chapter).

Step 3: Identify the Customers who receive the Outputs of the Process. To identify all the Customers, it is useful to ask

- Who pays for the process?
- Who represents the VOC?
- Who are the relevant end users for the process or product?
- Who represents the Voice of the Business (VOB)?
- Who are the relevant internal Customers who use the product or process downstream?
- Who typically pays for the product or service?
- Which external companies, such as vendors, does the process impact?

Step 4: Identify the *Inputs* required for the Process to function properly. Inputs are written as nouns and are usually

- Physical objects
- Information

Factors that influence the process.

If applicable, consider inputs from both internal and external Customers; for example, in an invoicing process an external Customer input might be a purchase and an internal Customer input might be pricing information.

Step 5: Identify the *Suppliers* of the *Inputs* that are required by the Process. Suppliers could be

- Any person or organization that provides an input to the Process
- Internal Suppliers
- Co-workers that provide inputs to the process in the same or different departments
- External Suppliers, such as vendors

Sometimes, Customers are also suppliers and in this case the Customer/Supplier becomes a vital partner in process success.[81]

Interpreting the Output

As mentioned before, the importance of the SIPOC is having the Team going through constructing it, not necessarily the physical end resulting SIPOC map itself.

Based on the SIPOC, the scope of the process (thus the scope of the project) should be clearly defined. Examination of the Outputs of the process should guide the Team to what *might* make sensible Big Ys for the process.

Do not try to read much more than this into it. The tendency is for Teams to want to base significant improvement decisions on the SIPOC result, when it is really only a tool to help frame the project. The SIPOC represents only current Team thinking; it does not include true Customer input—that comes later.

42: Spaghetti (Physical Process) Map

Overview

A Spaghetti Map or Physical Process Map is the simplest Lean Sigma tool. It demonstrates the physical flow of an entity or multiple entity types (product, patient, information, order, and so on) and the associated travel distance for a single cycle of a process. It is a graphical representation of travel distance and travel patterns.

[81] More common in Transactional and Service processes than in Manufacturing processes.

The Spaghetti Map is a particularly useful tool when there is excessive movement of an entity or entities through a process. A highly simple, visual tool, it can help streamline a process and is part of the standard toolkit used when running a kaizen event. In that context, it can show the existing problems in a process and also communicate the potential benefit of change to a new layout or flow, as in Figure 7.42.1.

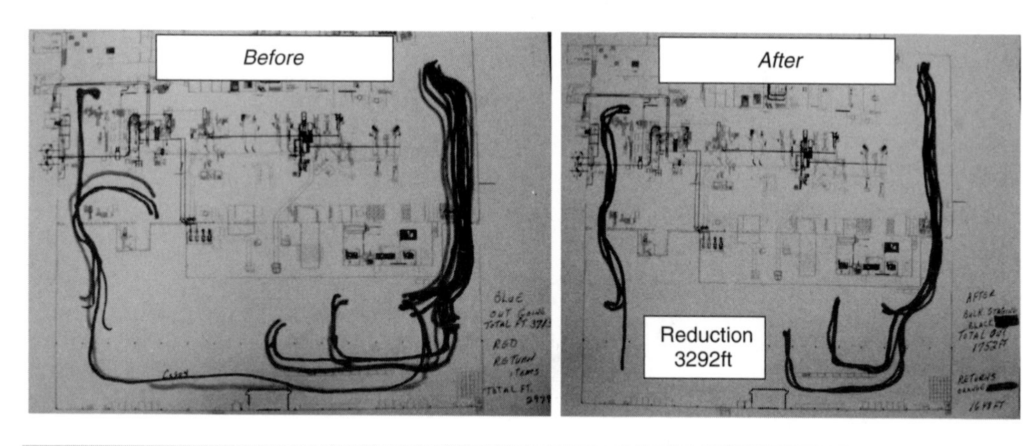

Figure 7.42.1 Spaghetti Map showing material movement before and after a kaizen event.

Roadmap

To construct the map, the roadmap is as follows:

Step 1: Determine the scope of the process in question (i.e., the start point, end point, and geographical boundaries of the process).

Step 2: Sketch or obtain a blueprint or CAD drawing of the facility/process layout as per the geographical boundaries identified in Step 1.

Step 3: Mark the process locations and steps onto the layout, as shown in the left side of Figure 7.42.2.

Step 4: Connect the dots in accordance with the actual travel or walk patterns for the entity. The commonest mistake here is to draw the line "as the crow flies," as shown in Figure 7.42.3. This is incorrect, because entities don't typically tend to fly through the air around a facility. Map the path as the entity actually travels, similar to that shown in Figure 7.42.1. The paths should be drawn for just a single cycle of the process.

Step 5: Calculate the distances traveled. This is done with a measuring device such as the wheel-on-a-stick contraption or an electronic pedometer. In the case of the kaizen event in Figure 7.42.1, the warehouse in question was quite large and most of the travel distance data came from setting the trip on the odometer of a forklift truck.

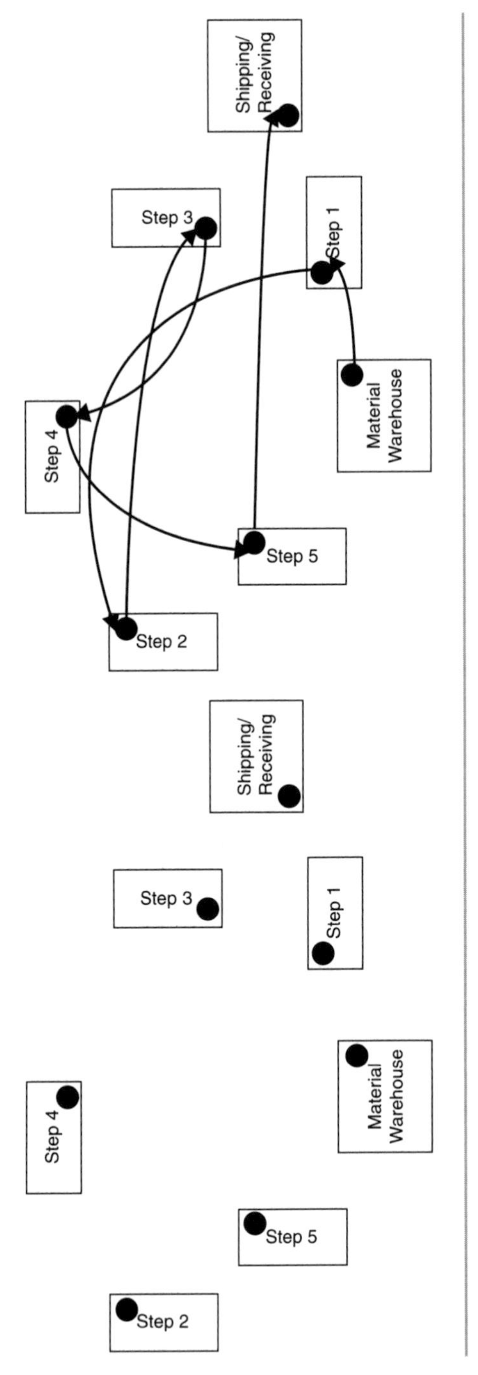

Figure 7.42.2 Construction of a Spaghetti Map.

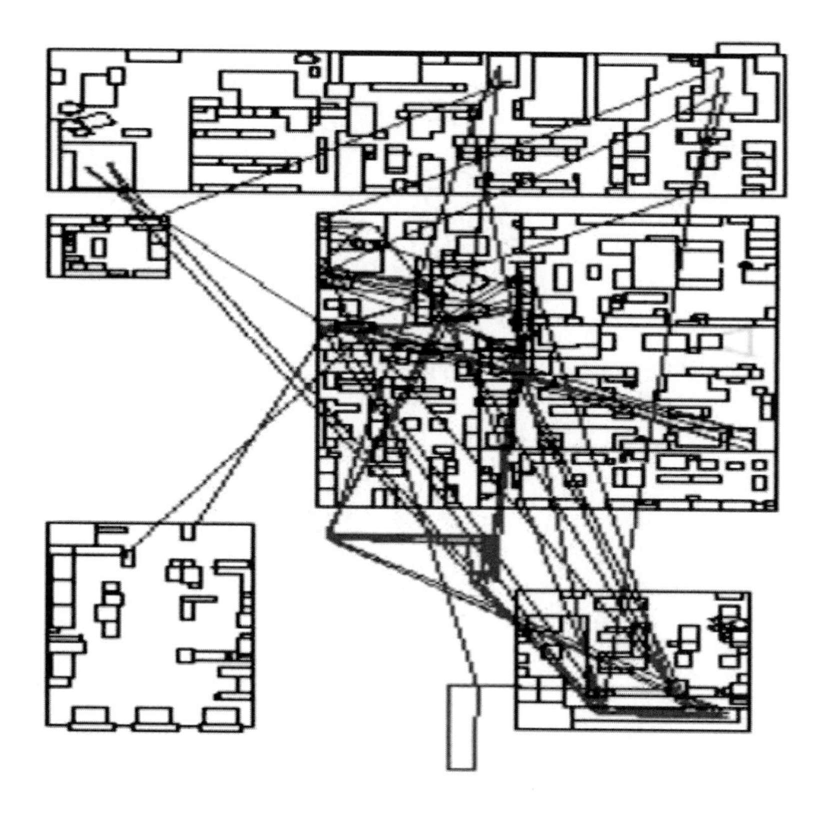

Figure 7.42.3 Incorrect use of mapping lines.

Step 6: After the Current State Map is completed and work is done to improve the layout of the process, a second, Future State Map, is constructed with an indication of the reduction in travel distances, as shown in the right side of Figure 7.42.1.

Interpreting the Output

The output of the Spaghetti Map typically requires no explanation and is a great communication tool for changes made or about to be made.

43: STATISTICAL PROCESS CONTROL (SPC)

OVERVIEW

Statistical Process Control[82] (SPC) is the use of *Control Charts* to help control a process. Most novice Belts confuse SPC with *Control Charts* and vice versa. *Control Charts* are used in a number of places in the Lean Sigma roadmap, for example:

- To test for stability before performing a Capability Study in the Measure Phase
- To test stability before performing a test of means in the Analyze Phase
- To validate stability after project completion

SPC on the other hand is solely a Control tool and appears in the Measure Phase (to check what is currently being controlled) and the Control Phase (to control what should be controlled henceforth).

The key to understanding SPC is an appreciation of the different types of variation in a process,[83] specifically:

- **Common Cause.** The inherent variation present in every process, produced by the process itself. This is effectively the background noise in the process and can only be removed or lessened by a fundamental change in the process, usually the process physics, chemistry, or technology. A process is Stable, Predictable, and In-Control when only Common Cause variation exists in the process.
- **Special Cause.** The unpredictable variation in a process caused by a unique disturbance or a series of them. Special Cause variation is typically large in comparison to Common Cause variation, but is not part of the underlying physics of the process and can be removed or lessened by basic process control and monitoring. A process exhibiting Special Cause variation is said to be Out-of-Control and Unstable.

Control Charts are used to find the signals (Special Cause variation attributable to assignable causes) in amongst all of the background noise (Common Cause variation).

SPC is placed on critical Xs in the process and uses *Control Charts* to detect when there are out of the ordinary events in amongst the regular background noise of the process. The Control element of SPC is that once an event is detected, action is taken to identify and remedy the cause. Without these controlling actions, someone accountable to make

[82] Use of Control Charts for Process Control dates back to the 1920's and Dr. Walter Shewhart of Western Electric.

[83] A wonderful reference here, written in plain language rather than Statspeak, is Donald Wheeler's book, *Understanding Variation—The Key To Managing Chaos.*

them and correct placement on the critical Xs, SPC does not exist. What exists instead, which is common in many misinformed groups, is a piece of paper with a graph on it.

Logistics

As mentioned in Chapter 5, "Control—Tools Used at the End of All Projects," SPC is part of the Control Plan; the group of all tools, physical changes, procedures, and documentation that is used to ensure that process performance consistently remains at the desired level. SPC is not placed on every single X, it is placed on critical Xs that cannot be designed out of the process or controlled by physical means or with mistake-proofing devices. It clearly also relies on the ability to measure the X and respond accordingly.

The Process Owner should own SPC on an on-going basis, with little to no Belt involvement. If a Belt cannot walk away from the process at the end of a project, then their project isn't complete and more time needs to be spent on a more robust Control Plan and handoff.

Control should be made as close to the process as possible with *Control Charts* being generated by the operators, special causes detected, and action taken at that level. There needs to be clear management commitment to do this.

Accountability for SPC is typically on a key operator or line supervisor and it should be written into both their role and appraisal criteria.

Roadmap

The roadmap to setting up SPC on a process is as follows:

Step 1: Identify the critical Xs or KPIVs for the process that are controlled using SPC. The biggest mistake here is to pull them out of thin air. The Xs should be chosen based on the following:

- They have been determined by analysis to be critical Xs in the process and drive a large percentage of the variability in the major performance characteristics, the Big Ys or KPOVs of the process.
- They cannot be designed out of the process.
- They cannot be controlled by physical methods.
- They cannot be controlled using mistake-proofing devices.
- They are measurable on a Continuous or Attribute scale.

Step 2: From the data type of the X (Continuous or Attribute) determine the specific type of *Control Chart* required (see "Control Charts" in this chapter).

Step 3: Use a sample of recent historical data to form a base chart. If none is available, then create a blank chart to be populated as data does become available. A common mistake here is to want to invent Control Limits. The process determines these.

Step 4: Determine the actions required if the process shows an out-of-control condition. This could have been already determined earlier in a *Failure Mode & Effects Analysis.*

Step 5: Train the Operators how to read the Chart and the actions required if the process goes out of control.

Step 6: Transfer ownership of the chart to the Operators or Line Supervisor, whoever is the most appropriate.

Step 7: Update all procedures and job descriptions to include accountability for the charts.

Step 8: Hold Operators or Supervisors (as per Step 6) accountable by whatever existing means the business is currently managed.

INTERPRETING THE OUTPUT

As mentioned previously, SPC relies on *Control Charts* to detect when an out of the ordinary event has occurred. See also "Control Charts" in this chapter. *Control Charts* typically take the form shown in Figure 7.43.1. Data is plotted over time across the x-axis, with the height on the y-axis representing the level of the X in question. From the data in the chart, "Control Limits" are calculated that represent the boundaries of reasonable behavior within the process. A point landing outside of these boundaries is considered special cause (out of the ordinary). The Control Limits are calculated from the process data itself using specific equations based on the data type. A statistical software package does this automatically.

The odds of a point lying outside of a Control limit are of the order of 300 to 500:1.[84] This is considered an unusual event. Obviously in a process that generates hundreds or thousands of entities, the occasional point falls outside the lines. Two in a row almost guarantees that something highly unusual has occurred or that the process has changed in some way.

[84] For an Individuals Chart used for charting normal data, the Control Limits are placed at ±3 Standard Deviations. Analysis of a Normal Distribution shows that approximately 99.73% of all data points should fall between these lines and hence falling outside is an event with probability 0.27%. See "Control Charts" in this chapter.

Statistical software generally also highlights other unusual points for instances such as:

- Points hugging the center line
- Points oscillating back and forth
- Points continuously rising or falling

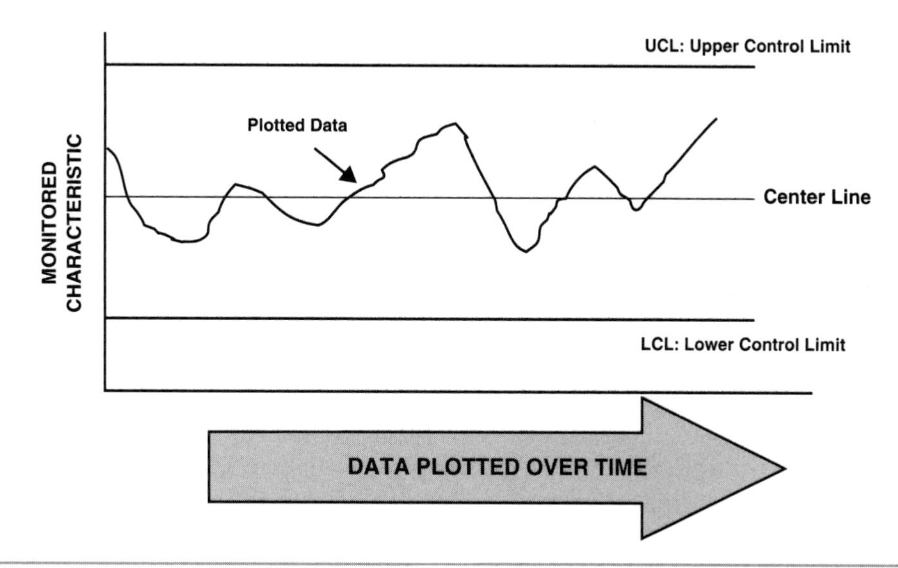

Figure 7.43.1 Structur of a Control Chart.

For each, statistics are used to determine patterns that occur with odds at least 300:1 against.

The biggest mistakes made with the use of *Control Charts* for SPC are

- Putting Product Specification limits on the Control Chart causes it to become just an inspection tool—it is no longer a Control Chart at this point. The Control Limits tend to be ignored and focus is only on the specifications, or even worse the chart is incomprehensible to the Operator and is ignored completely. Understanding how a process performs against specification (known as Process Capability) is important, however, not in this application.
- Treating the Control Limits as specification limits. The Control Limits are not directly tied to customer defects. If a point goes out of control it does not necessarily mean that the entity concerned is defective; it just means that the process has changed.
- Flooding the system with *Control Charts*, and then not taking action on the data.
- Not following up on unusual negative events to remedy them by determining and eliminating the root causes. In this case there is no "Control" in Statistical Process Control.

- Not following up on unusual positive events to learn from them and capture the improvement. Sometimes the process suddenly gets better and the change is statistically unlikely; for example, the data points might suddenly start hugging the centerline. Often the mindset is to feel good about the process improvement, but to do little to understand why the process has improved. It is important in this case that Operators and Supervisors seek to capture the improvement and set this as the new standard going forward.

44: SWIMLANE MAP

OVERVIEW

The Swimlane Map is actually two different forms of maps, more formally known as the Cross-Functional Map and the Cross-Resource Map. Both are a simple extension of the detailed *Value Stream Map (VSM)* and are created from it.[85]

The simpler of the two, the Cross-Functional Map, is one that conveys process handoffs between functions or departments, by rearranging *VSM* process steps into lanes of functions. The lanes run horizontally (or sometimes vertically) and the rows (or columns if vertically oriented) represent different functions in the business.

Therefore, all activities from a particular function fall in appropriate swimlanes as they are known. Time runs along the horizontal axis of the Swimlane Map; thus, any concurrent activities appear vertically aligned with one-another. An example of the structure is shown in Figure 7.44.1.

The Cross-Resource Map is an extension to the Cross-Functional Map and takes the level of detail in the lanes one step lower. Where appropriate, Function Lanes are broken down to become individual Resource lanes. This is only appropriate where a specific Resource (a role, not a person) is always solely responsible for doing a particular task or activity. Functions that use pooled resources (anybody in that Function can pick up the task) are not split. An example of the structure is shown in Figure 7.44.2. It is unlikely in a Cross-Resource Map to see every single Function Lane broken down to the resource level, so the map is more of a hybrid than anything.

LOGISTICS

The Swimlane Map (either Cross-Function or Cross-Resource) is a quite a simple extension of the detailed *VSM*. It certainly requires the whole Team to construct it and should

[85] The Swimlane Map can also be created from any pre-existing Flowchart or Block Diagram.

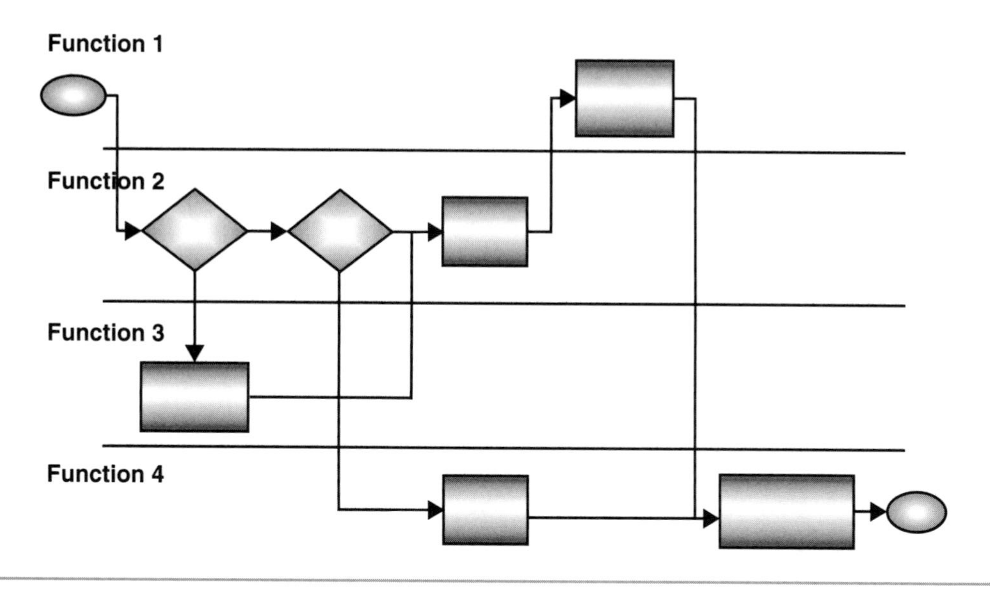

Figure 7.44.1 Cross-Functional Map Structure.

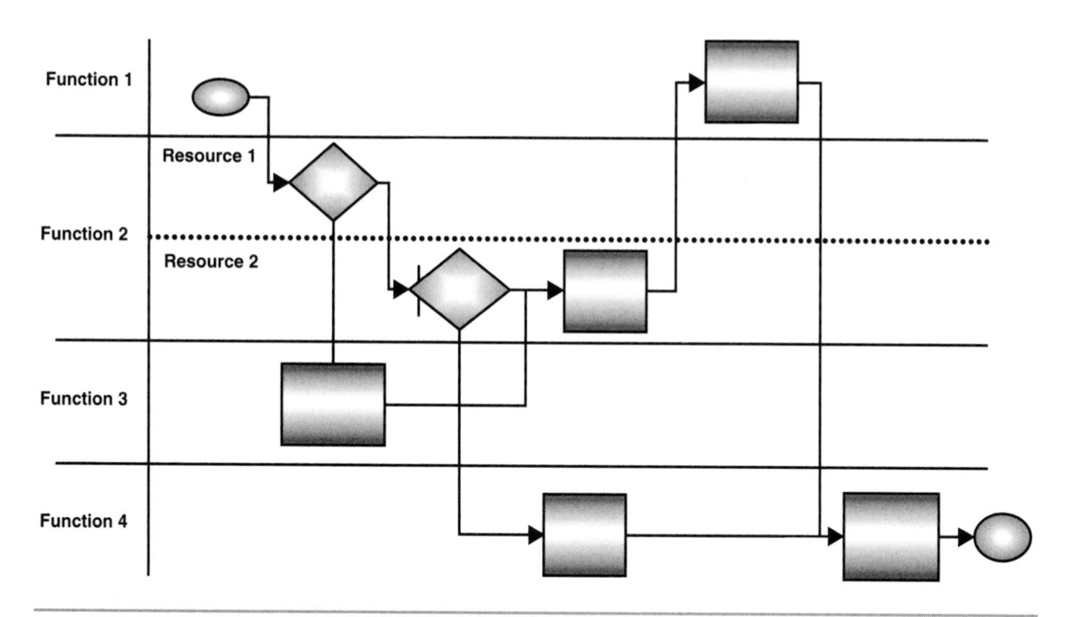

Figure 7.44.2 Cross-Resource Map Structure.

not be attempted by the Belt alone. Constructing the Swimlane Map from the *VSM* can take 30 minutes to an hour (depending on the size an complexity of the map) and can be done using the same sticky notes and repositioning them. It can, in theory, be done directly after the *VSM* is complete, but it is important to make sure the *VSM* is captured in electronic form first, so it is usually best to create the Swimlane Map at the next Team meeting to allow some transcription time to get the *VSM* into a flowcharting software package.

As with the *VSM*, it is best to create the Swimlane Map as a Team by hand using sticky notes versus having one person entering the Map directly into software as the Team looks at the projected image on the wall. The latter approach tends to lead to the Team disengaging and errors creeping in.

ROADMAP

The roadmap to converting the detailed *VSM* into a Swimlane Map is as follows:

Step 1: Prepare the work area. Construction requires a large wall space to accommodate the Map (typically 12–15 feet). Place a blank length of blank paper on the wall to create the Swimlane Map on. Place the existing *VSM* either directly above or directly below the blank sheet.

Step 2: From the *VSM*, identify the Functional groups that the process crosses and record them on each on a sticky note on left side of page. Don't write them directly onto the page, as they invariably have to get moved around early in the construction. Draw faint pencil lines to represent the lanes between functional groups. About two to three times the height of a sticky note works well for the lane separation at this point.

Step 3: Start on left side of map; place the first step from *VSM* in the appropriate functional lane decided by who actually does the step *currently*. Continue to place process steps according to following guidelines:

- Time increases from left to right across the map
- Steps that occur simultaneously should appear one directly above the other

The *VSM* itself was based around the Primary Entity, so there is likely additional sticky notes added for activities not captured on the *VSM*. This is normal. Do not draw in connectors at this point; they come later.

Step 4: After the sticky notes are in place functionally, examine each Functional lane carefully to see if it makes sense to split the lane down to the Resource level. If so, then create new lane titles again on sticky notes and separate the notes in the Functional lane into separate Resource lanes. This should be straightforward without full-scale rearrangement because the initial line spacing in Step 2 was broad. Repeat for all Functional lanes, remembering that it is only appropriate to split a few of them by Resource.

Some shuffling of sticky notes might be required to gain space.

Step 5: Draw on the lines in black marker pen. Swimlanes are usually 1.5–2 sticky notes in height. First draw on lines between Functions as solid (continuous) lines. Lines between Resources within a Function are then drawn on as dashed lines.

Step 6: Organize the sticky notes and then draw on the connectors between the activities. Arrows typically exit on right and enter on left sides of process symbols, unless steps are at same time. A section of a completed example Swimlane Map is shown in Figure 7.44.3. The example includes both Functional and Resource lanes.

Interpreting the Output

The Swimlane Map should be understandable to anyone looking at it, even for the first time. The key value that the Map brings is to understand

- The activities that Functions should be focusing on and the role of each Function in the process.
- The number of handoffs occurring in the process and where they are occurring. Whenever a connector line crosses horizontal lanes, a handoff has occurred. Every handoff carries a risk and so the necessity of each should be questioned.
- The validity of ownership of a task by a certain Resource or Function. After the Map is complete for the current situation, it is worth examining every step in the Map and determining its "license." A license for an activity means that only the specific Function (or Resource) can do that particular activity. For each licensed activity, place a colored dot on the sticky note indicating the fact that the activity can be done only by that Function or Resource. All notes without licenses (dots) can, in theory, be moved around to reduce the number of handoffs and balance the workload more evenly.

Other Options

Both the *VSM* and Swimlane Map bring large value in understanding to the project. Sometimes Teams can combine the two maps to create a hybrid VSM-Swimlane Map. This is a strange beast, but can be useful to represent everything on a single Map.

The simplest way to construct the map is to actually construct the Swimlane Map from the *VSM* as before and then add a separate Swimlane on the Map (at either the very top or bottom) for the Primary Entity activity. There are no line connections to the other lanes, but the activities in those lanes are aligned chronologically based on the activities in the Value Stream, see Figure 7.44.4. The Map is useful because it shows who does what activity during the delays along the Value Stream, where and why the Primary Entity it stalled.

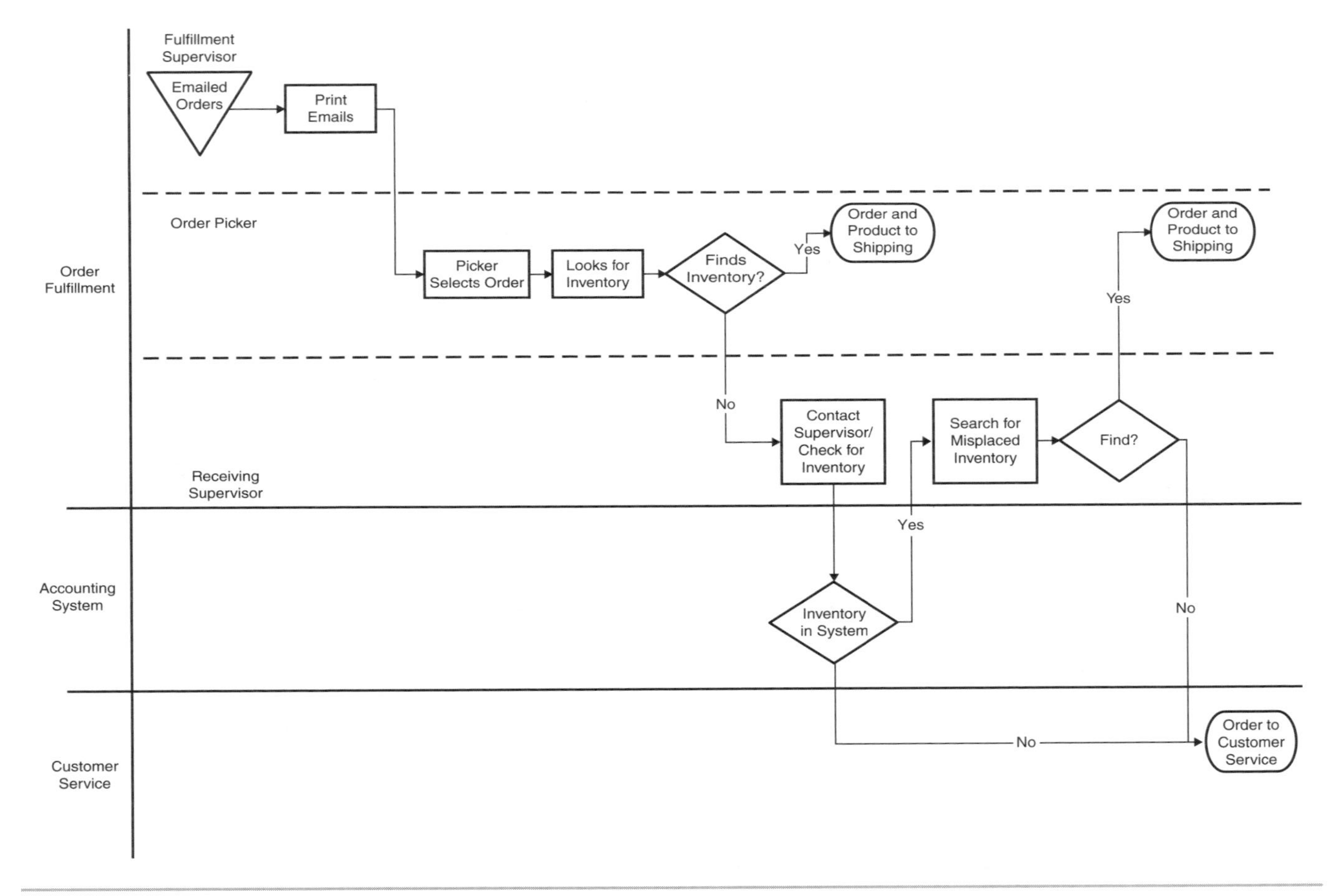

Figure 7.44.3 Section of a completed example Swimlane Map for order processing.[86]

[86] Adapted from SBTI's Transactional Process Improvement Methodology training material.

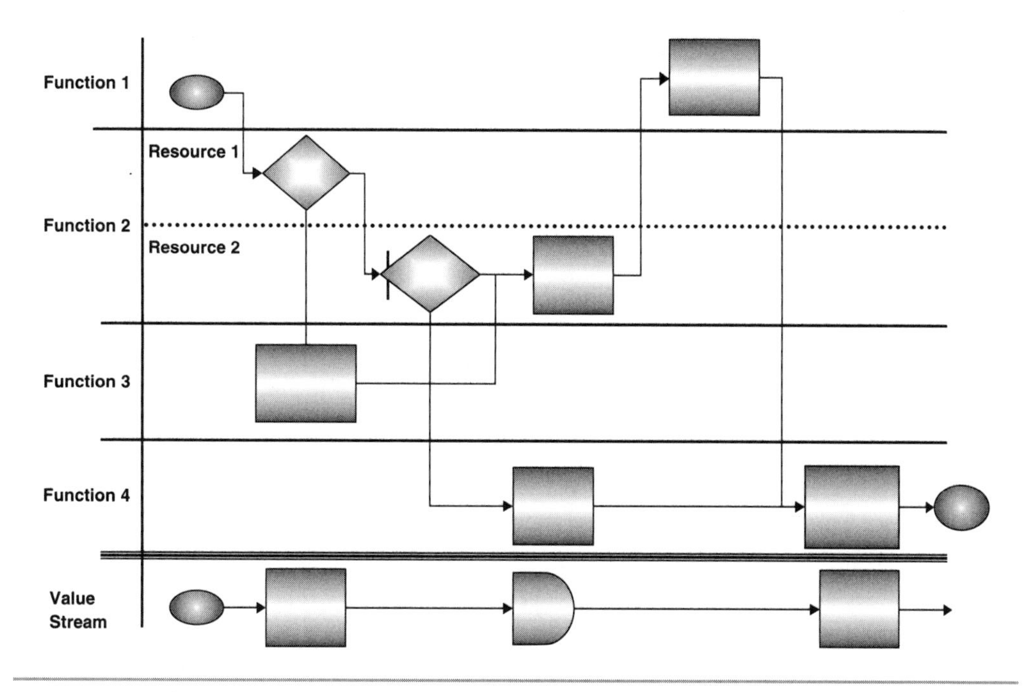

Figure 7.44.4 Structure of a hybrid VSM-Swimlane Map.

45: TEST OF EQUAL VARIANCE

OVERVIEW

The Test of Equal Variance is used to compare two sample variances against each other. For example, a Team might need to determine if two operators have the same amount of variation in the time they take to perform a task. For example, a data sample would be taken of 25 points from each operator to make the judgment, and the result would be the likelihood that the variation in the operator's task time (as work continues) is the same.

Thus, a sample of data points (lower curves) is taken from the two processes (the populations of all data points, upper curves), as shown in Figure 7.50.1. From the characteristics of the samples (standard deviation s and sample size n), an inference is made on the size of the population variances σ relative to the each other. The result would be a degree of confidence (a p-value) that the samples come from populations with the same variance.

ROADMAP

The roadmap is as follows.

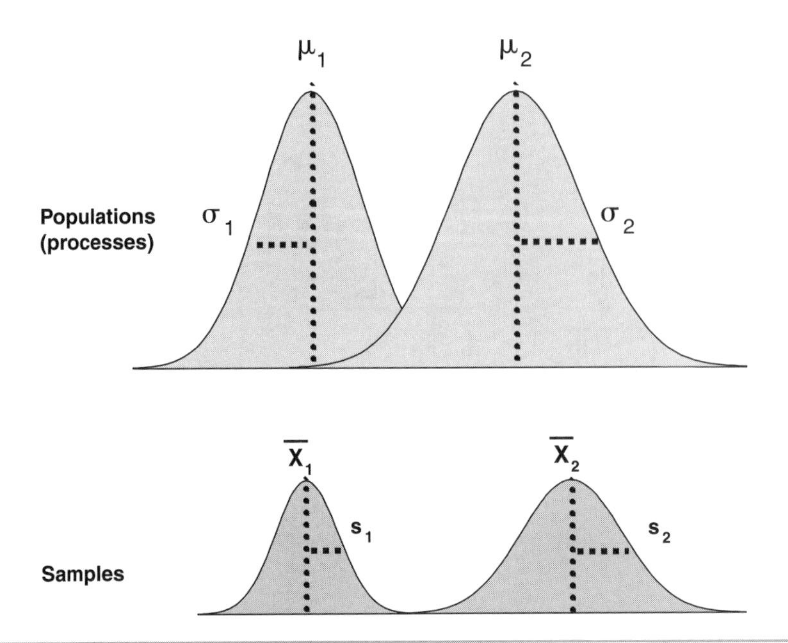

Figure 7.45.1 Graphical representation of a Test of Equal Variance.

Step 1: Identify the metric and levels to be examined (two operators or such like). Analysis of this kind should be done in the Analyze Phase at the earliest, so the metric should be well defined and understood at this point (see "KPOVs and Data" in this chapter).

Step 2: Collect two samples, one from each population (process) following the rules of good experimentation.

Step 3: Examine stability of both sample data sets using a *Control Chart* for each, typically an Individuals and Moving Range Chart (I-MR). A *Control Chart* identifies whether the processes are stable, that is having

- Constant mean (from the Individuals Chart)
- Predictable variability (from the Range Chart)

This is important because if the processes do not have predictable variation then it is impossible to sensibly make the call as to whether their variances are the same or not.

Step 4: Examine normality of the sample data sets using a *Normality Test* for each. The Test of Equal Variance uses a different statistic depending on normality.

Step 5: Perform a Test of Equal Variance on the sample data sets. The Test of Equal Variance has hypotheses:

- H_0: Population (process) $\sigma_1^2 = \sigma_2^2$ (variances equal)
- H_a: Population (process) $\sigma_1^2 \neq \sigma_2^2$ (variances not equal)

If both sample data sets are normal, then look to the Bartlett's or F-Test. If either or both sample data sets are non-normal, then look to the less powerful Levene's Test.

INTERPRETING THE OUTPUT

The Test of Equal Variance[87] compares the sample data sets' characteristics (standard deviation s and sample size n) to a reference distribution, to determine whether the sample data sets indicate that the populations variances are statistically different or not. Amongst other things the test returns a p-value, the likelihood that for the samples a difference in variances this large could have occurred purely by random chance even if the populations had the same variation.

Based on the p-values, statements can be generally formed as follows:

- Based on the data, I can say that there is a difference in variances and there is a (p-value) chance that I am wrong
- Or based on the data, I can say that there is an important effect on the variance and there is a (p-value) chance the result is just due to chance

The output of an example Test of Equal Variance is shown in Figure 7.50.2. Depending on whether the data is normal or non-normal affects which test results to examine. As stated previously, if both sample data sets are normal, then look to the Bartlett's or F-Test. If either or both sample data sets are non-normal then look to the less powerful Levene's Test. Both tests return a p-value that is interpreted in the usual way:

- p less than 0.05—reject H_0 and conclude that the variances are different
- p greater than 0.05—accept H_0 and conclude that the variances are the same

For the example of Bob and Jane shown in Figure 7.50.2, both data sets had previously been determined to be normal, so looking to the F-Test (with a p-value well above 0.05) the conclusion should be that Bob's variance cannot be differentiated from Jane's.

[87] The technical details of a Test of Equal Variance are covered in most statistics textbooks; *Statistics for Management and Economics* by Keller and Warrack makes it understandable to non-statisticians.

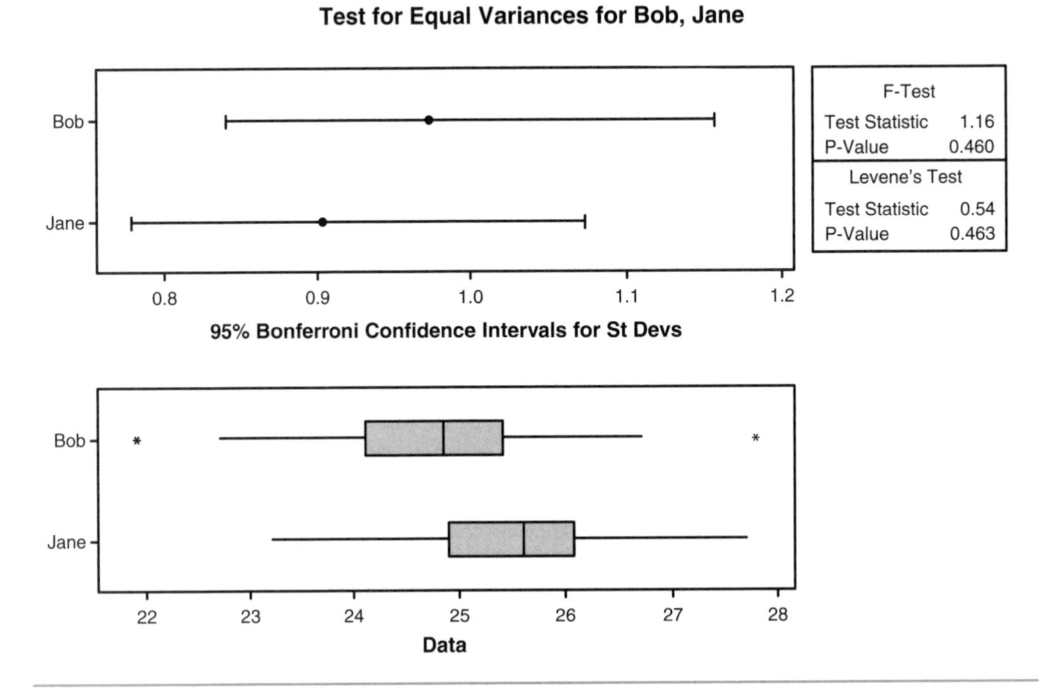

Figure 7.45.2 Test of Equal Variances for two operators: Bob and Jane (output from Minitab v14).

46: TIME—GLOBAL PROCESS CYCLE TIME

NOTE

As with other definitions described in "Time—Individual Step Cycle Time," "Time—Process Lead Time," "Time—Replenishment Time," and "Time—Takt Time" in this chapter, there is disagreement between different Process Improvement camps. To be frank, it really doesn't matter which naming convention you select, as long as you measure the right thing from a project perspective and do the right things based on it. In the subsequent description, focus on *what to measure*, as opposed to *what it's called*.

The naming convention used here is chosen for its clarity and practicality. If a naming convention is already established in your business, then use that; however, a little digging usually uncovers a lot of inconsistency and misuse.

In reports, always list the metric and how you define it; that way confusion and inevitable fruitless debate are reduced.

It is useful to read "Time—Individual Step Cycle Time," "Time—Process Lead Time," "Time—Replenishment Time," and "Time—Takt Time" in this chapter prior to proceeding with data capture to ensure comprehension.

Overview

(See also "Time—Individual Step Cycle Time" in this chapter.)

Cycle Time, when applied to the whole process, is the time between entities being processed (not the time taken for a single entity to traverse the whole process, which is known as the Process Lead Time).

Thus, the Global Process Cycle Time is measured by standing at the end of the process and timing between entities as they exit the process, as shown in Figure 7.45.1.

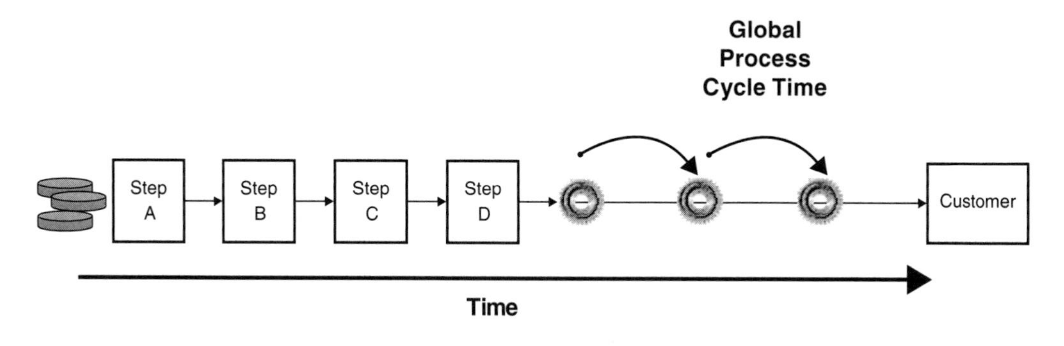

Figure 7.46.1 Graphical representation of Global Process Cycle Time.

Global Process Cycle Time and Process Lead Time can be the same if the process is always empty and processes only one entity at a time (i.e., to get an entity to fall out of the end of the process, that entity has to traverse the process from beginning to end).

Clearly the Global Process Cycle Time is driven both by the Individual Step Cycle Times and the positioning of inventory (Work in Process) in the process. The Global Process Cycle Time can be reduced significantly by having the correct Work in Progress (WIP) in the system.

There is also an obvious relationship here between Global Process Cycle Time and Takt Time (the pace of customer demand). Global Process Cycle Time should be balanced with Takt Time to go quickly enough to meet customer demand and yet not so quickly that you generate excess inventory. Ideally, the Global Process Cycle Time is set up to be slightly less than Takt Time, typically 95% (i.e., quicker) to retain flexibility and robustness to variation in demand.

Logistics

Calculating the average Global Process Cycle Time can usually be done from data captured over a period of a week or so. To calculate variation in the Cycle Time, a longer study is needed to allow normal variation to appear in the process; for example, if there is weekly variability, then data has to be captured over a few weeks to be able to see it.

It takes approximately one hour of a preparatory meeting to structure a data capture method. It is always useful to do a dry run data capture, to ensure the correct data is being captured. For more detail regarding the data capture, see "KPOVs and Data" in this chapter.

Global Process Cycle Time inevitably varies by entity type. It is useful, therefore, to capture either data for just one entity type or preferably data for all (or multiple) entity types and then stratifying the Time by type during the data analysis.

ROADMAP

It is possible to calculate the Global Process Cycle Time from as few as 7–10 data points. By taking the truncated mean (discarding the highest and lowest points and averaging the rest) a good approximation can be made. If understanding the variation in the Global Process Cycle Time is required, then 25–30 data points are required as a minimum.

If multiple entity types are examined, seven data points per type allow calculation of an average for each type as well as the whole. If an understanding of variation is required *by type*, then 25–30 data points *per type* are required.

Step 1: Determine the timing point for an entity. Usually this is the exact time at which an entity exits the process, but there should be a clear operational definition for this, so that timing is consistent during the data capture. At this point for an entity, the clock starts and when the subsequent entity meets the same timing point the clock stops.

Step 2: If the Global Process Cycle Time is less than one minute, then it is probably best to use a stopwatch, starting it when the first entity exits the process and stopping it when the second entity exits the process. This is the Global Process Cycle Time for the second entity, so if the entity type varies, then record the entity type of the second entity (not the first).

If the step takes significantly longer than a minute then it probably is best to write down the start and stop times and then calculate the Global Process Cycle Time later. Record the times and also the entity type being processed to allow stratification by type later during the data analysis.

It is sometimes useful, if the Global Process Cycle Time is short, to time three or more *consecutive* entities passing through the step to get a better reading on the average time.

Step 3: From the captured data, calculate mean and variation (standard deviation) in the Global Process Cycle Time.

If data for multiple entity types was captured, analyze the data as a whole and then conduct separate analyses stratifying by entity type.

Interpreting the Output

Interpreting the Global Process Cycle Time should not be done in isolation from other elements. The trick is to understand the relationship between the internal times relating to our process:

- Process Lead Time—Total time for an entity to progress through the process
- Replenishment Time to Customer—Time to fill the Customer's "in-box" when they request it be refilled
- Global Process Cycle Time—Rate (expressed in time) at which the process can generate entities

And to understand the relationship with the external Customer or Market related times:

- Requested Delivery Time—How quickly the Customer needs delivery of an entity when they request it
- Takt Time—The pace of usage of entities by the Customer, expressed in time

If the Global Process Cycle Time is greater than the Takt Time, then the process cannot meet Customer demand and options are

- Shedding load from the process to speed it up
- Eliminating work content (specifically NVA)
- Providing the correct Work In Process inventory to ensure no steps are starved of entities to work on
- Adding more resources to the process, specifically on the bottleneck steps
- Using multiple lines in parallel

47: Time—Individual Step Cycle Time

Overview

Simply put, the Individual Step Cycle Time is the time required, on an ongoing basis, by an individual step to process an entity. If the process step has a five-minute Cycle Time (if there is an entity ready and waiting to be processed in front of the process step and there is the appropriate resource to *start* processing the entity), then it takes five minutes for the resource to process the entity through this step.

The assumptions in parentheses here are key—Individual Step Cycle Time takes no account of lack of an entity to work on, it is measured from the time an entity begins this step to the time that the step ends after processing. Neither does it take into account the availability of resources to commence the step, because it is measured from when the step starts. It does, however, take into account availability of resources *during* the processing of the entity (i.e., if resources become unavailable during processing (after start), then the Cycle Time is increased.

If the assumptions are met, then by standing at the end of the process step it should be possible to see an entity exiting the process step every Cycle Time interval as per Figure 7.46.1.

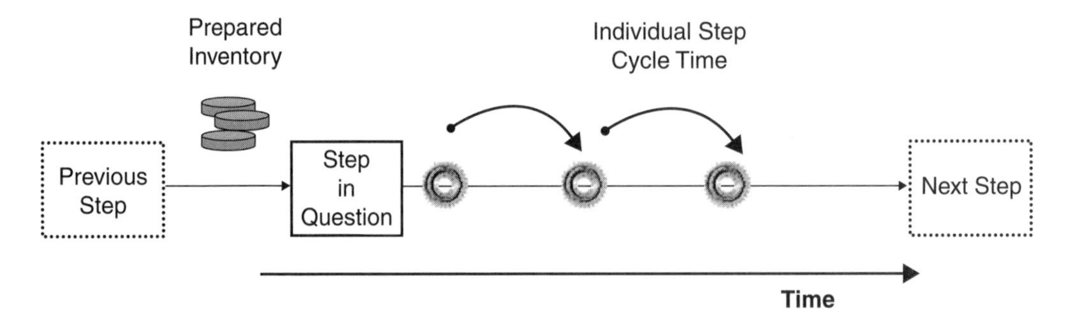

Figure 7.47.1 Graphical representation of Individual Step Cycle Time.

Logistics

Calculating an Individual Step Cycle Time can be done in a relatively short period of time, because Cycle Time at this level tends to be a lot more consistent than across a process as a whole.

To capture average Cycle Time, data can usually be captured over a period of a day or so, with a week being more than sufficient. To calculate variation in the Cycle Time, a longer study is needed to allow normal variation to appear in the process; for example, if there is weekly variability, then data has to be captured over a few weeks to be able to see it.

While the data is collected, the step must never be starved of something to work on, so it is important to only time the step when an input entity is available.

Recording data across multiple steps in the process in the same study is known as a *Multi-Cycle Analysis*.

It takes approximately one hour of a preparatory meeting to identify exact timing points for the step, preferably using an already constructed *Value Stream Map*, and then structure a data capture method. It is always useful to do a dry run data capture, to ensure the correct data is being captured and no biases are creeping in, but inevitably some Hawthorne Effect occurs. For more detail regarding the data capture, see "KPOVs and Data" in this chapter.

Individual Step Cycle Time often varies by entity type. It is useful, therefore, to capture either data for just one entity type or preferably data for all (or multiple) entity types and then stratifying the Time by type during the data analysis.

Roadmap

Due to the relative consistency in times, it is possible to calculate Individual Step Cycle Time from as few as 7–10 data points. By taking the truncated mean (discarding the highest and lowest points and averaging the rest) a good approximation can be made. If understanding the variation in the Cycle Time is required, then 25–30 data points are required as a minimum.

If multiple entity types are examined, seven data points per type allows calculation of an average for each type as well as the whole. If an understanding of variation is required *by type* then 25–30 data points *per type* is required.

Step 1: Determine the scope of the Process Step Time, by identifying the beginning point for the entity (when the clock starts) and the beginning point of the next entity (when the clock stops). The step Cycle Time must include all the work done to process an entity, including the setup, and so on; so measuring from beginning to beginning (or end to end) captures a full Cycle Time.

Step 2: If the Cycle Time is less than one minute, then it is probably best to use a stopwatch, starting it when the step begins, processing the first entity, and stopping it when the step begins processing the subsequent entity.

If the step takes significantly longer than a minute, then it probably is best to write down the start time and the end time and then calculate the step Cycle Time later. Record the times and also the entity type being processed to allow stratification by type later during the data analysis.

It is sometimes useful, if the step time is short, to time three or more *consecutive* entities passing through the step (making sure there is always entity inventory available), to get a better reading on the average time.

Step 3: From the captured data, calculate the Individual Step Cycle Times by subtracting the start time from the stop time for each cycle. Then calculate the mean and variation (standard deviation) in Individual Step Cycle Times.

If data for multiple entity types was captured, analyze the data as a whole and then conduct separate analyses stratifying by entity type.

Interpreting the Output

Interpreting the Individual Step Cycle Times should not be done in isolation from other elements. The trick is to understand the relationship between the internal times relating to our process:

- Process Lead Time—Total time for an entity to progress through the process
- Replenishment Time to Customer—Time to fill the Customer's "in-box" when they request it be refilled
- Global Process Cycle Time—Rate (expressed in time) at which the process can generate entities

And to understand the relationship with the external Customer or Market related times:

- Requested Delivery Time—How quickly the Customer needs delivery of an entity when they request it
- Takt Time—The pace of usage of entities by the Customer, expressed in time

By looking at the Individual Step Cycle Times versus the Takt Time, it is possible to ascertain whether the steps in the process will be able to keep up with the pace of Customer demand and also which step, if any, is the bottleneck.

See "Load Chart" in this chapter.

48: Time—Process Lead Time

Note

As with other definitions described in "Time—Global Process Cycle Time," "Time—Individual Step Cycle Time," "Time—Replenishment Time," and "Time—Takt Time" in this chapter, there is disagreement between different Process Improvement camps. To be frank, it really doesn't matter which naming convention you select, as long as you measure the right thing from a project perspective and do the right things based on it. In the subsequent description, focus on what to measure, as opposed to what it's called.

The naming convention used here is chosen for its clarity and practicality. If a (different) naming convention is already established in your business, then use that; however, a little digging usually uncovers a lot of inconsistency and misuse.

In reports always list the metric and how you define it; that way confusion and inevitable fruitless debate is reduced. It is useful to read "Time—Global Process Cycle

Time," "Time—Individual Step Cycle Time," "Time—Replenishment Time," and "Time—Takt Time" in this chapter prior to proceeding with data capture to ensure comprehension.

Overview

Simply put, the Process Lead Time is the time taken for a single entity to traverse the whole process from beginning to end.

The way to think of this is to consider yourself being stapled to the entity; starting a stopwatch at commencement of the first step in the process and stopping the stopwatch after completing the last step in the process. Process Lead Time takes into account all the delays throughout the process (transportation, and so on), including the entity sitting in inventory between steps. See Figure 7.47.1.

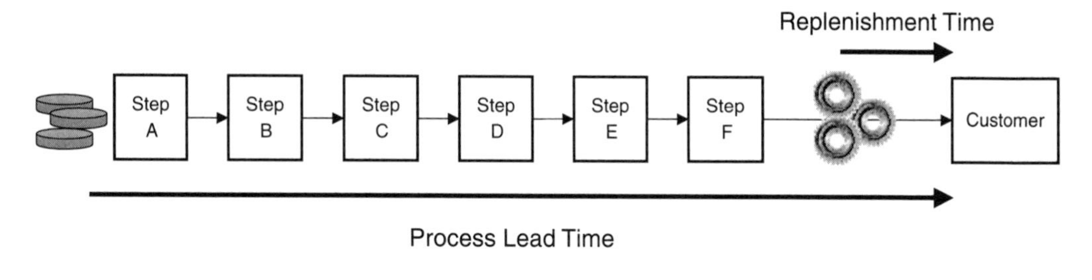

Figure 7.48.1 Graphical representation of Process Lead Time (versus Replenishment Time).

Thus, the Process Lead Time is the sum of all the Individual Step Cycle Times plus all the delays.

Logistics

Calculating the Process Lead Time requires involvement from the process operators, to both set up and capture the data. Data should be captured over a representative time period to take into account variation due to noise factors in the process.

It takes approximately 1–2 hours of a preparatory meeting to identify exact timing points through the process, preferably using an already constructed *VSM*, and then structure a data capture method. It is always useful to do a dry run data capture, to ensure the correct data is being captured and no biases are creeping in, but inevitably some Hawthorne Effect does occur. For more detail regarding the data capture, see "KPOVs and Data" in this chapter.

Process Lead Time often varies by entity type. It is useful, therefore, to capture either data for one entity type or preferably data for all (or multiple) entity types and then stratifying the Process Lead Time by type during the data analysis.

ROADMAP

Calculating the Process Lead Time is usually part of a *Multi-Cycle Analysis*. Ten data points suffice for calculating an average Process Lead Time, but 30–50 points are required to understand the variation in the metric.

Step 1: Determine the scope of the Process Lead Time, by identifying the beginning point (when the clock starts in the process) and the end point (when the clock stops).

Step 2: It is usually best to use a piece of paper that accompanies the entity through the process, known as a "rider." At the beginning point (as defined in Step 1), the start time is recorded on the rider. The entity progresses along the length of the process and at the end point (as defined in Step 1) the stop time is recorded onto the rider.

It is a good idea at this stage to capture other key timing points along the process (extra boxes to capture data on the rider), to understand how the Process Lead Time is distributed across the process.

Step 3: From the captured data, calculate the Process Lead Times by subtracting the start time from the stop time for each rider. Then calculate the mean and variation (standard deviation) in Process Lead Times.

INTERPRETING THE OUTPUT

Interpreting the Process Lead Time should not be done in isolation from other elements. The trick is to understand the relationship between the internal times relating to our process:

- Process Lead Time—Total time for an entity to progress through the process
- Replenishment Time to Customer—Time to fill the Customer's "in-box" when they request it be refilled
- Global Process Cycle Time—Rate (expressed in time) at which the process can generate entities

And to understand the relationship with the external Customer or Market related times:

- Requested Delivery Time—How quickly the Customer needs delivery of an entity when they request it
- Takt Time—The pace of usage of entities by the Customer, expressed in time

By looking at the Process Lead Time versus the requested Delivery Lead Time, it becomes apparent whether an entity can be processed from scratch to meet an order, or whether inventory is needed throughout the process to maintain an effective Replenishment Time less than the Process Lead Time.

For some processes, a Replenishment Time shorter than the Process Lead Time is not possible (every entity must be processed from scratch when an order arrives); thus, the only solution in this case is to shorten the Process Lead Time, by eliminating NVA activity throughout the process.

49: Time—Replenishment Time

Overview

The Replenishment Time is based around timing from the perspective of a location. It is the time required to receive replacement of an entity after there is an identified need for replacement. Thus, consider yourself as the receiving location itself; Replenishment Time is the time from when you *ask* to be refilled to when you are *actually* refilled.

Replenishment Time is counted in *actual* elapsed days (or hours/minutes depending on the process), not just working days, so a Replenishment Time from Friday across a non-working weekend to the Monday is 3 days not 1.

Replenishment Time typically includes many of the following:

- The need occurs
- The need is recognized
- The request is made
- The request "travels" to supplier
- The supplier processes the request
- The supplier generates replacement entity or retrieves it from storage
- The replacement entity is transported
- The replacement entity is received
- The need is fulfilled
- The need fulfillment is recognized

Replenishment Time is often confused with Process Lead Time. Replenishment Time is purely based from the perspective of the receiving location, whereas Process Lead Time is entity-based and tracks the time as the entity progresses through the process. The two are equivalent only if the entity has to be processed entirely from scratch; thus, to replenish, the entity has to progress the full length of the process.

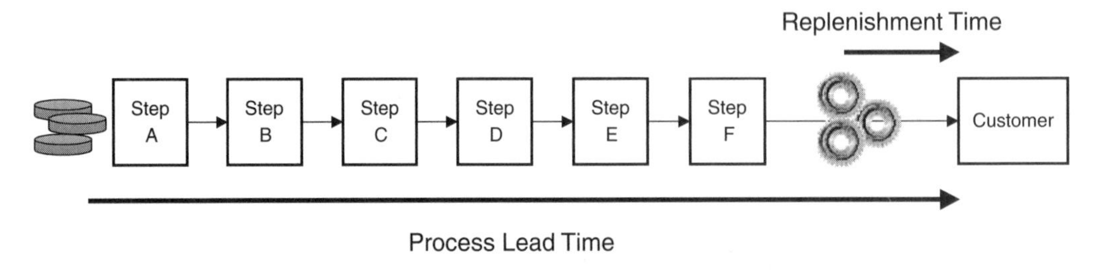

Figure 7.49.1 Graphical representation of Replenishment Time (versus Process Lead Time).

Logistics

Calculating Replenishment Time requires data collection at the receiving location (typically a Customer site), not from supplying process. Also, given the infrequent nature of replenishment in some processes, data can be in short supply. These two factors can cause great difficulty in getting a usable value for Replenishment Time. To calculate average Replenishment Time, 7–10 data points are required. Calculating variation in Replenishment Time requires some 25–30 data points and might prove to be impossible within the time constraints of a project unless viable historical data is available (which is highly unlikely).

In addition to this, Replenishment Time can vary by entity type, in which case data might be required to stratify by type (i.e., 7–10 points for each and every entity type). In most cases, however, Replenishment Time is independent of type or can at least be grouped into a few small categories, for example:

- Short term—Available to deliver from a finished goods inventory
- Medium term—Available to deliver from a responsive process
- Long term—Requires special ordering and is not produced on a consistent basis

Thus, Replenishment Time can be calculated for short, medium, or long-term groups of entity types.

Roadmap

Step 1: Determine the scope of the Replenishment Time, by identifying the beginning point (when the clock starts in the process) and the end point (when the clock stops). This is generally defined as from "when the need is identified" to "when the need is fulfilled." If an earlier point in time is available, such as "when the need arises," then time from that point to the fulfillment point instead.

Step 2: Data capture can be done in two pieces:

- At the Customer location, record a unique identifier for the entity and the time of the beginning of the process as defined in Step 1. After the end is reached (the need is fulfilled), then record the end time. The Replenishment Time is the difference in the two times recorded.
- (Optional, but recommended) At the supplying location it is possible to use a piece of paper that accompanies the entity through the process, known as a "rider" or "traveler." From the point that the request is received, the time is recorded on the rider. As the entity progresses through the replenishment process, capture other key timing points along the process (extra boxes to capture data on the rider), and at the point that the entity leaves the supply location to the Customer location, the stop time is recorded onto the rider. This gives an understanding of the timing of events within the Replenishment Time to understand at a high level where time is lost.

Step 3: From the captured data, calculate the mean and variation (standard deviation) in Replenishment Times. If the optional data capture was done through the supplying process, then calculate mean durations of steps through the process.

Interpreting the Output

Interpreting the Replenishment Time should not be done in isolation from other elements. The trick is to understand the relationship between the internal times relating to our process:

- Process Lead Time—Total time for an entity to progress through the process
- Replenishment Time to Customer—Time to fill the Customer's "in-box" when they request it be refilled
- Global Process Cycle Time—Rate (expressed in time) at which the process can generate entities

And to understand the relationship with the external Customer or Market related times:

- Requested Delivery Time: how quickly the Customer needs delivery of an entity when they request it
- *Takt Time*: the pace of usage of entities by the Customer, expressed in time

By looking at the Replenishment Time versus the requested Delivery Lead Time, it becomes apparent whether an entity can be delivered in time to meet an order. If not, and entities are processed from scratch to meet an order, first look to the short-term solution of holding inventory at key points along the process (i.e., make the Replenishment Time less than the Process Lead Time).

For some processes, a Replenishment Time shorter than the Process Lead Time is not possible (every entity must be processed from scratch when an order arrives), and the solution in this case is to shorten the Process Lead Time, by eliminating NVA activity throughout the process.

One subtlety (and often in fact a big opportunity) here is that Replenishment Time is measured from the need occurring, not when the actual request trigger is placed. By mapping the process from *need occurring* to *request trigger placed* often some earlier trigger points become apparent, thus extending the effective Delivery Lead Time.

50: TIME—TAKT TIME

OVERVIEW

Takt (a real word, not an abbreviation) is the German word for rhythm or cadence. A common mistake here is to confuse it with TACT (Total Activity Cycle Time) or similar, which is an entirely different thing.

Takt is defined as "The rate at which the end product or service must be produced and delivered in order to satisfy a defined customer demand within a given period of time." Simply put, it is the drumbeat of the market demand based on our working hours. Takt Time is calculated as

$$\text{Takt Time} = \frac{\text{Available Work Time in Period}}{\text{Average Demand in Period}}$$

If a process were perfectly balanced with the market demand, then every Takt increment an entity is processed and used by the market. See Figure 7.49.1. For example, if a process runs 24 hours a day and the market demand is 240 entities per day then

$$\text{Takt Time} = \frac{(24*60)\text{ minutes}}{240\text{ units}} = 6\text{ minutes/unit}$$

If an entity is not processed (on average) each and every six minutes, the process falls behind Customer demand. In Lean Sigma terms, the processing time is known as the Global Process Cycle Time. Thus, if the Global Process Cycle Time is above the Takt Time, the process falls behind Customer demand. Likewise if the Global Process Cycle Time is less than the Takt Time of, for example, five minutes, then the process is cycling faster than Customer demand and building inventory or spending one minute in every six waiting, to avoid creating unused inventory.

Note that there is no mention in Takt of how quickly the process can possibly go or how quickly it is actually going, only how fast it should go to meet demand.

Clearly if the work period (shift time) is less than 24 hours per day, then the process has to go proportionately faster during those times that work is actually done to meet the daily market demand. For example, if the work period is 12 hours per day in the preceding example, then the new Takt Time is now three minutes, because there is only half the available work time and the process has to cycle twice as fast. So, if the Global Process Cycle Time is not three minutes, the process is not balanced with the demand.

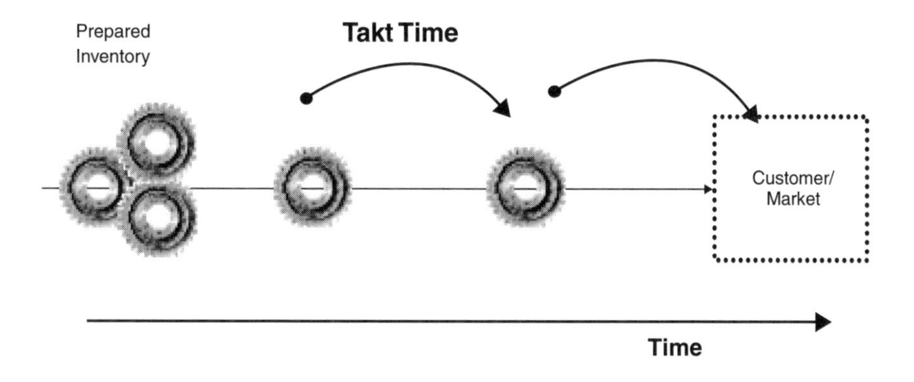

Figure 7.50.1 Graphical representation of Takt Time.

Most processes deal with more than one entity type, and if this is the case, then the Takt Time must be adjusted to take this into account. For example, if a process has two main entity types:

- Product "A" currently has demand for 350 a month
- Product "B" currently has demand for 525 a month

Then the *total* demand is 875/month.

If the business runs a two-shift operation, 5 days/week, 4.2 weeks/month

- 5 days × 4.2 weeks = 21 available days
- 21 days × 16 hours = 336 hours

And so monthly available time is 336 hours × 60 minutes = 20160 minutes

$$\text{Takt Time } = \frac{\text{Avilable Time}}{\text{Demand}} = \frac{\text{20160 minutes}}{\text{875 units}} \simeq 23 \text{ minutes/unit}$$

There is a strong caution here that Takt is a metric based on *average* demand and does not take into account the variation in demand. For this you should also consider a tool such as *Demand Segmentation*.

Takt is Customer or Market dependent; thus, as the demand changes, then the Takt has to be recalculated. This is usually done on a monthly, or more typically, quarterly basis as a matter of course. Takt, however, should not be used as a reactive measure to tweak daily operations planning; it is a longer term guiding metric for determining how a process should be structured. Based on the Takt, the required Global Process Cycle Time can be determined and from this the number of operators or lines in the process.

Governing a process by Takt only makes sense when demand is reasonably consistent, either for a single entity type or for the total output across multiple entity types.

LOGISTICS

On a quarterly basis, the Process Owners should calculate the Takt Times for all the processes under their control. It is not particularly a team sport to do this and all the data to make the calculation should be readily available historically.

ROADMAP

The roadmap typically requires no fresh data capture and is as follows:

Step 1: Identify the process(es) for which the Takt Time is calculated. Takt is process-specific and should not be aggregated across multiple lines unless they generate the same entity types.

Step 2: Select the time period for which the Takt is calculated—usually a week or a month. For the time period selected, identify the true time available to produce units within that period.

If using Monthly demand, then the time available per month is

(#Days/Month × #Shifts/Day × #Hours/Shift × #Min/Hour)

If using Weekly demand, then the time available per week is

(#Shifts × #Days/Week × #Hours/Shift × #Min/Hour)

The available work time should be calculated based on the full-shift time and not a depleted time that might have deducted the following:

- Wait for information or material time
- Rework time
- Equipment breakdown time
- Regular labor break time
- Fatigue/rest time

It is done this way because the preceding elements are NVA activity in the process. If the Takt is calculated around them, then they tend to be overlooked as opportunities for capacity increase. The process still has to be generating entities every Takt increment and thus the drive should be to discover how to actively manage that.

Step 3: Identify which entities likely flow through the process during the next time period. Unless absolute booked work is available, then this can be taken from a historical perspective; the assumption being that the pace of the process in the coming period should be close to that of the last period.

Step 4: Calculate the total demand for this process for all of the entities identified in Step 3. This is a unit count of entities. As in the preceding example, if the demand is for 350 units of Product A and 525 units of Product B, then the total demand is 875 units.

Step 5: Calculate the Takt Time from the work time calculated in Step 2 and the demand calculated in Step 4 as

$$\text{Takt Time} = \frac{\text{Available Work Time in Period}}{\text{Average Demand in Period}}$$

INTERPRETING THE OUTPUT

Interpreting the Takt Time should not be done in isolation from other elements. The trick is to understand the relationship between the internal times relating to the process:

- Process Lead Time—Total time for an entity to progress through the process
- Replenishment Time to Customer—Time to fill the Customer's "in-box" when they request it be refilled

- Global Process Cycle Time—Rate (expressed in time) at which the process can generate entities

And to understand the relationship with the external Customer or Market related times:

- Requested Delivery Time—How quickly the Customer needs delivery of an entity when they request it
- Takt Time—The pace of usage of entities by the Customer, expressed in time

If the Global Process Cycle Time is greater than the *Takt Time*, then the process cannot meet Customer demand and options are

- Shedding the load from the process to speed it up
- Eliminating work content (specifically NVA)
- Providing the correct Work In Process inventory to ensure no steps are starved of entities to work on
- Adding more resource to the process, specifically on the bottleneck steps
- Using multiple lines in parallel

51: TOTAL PRODUCTIVE MAINTENANCE

OVERVIEW

After determining the *Overall Equipment Effectiveness (OEE)* of a machine or process, clearly the next step is to determine why it isn't 100% and work towards the 100% goal. OEE is made up of three elements as per the equation:

$$OEE = \%Uptime \times \%Pace \times \%Quality$$

After the level of each of Uptime, Pace, and Quality are known then focused Lean Sigma projects can be applied to each stream. This can have immediate breakthrough improvement in each area, but is sometimes difficult to sustain without embedding a fundamental change into the way that the process is maintained. To gain a lasting result it is highly recommended that a Total Productive Maintenance[88] initiative be undertaken.

[88] For more information, *see Inspection and Training for TPM* by Terry Wireman, published by Industrial Press Inc.

This is a large and complex subject area and is not dealt with in great detail here. However, there are some commonly held misconceptions, which hopefully can be dispelled. The biggest is that TPM stands for Total *Preventative* Maintenance and is just doing a bit of oiling and greasing of a machine like it says to in the handbook. TPM stands for Total *Productive* Maintenance and is much more.

To understand TPM better, it is useful to first consider the major causes of equipment breakdowns:

- Failure to maintain fundamental machine requirements, such as housekeeping, oiling, tightening bolts, and so on
- Failure to maintain correct machine operation conditions, such as temperature, vibration, pressure, speed, torque, and so on
- Lack of operator skills causing incorrect machine operation
- Lack of maintenance crew skills causing maintenance errors
- Physical deterioration of bearings, gears, fixtures, and so on
- Design deficiency of the equipment in terms of material construction (the machine just isn't strong enough to survive daily use), dimension (the machine works well within certain parameters, but is less robust outside certain settings), and so on
- Misapplication of equipment, such as using light duty cycle equipment for a medium/heavy duty cycle task.

To be successful, TPM needs to address all these issues, which simple "Preventative Maintenance" certainly cannot.

Logistics

As stated in "Overview" in this section, TPM should be undertaken as an operations-wide initiative and should be resourced as such. It should have an Executive Champion, usually a Vice President of Operations or similar and an accountable hierarchy of resources to plan, structure, and implement change. Commonly multiple TPM Teams are active in a business at any one time working systematically through key equipment issues to raise the OEE of the facility.

Roadmap

The high-level steps in a TPM timeline for critical pieces of equipment are as follows:

Step 0: It is the bare minimum responsibilities of any operators and maintenance crew to conduct cleaning, inspecting, and repairing the area and equipment. Items that are not fixed immediately should be Red Tagged. Create a log of who is responsible, what the action is, and when it is due.

Step 1: Preventive Maintenance—This is typically what the newcomer assumes TPM represents; preventing breakdowns from occurring through a regular maintenance program that fills fluid levels, replace filters, and so on. Create cleaning and lubrication standards and a schedule with identified responsibilities for operators and maintenance personnel.

Step 2: Corrective Maintenance—Making Preventive Maintenance easier and user-friendly by eliminating problem sources, such as leaks, and improving access to equipment to be able to clean, inspect, and repair it. Central lubrication systems are a common approach here.

Step 3: Breakdown Maintenance—Identifying the primary causes of machine downtime to be able to create recovery plans that reduce the amount of time it takes to repair a machine after a breakdown. It is often useful to use a *Rapid Changeover (SMED)* approach here. The assumption is that the equipment goes down at some point (for whatever reason), and so it makes sense to have a standard approach to rapidly bring it back on-line.

Step 4: Maintenance Prevention—Buying or designing equipment so that it needs minimal maintenance. Identifying the key maintenance risks and working to eliminate them with (typically) more advanced technology, such as replacing standard bearings with sealed bearings or traditional cutting heads with more exotic alloys or better still lasers.

Step 5: Predictive Maintenance—Identifying the key Xs that cause machine failure using the Lean Sigma $Y = f(X_1, X_2,..., X_n)$ approach, then using technology to track those Xs to predict machine failure. Xs typically are one or more of heat analysis, vibration, electrical tests, measurements, and so on.

Step 6: Scheduled Maintenance—The previous steps give a detailed understanding of the Xs that drive failure and the systematic methods to bring equipment back online after failure or during routine maintenance. Thus, it is possible with this information to regularly schedule machine downtime with the production department.

To reiterate, TPM is certainly not a simple undertaking, but nonetheless is crucial to most Operational Excellence initiatives. The preceding roadmap is not nearly enough to commence such an initiative, but hopefully gives an indication of what is involved and hopefully the value of such an endeavor.

52: T-TEST—1-SAMPLE

OVERVIEW

The 1-Sample t-Test is used to compare a sample mean against a specific target value. For example, a Team might need to determine if an operator is taking a certain amount of time to perform a task. The target value would be the nominal task time, and a data sample would be taken of, for example, 25 points to make the judgment. The result is the likelihood that the average operator task time (as work continues thenceforth) is the same as the nominal value.

Thus, a sample of data points (lower curve) is taken from the process (the population of all data points, upper curve), as shown in Figure 7.52.1. From the characteristics of the sample (mean $\bar{X}$, standard deviation s, and sample size n), an inference is made on the location of the population mean μ relative to the target value. The result would be a degree of confidence (a p-value) that the sample came from a population with mean μ the same as the target.

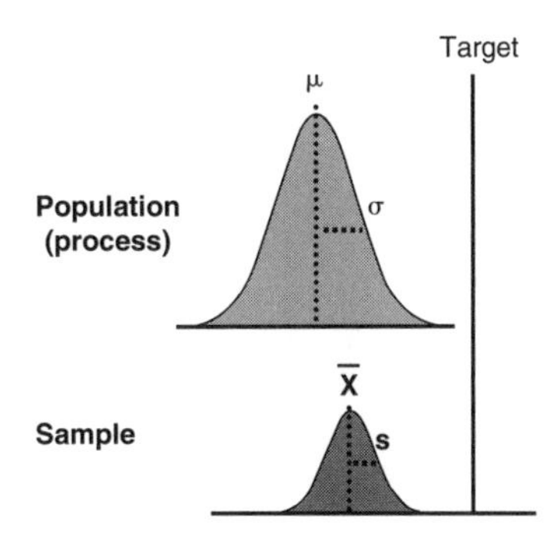

Figure 7.52.1 Graphical representation of a 1-Sample t-Test.

ROADMAP

The roadmap of the test analysis itself is shown graphically in Figure 7.52.2.

Step 1: Identify the metric and target value. Analysis of this kind should be done in the Analyze Phase at the earliest, so the metric should be well defined and understood at this point (see "KPOVs and Data" in this chapter).

Step 2: Determine the sample size. This can be as simple as taking the suggested 25 data points or using a sample size calculator in a statistical package. These rely on an equation relating the sample size to

- The standard deviation (the spread of the data). This would have to be approximated from historical data.
- The required power of the test (the likelihood of the test seeing a difference between the mean and the target value if there truly was one). This is usually set at 0.8 or 80%. The power is actually $(1 - \beta)$, where β is the likelihood of giving a false negative and might need to be entered as a β of 0.2 or 20%.
- The size of the difference δ between the mean and the target value that is desired to be detected (i.e., the size of deviation from the target that would lead the Team to say that the two values were different).
- The alpha level for the test (the likelihood of the test giving a false positive) usually set at 0.05 or 5% and represents the cutoff for the p-value (remember if p is low, H_0 must go).
- Whether the test is one-tailed or two-tailed see "Other Options" in this section.

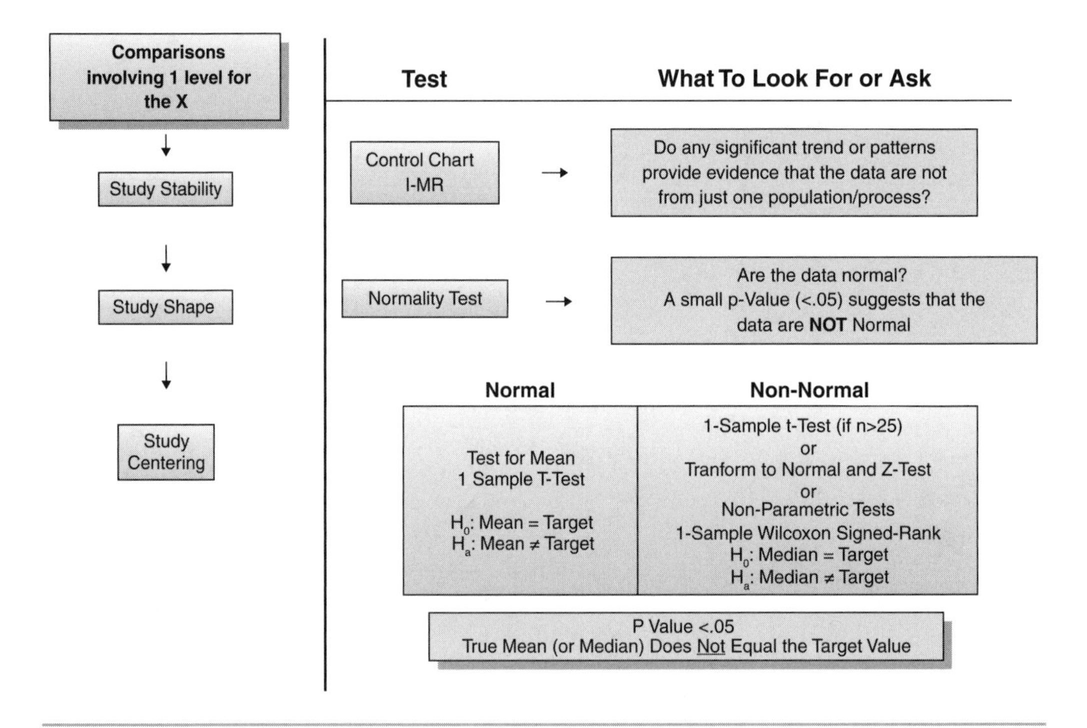

Figure 7.52.2 1-Sample t-Test Roadmap.[89]

[89] Roadmap adapted from SBTI's Process Improvement Methodology training material.

Step 3: Collect the sample following the rules of good experimentation.

Step 4: Examine stability of the sample data using a *Control Chart*, typically an Individuals and Moving Range Chart (I-MR). A *Control Chart* identifies whether the process is stable, that is having

- Constant mean (from the Individuals Chart)
- Predictable variability (from the Range Chart)

This is important because if the process is moving around, it is impossible to sensibly make the call as to whether it is on target or not.

Step 5: Examine normality of the sample data using a *Normality Test*. This is important because the statistical test in Step 6 relies on it, but in simple terms if the sample curve in Figure 7.52.1 were a strange shape it would be difficult to determine if the middle of it were off target. In fact if the data becomes skewed, then the mean is probably not the best measure of center (a t-Test is a mean-based test), and a median-based test is probably better. The longer tail on the right of the curve drags the mean to the right; however, the median tends to remain constant, see Figure 7.52.3. Medians-based tests can, in theory, be used for everything as a more robust test, but they are less powerful than their means-based counterparts, and, hence, the desire to go with the mean.

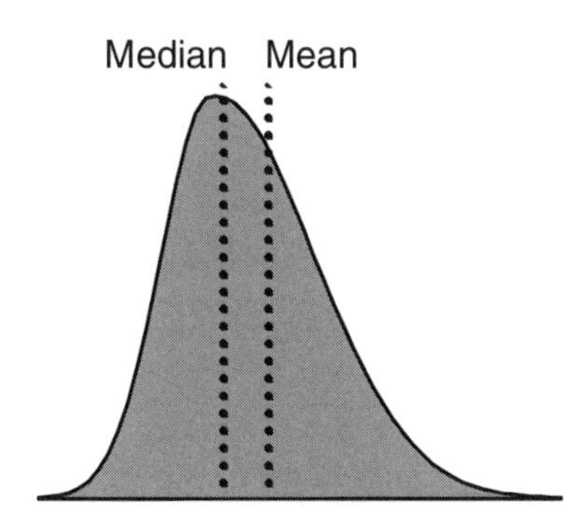

Figure 7.52.3 Measures of Center.

Step 6: Perform the 1-Sample t-Test if the sample data was normal. The hypotheses in this case are

- H_0: Population (process) Mean = Target
- H_a: Population (process) Mean ≠ Target

If the data were non-normal, then as per Figure 7.52.2:

- Continue unabated with the 1-Sample t-Test if the sample size is large enough (>25)
- Transform the data and perform the analysis using a 1-Sample Z-Test
- Perform the median-based equivalent test, a Wilcoxon Test (if the data is symmetrical) or a Signed-Rank Test if not

The last option sounds complicated, but the medians tests look identical in form to the means test and both return the key item, a p-value (the p-values are unlikely to be the same though).

INTERPRETING THE OUTPUT

The 1-Sample t-Test[90] compares the sample characteristics (mean $\bar{X}$, standard deviation s, and sample size n) to a reference distribution, the t-distribution, to determine whether the sample indicates that the population is statistically different from the target or not. Amongst other things, it returns a p-value; the likelihood that for the sample a difference from a target this large can have occurred purely by random chance even if the population were on target.

Based on the t-Test and the p-values, statements can be generally formed as follows:

- Based on the data, I can say that there is a statistically significant difference and there is a (p-value) chance that I am wrong
- Or based on the data, I can say that there is an important effect and there is a (p-value) chance the result is just due to chance

Output from a 1-Sample t-Test is shown in Figure 7.52.4.

One-Sample T: Bob

Test of mu = 25 vs not = 25

Variable	N	Mean	StDev	SE Mean	95% CI	T	P
Bob	30	24.8482	0.8693	0.1587	(24.5236, 25.1728)	−0.96	0.347

Figure 7.52.4 Test results for a comparison of a sample of Bob's performance data versus a target value of 25.

[90] The technical details of a t-Test are covered in most statistics textbooks; *Statistics for Management and Economics* by Keller and Warrack makes it understandable to non-statisticians.

From the example results

- They are based on hypotheses $\mu = 25$ (H_0) versus $\mu \neq 25$ (H_a)
- A sample of 30 data points was taken
- The mean of the sample was 24.8482
- The standard deviation of the sample was 0.8693
- There is a 95% likelihood that the mean of the population lies between 24.5236 and 25.1728
- The mean of the sample lies 0.96 Standard Errors (0.1587) below the target (hence the minus sign)
- The likelihood of seeing a sample mean $\bar{X}$ this far from the target value (if the population were actually perfectly on target) is 34.7% (p-value), which is above 0.05; thus, a conclusion that Bob's performance cannot be differentiated from target given the sample data.

Other Options

The test described previously has hypotheses defined as

- H_0: Population (process) Mean = Target
- H_a: Population (process) Mean ≠ Target

Therefore, this test is known as a two-tailed test, because (by the "≠" in H_a) it is not known whether the mean is above or below target. Therefore, the test needs to cover both sides (tails).

A one-tailed test, on the other hand, is used when it can be stated up front whether the mean is above target or below target. The hypotheses in this case would either be

- H_0: Population (process) Mean = Target
- H_a: Population (process) Mean > Target

or

- H_0: Population (process) Mean = Target
- H_a: Population (process) Mean < Target.

A one-tailed test (greater than or less than) can detect a smaller difference between the mean and the target value than a two-tailed test[91] (not equal to). Given the choice, go with the one-tailed test if there is data to show which side of the target value the mean is sitting.

53: T-TEST—2-SAMPLE

OVERVIEW

The 2-Sample t-Test is used to compare two sample means against each other. For example, a Team might need to determine if two operators are taking the same amount of time to perform a task. A data sample would be taken of, for example, 25 points for each operator to make the judgment, and the result is the likelihood that both operators' average task time (as work continues) is the same.

Thus, a sample of data points (lower curves) is taken from the two processes (the populations of all data points, upper curves), as shown in Figure 7.53.1. From the characteristics of the samples (mean $\overline{X}$, standard deviation s, and sample size n), an inference is made on the location of the population means μ relative to the each other. The result would be a degree of confidence (a p-value) that the samples come from populations with the same mean.

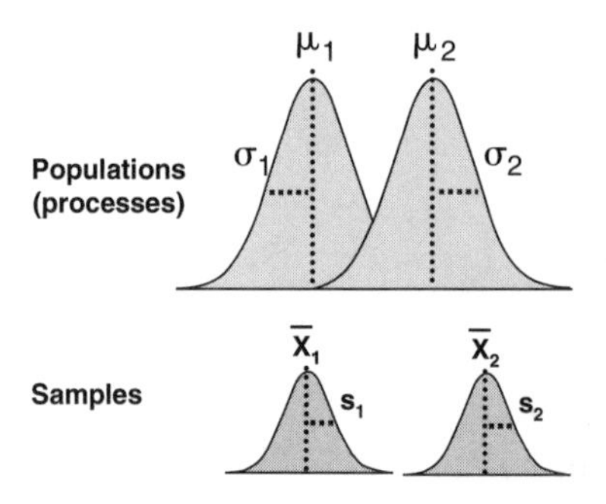

Figure 7.53.1 Graphical representation of a 2-Sample t-Test.

Caution: It should be noted that there is a subtly different test known as a Paired t-Test that is often confused with a 2-Sample t-Test. A 2-Sample t-Test for the preceding example looks at the mean of the sample for Operator 1 versus the mean of the sample for Operator 2. The Paired t-Test, on the other hand, is used to compare two samples against each other, but

[91] A detailed explanation of the reason for this can be found in most statistics textbooks; *Statistics for Management and Economics* by Keller and Warrack is useful here.

where a data point from each set needs to be considered together in pairs. For example, a Team might need to determine if two operators are quoting the same price for packages of products or services. It makes no sense to compare the average of all the quotes for Operator 1 to the average of all the quotes for Operator 2. Each package quote needs to be examined separately, making paired comparisons of Operator 1's quote for the package and Operator 2's quote for the package. In this case refer to "t-Test—Paired" in this chapter.

Roadmap

The roadmap of the test analysis itself is shown graphically in Figure 7.53.2.

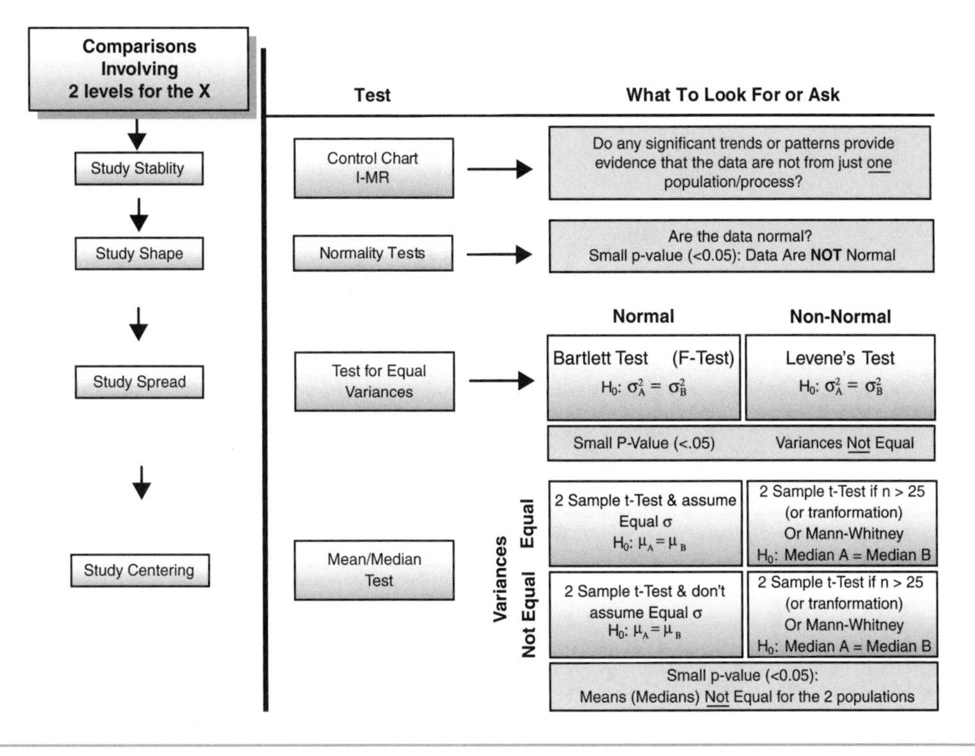

Figure 7.53.2 2-Sample t-Test Roadmap [92]

Step 1: Identify the metric and levels to be examined (two operators or such like). Analysis of this kind should be done in the Analyze Phase at the earliest, so the metric should be well defined and understood at this point (see "KPOVs and Data" in this chapter).

[92] Roadmap adapted from SBTI's Process Improvement Methodology training material.

Step 2: Determine the sample size. This can be as simple as taking the suggested 25 data points or using a sample size calculator in a statistical package. These rely on an equation relating the sample size to the following:

- The standard deviations (the spread of the data) of each population. This would have to be approximated from historical data.
- The required power of the test (the likelihood of the test identifying a difference between the means if there truly was one). This is usually set at 0.8 or 80%. The power is actually $(1 - \beta)$, where β is the likelihood of giving a false negative, and might need to be entered in the software as a β of 0.2 or 20%.
- The size of the difference δ between the means that is desired to be detected (i.e., the distance between the means that would lead the Team to say that the two values were different).
- The alpha level for the test (the likelihood of the test giving a false positive) usually set at 0.05 or 5% and represents the cutoff for the p-value (remember if p is low, H_0 must go).
- Whether the test is one-tailed or two-tailed see "Other Options" in this section.

Step 3: Collect two sample data sets, one from each population (process), following the rules of good experimentation.

Step 4: Examine stability of both sample data sets using a *Control Chart* for each, typically an Individuals and Moving Range Chart (I-MR). A *Control Chart* identifies whether the processes are stable, having

- Constant mean (from the Individuals Chart)
- Predictable variability (from the Range Chart)

This is important because if the processes are moving around, it is impossible to sensibly make the call as to whether they are the same or not.

Step 5: Examine normality of the sample data sets using a *Normality Test* for each. This is important because the statistical tests in Step 6 and 7 rely on it, but in simple terms if the sample curves in Figure 7.53.1 were strange shapes it would be difficult to determine if the middles were aligned. In fact if data becomes skewed, then the mean is probably not the best measure of center (a t-Test is a mean-based test), and a median-based test is probably better. The longer tail on the right of the example curve in Figure 7.53.3 drags the mean to the right, however, the median tends to remain constant. Medians-based tests could in theory be used for everything as a more robust test, but they are less powerful than their means-based counterparts and hence the desire to go with the mean.

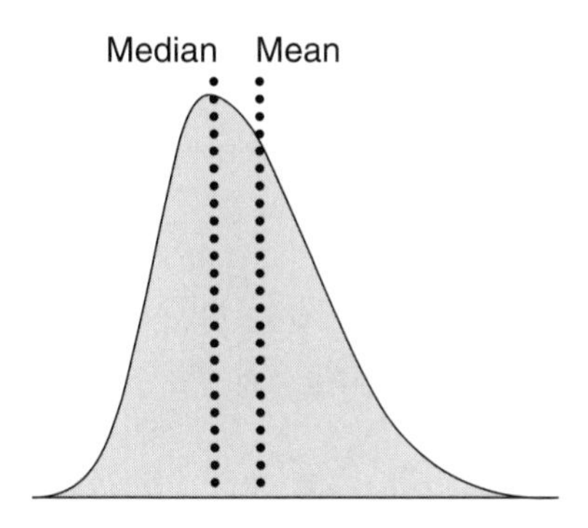

Figure 7.53.3 Measures of Center.

Step 6: Perform a Test of Equal Variance on the sample data sets. In simple terms, any test of centering (Step 7) looks to measure the distance from μ_1 to μ_2 in units of standard deviation. Figure 7.53.4 highlights the problem that if the variances of each population were different, then measuring the difference between them in units of standard deviation would be different using σ_1 versus σ_2. If the standard deviations are different, then the test should use a composite value known as the pooled standard deviation.

The Test of Equal Variance uses the sample data sets and has hypotheses:

- H_0: Population (process) $\sigma_1^2 = \sigma_2^2$ (variances equal)
- H_a: Population (process) $\sigma_1^2 \neq \sigma_2^2$ (variances not equal)

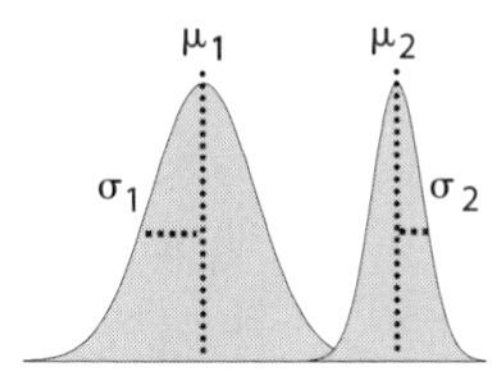

Figure 7.53.4 The impact of different variances.

Step 7: Perform the 2-Sample t-Test if both of the sample data sets were determined to be normal in Step 5. The hypotheses in this case are

- H_0: Population (process) $\mu_1^2 = \mu_2^2$ (means equal)
- H_a: Population (process) $\mu_1^2 \neq \mu_2^2$ (means not equal)

The output of Step 6 also needs to be included in the test. Most statistical software packages include a checkbox or similar to select if the variances are equal or not.

If the data in either or both of the samples were non-normal, then as per Figure 7.53.2:

- Continue unabated with the 2-Sample t-Test if the sample size is large enough (>25)
- Transform the data first and then perform the analysis using the 2-Sample t-Test[93]
- Perform the median-based equivalent test, a Mann-Whitney Test

The last option often worries Belts, but the medians tests look identical in form to the means test and both return the key item, a p-value (the p-values for a means test and a medians test on the same data are unlikely to be the same though).

INTERPRETING THE OUTPUT

The 2-Sample t-Test[94] compares the sample characteristics (mean $\bar{X}$, standard deviation s, and sample size n) to a reference distribution, the t-distribution, to determine whether the sample data sets indicate that the populations are statistically different or not. Amongst other things it returns a p-value, the likelihood that for the sample a difference between means this large could have occurred purely by random chance even if the populations were aligned.

Based on the t-Test and the p-values, statements can be generally formed as follows:

- Based on the data, I can say that there is a difference and there is a (p-value) chance that I am wrong
- Or based on the data, I can say that there is an important effect and there is a (p-value) chance the result is just due to chance

Output from a 2-Sample t-Test is shown in Figure 7.53.5.

From the example results

- A sample of 100 data points was taken for each operator
- Bob's sample mean was 24.811, Jane's was 25.525
- Bob's sample standard deviation was 0.973, Jane's was 0.904
- The test is based on the hypotheses: $\mu_{Bob} - \mu_{Jane} = 0$ (H_0) versus $\mu_{Bob} - \mu_{Jane} \neq 0$ (H_a)

[93] Transformation of data is considered beyond the scope of this book.

[94] The technical details of a t-Test are covered in most statistics textbooks; *Statistics for Management and Economics* by Keller and Warrack makes it understandable to non-statisticians.

- There is a 95% likelihood that the difference between the population means lies between –0.975938 and –0.452062
- The means of the samples are –5.38 Standard Errors apart (the t-value)
- The likelihood of seeing sample means $\bar{X}$ this far apart (if the populations were perfectly aligned) is 0.0% (p-value), which is below 0.05; thus, the conclusion is that Bob's is performing significantly differently from Jane given the sample data.

Two-Sample T-Test and CI: Bob, Jane

Two-Sample T for Bob vs Jane

	N	Mean	St Dev	SE Mean
Bob	*100*	*24.811*	*0.973*	*0.097*
Jane	*100*	*25.525*	*0.904*	*0.090*

Difference = *mu* (*Bob*) – *mu* (*Jane*)
Estimate for difference: – 0.714000
95% CI for difference: (– 0.975938, – 0.452062)
T – Test of difference = *0* (*vs not* =): *T-Value* = – 5.38 *P-Value* = 0.000 *DF* = 196

Figure 7.53.5 Test results for a comparison of sample of Bob's versus Jane's performance (output from Minitab v14).

OTHER OPTIONS

The preceding test with the hypotheses defined as

- H_0: Population (process) $\mu_1 = \mu_2$
- H_a: Population (process) $\mu_1 \neq \mu_2$

The test is known as a two-tailed test, because (by the "$\neq$" in H_a) it is not known whether the mean of population 1 is above or below the mean of population 2, and, hence, the test needs to cover both sides (tails).

A one-tailed test on the other hand is used when it *can* be stated up front whether the mean of population 1 is above or below the mean of population 2. The hypotheses in this case is either:

H_0: Population (process) $\mu_1 = \mu_2$
H_a: Population (process) $\mu_1 > \mu_2$

or

H_0: Population (process) $\mu_1 = \mu_2$
H_a: Population (process) $\mu_1 < \mu_2$.

A one-tailed test (greater than or less than) can detect a smaller difference between the means than a two-tailed test[95]. Given the choice, go with the one-tailed test if there is data to show which of the two population means is greater.

54: T-TEST—PAIRED

OVERVIEW

The Paired t-Test is used to compare two samples against each other, where a data point from each set need to be considered together in pairs. For example, a Team might need to determine if two operators are quoting the same price for packages of products or services. It makes no sense to compare the average of all the quotes for Operator 1 to the average of all the quotes for Operator 2. Each package quote needs to be examined separately, hence making paired comparisons of Operator 1's quote for the package and Operator 2's quote for the package.

A data sample is taken for each Operator of, for example, 25 points to make the judgment and the result would be the likelihood that both operators' quotes (as work continues) are the same. It is interesting to see that this is not really like a 2-Sample t-Test at all, it is only a 1-Sample t-Test in disguise.

A sample of data points (lower curve) is generated from the two processes (the populations of all data points, upper curves) by taking the difference between them, as shown in the left side of Figure 7.54.1. If the two parent populations are identical, then the differences between the pairs are all zero. In reality, there is always some variation, but the mean of the population of differences would still be centered on zero. Thus a 1-Sample t-Test is applied to the sample of differences (right side of Figure 7.54.1) and from the characteristics of the samples (mean $\bar{X}$, standard deviation s, and sample size n), an inference is made on the location of the population mean μ relative to the target zero. The result is a degree of confidence (a p-value) that the sample had come from a population with a mean μ of zero.

95 An explanation for this can be found in most statistical textbooks; *Statistics for Management and Economics* by Keller and Warrack is useful in this case.

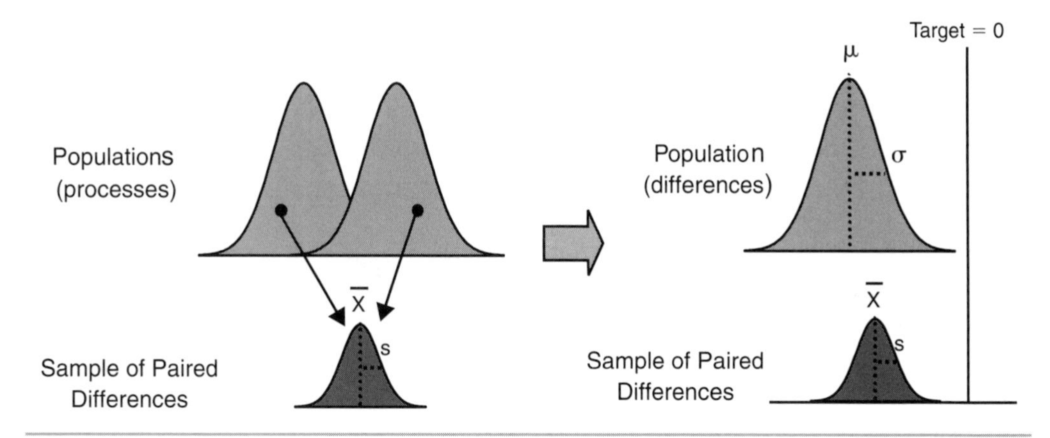

Figure 7.54.1 Graphical representation of a Paired t-Test.

ROADMAP

The roadmap here is exactly the same as the 1-Sample t-Test, but applied instead to the differences between each pairing in the sample data sets. The Target Value in this case is zero. The hypotheses are

- H_0: Population of Differences Mean = 0
- H_a: Population of Differences Mean ≠ 0

Go to "t-Test—1-Sample" in this chapter.

55: VALUE STREAM MAP

OVERVIEW

The Value Stream Map (VSM) is a key Lean Sigma tool in any project involving an overly complex process, or a desire to accelerate a process. Value Stream Mapping is used heavily in Lean projects[96] and Kaizens. Here the tool is similar, but there are some subtle differences, primarily in the detail of the map and its look.

VSM in Lean Sigma is similar to the traditional flowcharting approach used in many other Process Improvement initiatives, with three key distinctions:

[96] A good reference here is *Learning To See* by Rother and Shook.

- The map is centered around the Primary Entity in the process. For example, if the Primary Entity is an order, then the main horizontal center flow of the map is for the order and what occurs to it.
- The activities on the map are evaluated for the value they bring to the end Customer.
- Measures of performance for the process are integrated into the map.

There are two versions of the VSM used in Lean Sigma:

- High-level VSM sketched out in the Define Phase to begin to understand the process and populated with metrics during the Measure Phase
- Detailed VSM created in the Measure Phase to get a detailed understanding of the process flow

An example of a high-level VSM is shown in Figure 7.55.1.

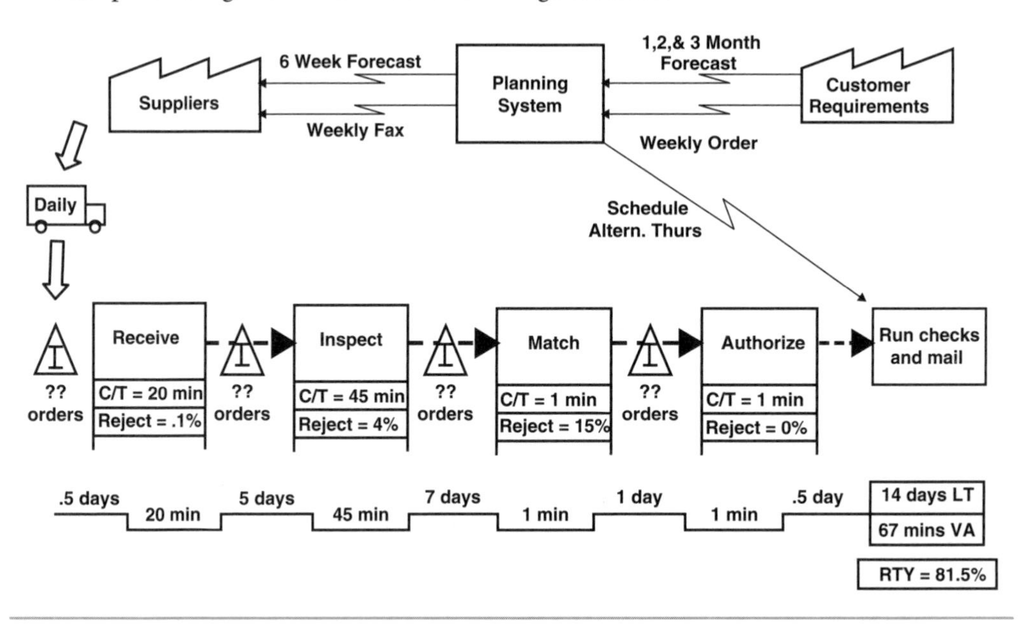

Figure 7.55.1 Example high-level Value Stream Map.[97]

The high-level map contains the major steps in the process, along with process metrics such as Cycle Times, Defect Rates, and Inventory levels. It also shows the control

[97] Taken from SBTI's Transactional Process Improvement Methodology training material, format adapted from *Learning To See* by Rother and Shook.

triggers, Suppliers, Inputs, and typically the Outputs and Customers (although not shown in Figure 7.55.1).

The detailed *VSM* in contrast breaks the process down into each component step as can be seen from Figures 7.55.2 and 7.55.3.

Figure 7.55.2 An example of a detailed Value Stream Map under construction.

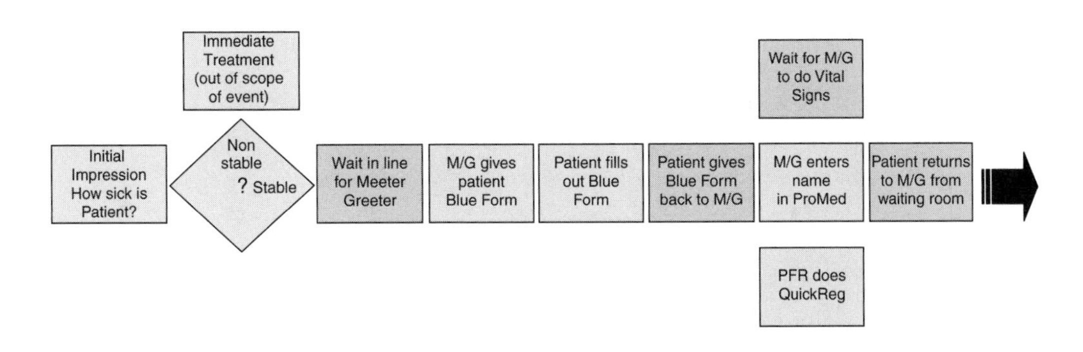

Figure 7.55.3 Magnified section of detailed Value Stream Map examining patient flow (prior to adding metrics).

The detailed map shows in microscopic detail exactly what happens to the Primary Entity as it progresses through the process. In Figure 7.55.2, associated procedures and documentation are attached to the VSM for Team reference. The magnified view of another VSM shown in Figure 7.55.3 highlights the difference in terms of value between steps in the process. Steps are either

- VA
- Non-Value Added but Required (NVAR) by the business or Required Waste
- NVA or Pure Waste

LOGISTICS

Construction of both the detailed and high-level VSM is a Team-based activity and absolutely should not be done in isolation by the Belt. Belts sometimes try to partially construct the VSM before a Team meeting to speed up the process, but this typically ends up taking more time and often Team members fail to buy-in to the end result.

Creating the high-level VSM from the P column in the *SIPOC* can take as little as 30 minutes, the detailed version however requires at least 3–4 hours of Team time. Sections of the process might have to be validated with other personnel, so they might have to be drawn in at appropriate times too.

The detailed VSM requires at least a 12–15 feet length of empty wall space and is typically done on butcher paper. Multiple packs of three-inch square sticky notes are required, preferably in different colors along with different colors of pens.

After completion, the maps are incredibly useful as training tools, and so on, but, for some reason, have a short lifespan (sticky notes tend to fall off, and so on), so transferal to electronic form using Flowcharting Software should be done promptly.

ROADMAP

The roadmap to creating a detailed VSM is as follows:

Step 1: Prepare the workspace as described in the preceding logistics section.

Step 2: Identify for the process:

- The Primary Entity
- Any secondary entities such as orders, information, and charts.
- The exact starting point(s) of the process
- The exact ending point(s) of the process

Step 3: Nominate one person on the Team to be the Primary Entity. They continue to ask, "Okay, what happens to me now?" This sounds a little crazy, but seems to work well. I am jokingly known within one client Team as "Ian the Order" after facilitating a VSM session.

Step 4: Step through the process, activity by activity just from the Primary Entity's perspective. This usually takes the form of the following:

- Arrive at site
- Wait
- First activity
- Wait
- Next activity
- Wait, and so on

Each activity (including Waits) should be written legibly, one per sticky note and placed on the paper in a stream horizontally down the center (this is the Value Stream; value is added to the Entity as it passes down this path). Some standard symbols are used, as shown in Table 7.55.1, so that anyone in the organization who picks up a map can immediately understand it.

Table 7.55.1 Value Stream Map Symbols and Meanings

Name	Description	Graphic
Start	Draws attention to beginning of the process.	Start →
Stop	Draws attention to end of process.	→ Stop
Flag	Shows connection to another page or point on the map; the number denotes where the connection links. It can also represent the input or output points of sub-processes.	→ 1
Activity	Also known as a Task, Step, or Operation, this represents points in the process where the entity is changed as the result of work being done. The label is typically *Verb Noun*, for example: • Enter Customer Name • Add Catalyst • Update General Ledger	→ Verb Noun → Metric 1 = ?? Metric 2 = ??

(continues)

Table 7.55.1 Value Stream Map Symbols and Meanings (Continued)

Name	Description	Graphic
	Metrics are added below the Task to reflect the current levels of • Cycle Time (C/T = ??) • Quality Rate (Reject = ??, Accuracy = ??) • Changeover Time (C/O = ??), and so on	
Trigger	Not to be confused with connectors or arrows between process steps, this symbol denotes a trigger from one step to another, generally in electronic form, for example: • Signal from surgery to preparation area to request next patient • Signal from operations area to warehouse for more materials	
Transport	Again, not to be confused with connectors or arrows between process steps, this symbol shows movement of the Primary (or a secondary) Entity and is used when the transport adds additional cycle time to the process or adds risk. The label is typically *Item Transported via Mechanism*, for example: • Send Quote via Mail • Send Patient via Wheelchair	Send "Item" via "Mechanism"
Decision	Identifies where a decision creates a branch in the process flow. The symbol label is the question itself, for example: • Are all fields complete? • Does the product meet specification? Each decision symbol can contain only *one* question. The decision should have two or more exit lines, each labeled with an answer to the question, for example: • Yes • No	Question ? Yes No Question ? Answer 1 Answer 2 Answer 3

(continues)

Table 7.55.1 Value Stream Map Symbols and Meanings (Continued)

Name	Description	Graphic
	Decisions start two or more mutually exclusive paths, but a single entity can only take *one* of the paths. The entity proceeds through path "A" *or* path "B," not both When a decision and, thus, a branch occurs during mapping, to reduce confusion it is best to follow and map one path to its conclusion before returning to the split and mapping the second path, and so on.	Question ? Answer 1 Answer 2 Answer 3 Answer 4
Store	Identifies a point at which an entity (or its parts) is physically stored. The entity is usually conveyed or placed in some sort of storage waiting for a trigger to release it. A store is not just a queue. The label is *Store Item at Location*, for example: • Store Hamburgers in Warmer. Burgers are made ahead of time waiting for the trigger of an order placed. • Store Parcels in Express Pickup Box. Mail waits in a box or on a shelf for pickup, the trigger being the arrival of the courier, for example.	Store "Item" at "Location"
Wait / Delay	Indicates that time is being consumed in the process by • A scheduled or planned delay • An activity or decision in another process The label is *Wait for Noun Verb*, for example: • Wait for Management Approval • Wait for Customer Review	Wait for "Action" "Noun"
Parallel Processing, Split and Join	Parallel processing occurs when two or more activities in a single process can be performed at the same time, thus reducing the overall time required to complete the process. Prerequisites for parallel processing: • Path "A" tasks must be independent of Path "B" tasks • A separate resource is available for each path	Path "A" Split Path "B" Path "A" Join Path "B"

(continues)

Table 7.55.1 Value Stream Map Symbols and Meanings (Continued)

Name	Description	Graphic
	The activity at which a parallel process is initiated is called a Split, at which point the entity is "cloned." Each of the clones then proceeds through different paths, for example, Order Faxed to Warehouse. This creates a cloned copy of the Order, one at the point of faxing, the other in the warehouse. The activity at which a parallel process is terminated is called a Join, at which point the pieces of an individual entity are recombined to form a new "whole" entity. Each piece brings with it the results of the activities through which it passed.	
Loop	Entities normally flow to the next activity in the sequence. Typically initiated by a decision, a Loop is the flow structure at which an entity flows *backward* to a previous activity. An example might be "Is the Entity Accurate?" If not, then the entity has to be reworked in some way.	

Step 5: Return to the start point of the process and map the path of the Secondary Entities in the process. These are not resources, such as staff, but objects, information, and so on, that supports the Primary Entity as it flows through the process.

Step 6: Return to the start point of the process, and, discussing each step in turn, determine if the steps are

- **Value Added.** Place a green dot on the sticky note. VA activities are those that
  - Change the size, shape, fit, form, or function of material or information (for the first time) to meet customer demands and requirements.
  - Or (specific to healthcare) contribute to a change in, or maintenance of a Patient's health status to meet customer (Patient/Physician) demands and requirements.
- **Non-Value Added but Required.** Place a blue dot on the sticky note. These activities do not bring value to the Customer, but are needed for some reason by the business or for legal or regulatory reasons. They could in theory be designed out if some process changes were made. Examples might include
  - Entering data into a system
  - Performing a quality check

- **Non-Value Added (Pure Waste).** Place a red dot on the sticky note. These are the activities in the process that are not required by anyone and would best be removed completely. There is often contention here, because people do not like to find out that a large portion of their work brings no value. It is important for Belts to really challenge the Team on each activity. After they get the hang of it and they understand that no one is casting aspersions on their abilities, there is typically at least one Team member that jumps in with gusto and puts the appropriate dots on the sticky notes. Examples of NVA activities include (some of these could arguably be required by the business, but challenge them anyway)
  - Waiting
  - Reworking entities
  - Scrapping entities
  - Transporting entities
  - Motion
  - Load/Unload
  - Secondary inspections
  - Storing

Step 7: Validate the basic VSM with other stakeholders in the process to ensure nothing is missing and the map is representative of the current process. Often disagreements arise; try to see these as opportunities—if people are doing the process differently, then that can lead only to an improvement if methods are aligned to be the best.

Step 8: Populate the steps in the VSM with process performance data. This might include

- Takt Times
- Process Lead Time
- Cycle Times
- Changeover Times
- % *Overall Equipment Effectiveness*
  - % Uptime
  - % Quality
  - % Pace
- Shift hours

For the process as a whole a Delay Ratio can be calculated as

$$\text{Delay Ratio} = \frac{\text{Process Lead Time}}{\text{Processing Time}}$$

For more detail on the Process Lead Time, see "Time—Process Lead Time" in this chapter. The standard Lean definition of Processing Time is all of the time in the process (whether it is VA, NVAR, or NVA) along the value stream (i.e., just related to the Primary Entity) minus the Waits. A preferable tweak to this is to consider the VA activity Processing Time for the Primary Entity, which gives a truer indication of the fat in the process:

$$\text{Value Ratio} = \frac{\text{Process Lead Time}}{\text{VA Processing Time}}$$

An impossibly perfect Lean process would have a Value Ratio of 1.0. Most processes run closer to 100.

INTERPRETING THE OUTPUT

The VSM is a key tool to identify unnecessary waste and risk in the process as follows:

- NVA activities should be the first target of any project Team. A few of them are Quick Hits and require little to no resource to remove. Others might take some reorganization of the process to remove. NVAR activities can also be a target—some can be removed quite quickly with a change of procedure or policy. Challenging a business, regulatory, or legal "requirement" often identifies that there is no requirement.
- Multiple Start points in a process can be a problem if the streams aren't quickly merged, or there are inconsistencies between streams. Seek to merge the multiple Starts into a single main stream as quickly as possible.
- Multiple Stop points in a process always seem to create problems, usually because there is no way to reconcile the end points. Seek to bring streams back together before the end of the process, or preferably don't have them separate in the first place.
- Improvement to Decisions is an obvious quick win in most projects. Decisions initiate branches and loops; thus, incorrect decisions initiate incorrect or unnecessary branches or loops. Each Decision should have the following elements:
  - A clearly articulated and VA Purpose
  - An accountable Decision Maker

  - Inputs necessary to make the decision, usually information and an entity
  - Criteria by which to make the decision
  - Alternative paths for the entity to take (clearly related back to the criteria)
  - An appropriate amount of Time to make the decision
- Parallel Processing seems at first glance to be a wonderful thing to do in any process. In reality the splitting or cloning of an entity brings its own problems, because as soon as any work is done on either clone, they are no longer clones, just parallel entities unless work is done to reconcile them. This reconciliation can be a huge undertaking and is often a major source of defects. Also, if the work content in the two paths is not balanced, then one clone ends up waiting at the join for another clone(s) to arrive.
- Loops always cause an increase in the amount of time required to complete the process and often disrupt the processing of other entities in the system. Consider focusing on reducing the number of entities going through the loop as a key part of any project.
- Triggers that allow the entity to move from step to step in the process are fertile ground for opportunities. Often an entity gets stalled at the sender when
  - The recipient doesn't know it is ready at the sender
  - The sender doesn't know that the entity is ready to send
  - The sender doesn't know that the recipient is ready to receive

Index

Symbols

1-sample T-tests
  conducting, 411, 414
  graphical representation, 411
  one-tailed, 416
  output, 414-415
  overview, 411
  two-tailed, 415
2-factor full factorials characterizing designs, 222
  data interaction, 225
  design matrices, 223
  generic model, 226
  graphical representation, 225
  runs, 224
2-level factorials, 213
2-sample T-tests
  conducting, 417-420
  graphical representation, 416
  one-tailed, 421
  output, 420-421
  overview, 416
  two-tailed, 421
5S
  assessments, 66
  baseline performance, 124
  events, 122-123
  overview, 122
  Shine, 128
  Sort, 124, 127
  Standardize, 129
  Store, 127
  Sustain, 129-130
5 Whys, 17, 119-120

A

abstract entities, 4
accounts payable, 93-95
accounts receivable, 95
  accuracy, 96-97
  customer non-payments, 97
  DSO, 96
  industry examples, 95
  measuring, 95-98
  payments, 97-98
  timeliness, 96

accuracy
  accounts receivable, 96-97
  deliveries, 24
  payments, 94-96
  problems, 29
  processes, 29
active listening, 180
activity symbol (VSMs), 427
administrative/service project examples, 5
affinity
  customer requirements, 131-133
  diagramming, 16, 117, 131-133
  discover, 117
  overview, 130
analysis. *See also* output
  characterizing designs, 228
  customer surveys, 192
  designs
    characterizing, 231, 234-235
    optimizing, 246-247
    screening, 216, 219-221
  multi-vari studies, 302
  optimizing designs, 241
  regression
    ANOVA example, 365
    conducting, 370
    fitted line plot example, 363
    full analysis example, 364
    linear, 371-372
    output, 370-371
    overview, 362-364, 367-368
    residual evaluation, 367-368
    signal-to-noise ratios, 366
Anderson-Darling tests, 309
ANOVA (one-Way analysis of variance), 133
  graphical representation, 134
  multi-way, 140-141
  output, 138-140
  overview, 133-135
  regression analysis, 365
  tables, 227
  test analysis, 135, 138
ARIMA (Autoregressive Integrated Moving Average), 89
attribute
  capability, 144
    calculating, 145
    DPMO, 145
    DPU, 144
    output, 146
  measures, 258
  MSA studies
    appraiser statistics, 279-281
    conducting, 278-279
    failures, 282-284
    kappa, 281
    output, 279, 282-284
    overview, 276-278
Autoregressive Integrated Moving Average (ARIMA), 89

B

backlogs
  orders, 20
  sales, 91-92
baseline capability studies
  accounts receivable, 96
  COV, 48
  cycle time, 100, 105
  delivery performance, 24
  DSO, 96
  entity types, 52
  failures, 40
  forecasts, 86
  headcount costs, 76
  inventory, 80

measurement systems, 61
pace, 37, 103
payments, 94
primary performance metrics, 30
processes
cycle time, 45
lead time, 42
processing rates, 45
sales, 90
scheduling variance, 55
setup/changeover times, 66
throughput, 28
unplanned maintenance, 69
uptime percentage, 34
yield, 84
baseline order backlogs, 92
baseline performance (5S), 124
Belbin, 14
between appraiser statistics, 281
bounds (continuous capability), 152
box plots
output, 143
overview, 141-142
brainstorming, 15, 116
breakdown maintenance, 410
broad inference, 208

C

C&E (Cause and Effect) Matrix, 31
entity types, 53
one-phase, 153-155
operations process scheduling, 59
output, 155
overview, 153
primary performance metrics, 31
product defects, 74
two-phase, 155-157
unplanned maintenance breakdowns, 71
cancellations, rescheduling, 58
capability
attribute, 144-146
calculating, 145
DPMO, 145
DPU, 144
output, 146
continuous, 146-152
bounds, 152
calculating, 148
Cp, 146
Cpk, 147
non-normal data, 151
output, 151
overview, 146-148
short-term versus long-term deviation, 149
single-sided specifications, 152
overview, 144-145
capability studies
finished goods out, 85
process losses, 85
processes, 111
raw materials in, 84
capacity (processes)
loss, 315
measuring, 27
OEE, 28
pace, 28
problems, 19
rolled throughput yield, 28
throughput, 28
too low, 25-27
uptime percentage, 28
categories of problems
identifying, 11
single process step, 21
whole process, 19-21

Cause & Effect matrix. *See* C&E matrix
center points
  characterizing designs, 237
  screening designs, 222
central composite designs, 242-244
central limit theorem, 170
changeover
  problems, 20
  rapid. *See* SMED
  times, 65
    industry examples, 65
    measuring, 65-67
    reducing, 56
    uptime percentages, 35
characterizing designs, 203
  2-factor full factorials, 222
    data interaction, 225
    design matrices, 223
    generic model, 226
    graphical representation, 225
    runs, 224
  analysis results, 228
  ANOVA tables, 227
  attribute Y data, 236
  center points, 237
  curvature, 237
  designing/analyzing, 231, 234-235
  Epsilon2, 229-231
  noise variables, 236
  overview, 222-225, 228-231
check sheets, 263
chemical manufacturing project examples, 5
Chi-square test
  applying, 160
  output, 161-162
  overview, 157-160
coded units, 223
Coefficient of Variation (COV), 47, 199
communication
  5S events, 123
  plans, 112
competences (processes), 112
concentration diagram check sheet, 264
configuring data points, 204-205
continuous capability
  bounds, 152
  calculating, 148
  Cp, 146
  Cpk, 147
  non-normal data, 151
  output, 151
  overview, 146-148
  short-term versus long-term deviation, 149
  single-sided specifications, 152
continuous improvement plans, 108
continuous measures (KPOVs), 258
continuous MSAs
  conducting, 289-290
  discrimination metric, 289
  gage R&R
    analytical results, 293
    components of variation plot, 293
    operator plot, 291
    operator-part interaction plot, 291
    by part plot, 292
    studies, 285-288
    Xbar-R charts, 290
  output, 290, 293-294
  overview, 284-289
  precision to tolerance ratio, 288
  %R&R, 289
  repeatability, 287
  reproducibility, 288
control charts
  central limit theorem, 170
  control limits, 163-164

data point distribution, 169
individuals chart, 164
moving range charts, 166
output, 168-169
overview, 163-164, 167
process variations, 169
range charts, 170
selecting, 167
stability, 165, 170
structure, 163
x-bar charts, 170
control limits, 163-164
control plans, 108
capability studies, 108
communication plans, 112
competences, 112
continuous improvement plans, 108
controls, building, 109
critical parameters, 108
customer information, 108
documenting, 111
effects analyses, 108
execution, 113
failure mode, 108
final reports, 113
final reviews, 113
implementation plans, 112
laying out, 110
maintenance, 108
methodology, 108
MSA, 108
process maps, 108
process triggers, 110
roles, 111
settings out, 111
signing off, 113
skills matrix, 112
SOPs, 108
standard work out, 110
summaries, 108
swimlane maps, 109
training
materials, 108
plans, 112
validation, 113
controls, building, 109
corrective maintenance, 410
costs of headcount, 76-78
COV (Coefficient of Variation), 47, 199
calculating, 199
measuring, 49
Cp, 146
Cpk, 147
critical parameters (control plans), 108
critical path analysis
applying, 172
output, 173
overview, 171
pace, 104
process lead time, 43
setup/changeover times, 67
cross-functional maps, 384
cross-resource maps, 384
current levels (time series analysis), 89
curvature
characterizing designs, 237
screening DOE, 222
customer
affinity diagram of customer requirements, 131-133
demand problems, 19
interviewing, 116, 174
conducting interviews, 179-182
output, 183
overview, 174
planning/preparation, 175-179

reviewing interviews, 182
validation, 182
matrix. *See* sampling plans
requirements
discovery, 116-118
processes, 15
requirements trees
creating, 184-185
discovery, 117
output, 186
overview, 183-184
surveys, 117
analysis plans, 192
conducting, 192
creating, 186-189, 192
designing, 187-189
data collection methods, 189
following-up, 192
objectives, 187
output, 192
overview, 186
sampling plans, 190-192
usage, measuring, 50-51
cycle time
global process
calculating, 394
output, 395
overview, 393
individual step
calculating, 397
output, 398
overview, 395
measuring, 27
processes, 21, 99
delivery performance, 25
measuring, 45
variations, 104-106

D

D-Study, 62
data collection
KPOVs
forms/tracking systems, 263-264
sampling, 262-263
testing, 266
plan summary sheets, 266
data integrity audits
output, 279, 282-284
reliability, 278-279
validity, 274-275
data points
distribution, 169
subgroups, 170
data sheets, 264
Days on Hand (DOH), 79
Days Sales Outstanding (DSO), 95
decision symbol (VSMs), 428
defects
defined, 144
processes, 19, 29
products, 73-75
Defects per Million Opportunities (DPMO), 145
Defects per Unit (DPU), 144
Define, Measure, Analyze, Improve, and Control (DMAIC), 7
deliverables, 4
deliveries
forecasting, 87
on-time delivery problems, 19
industry examples, 23
measuring performance, 23-26
demand
buckets, 198
delivery variations, 25

inventory, 80
profiling, 25
applying, 194
customer usage, 50
delivery performance, 25
inventory levels, 81
output, 194-196
overview, 193-194
Profile tool, 26
statistical analysis of plot data, 196
variable demands, 49
setting inventory, 81-82
segmentation
calculating, 198-199
COV, 48
delivery performance, 25
entity types, 52
forecasting, 87-88
output, 199-201
overview, 196-197
setting inventory levels, 81
variable, 47
customer usage, 50-51
industry examples, 47
measuring, 47-49
Design of Experiments. *See* DOE
destructive testing, 62
DISC, 14
discovery, 115
customer requirements, 116-118
project charter, 115
SIPOC, 116
discrete manufacturing project examples, 5
discrimination metric, 289
DMAIC (Define, Measure, Analyze, Improve, and Control), 7
documenting processes, 111
DOE (Design of Experiments), 32, 202
broad inference, 208
characterizing, 203
2-factor full factorials, 222-226
analyzing, 228, 231, 234-235
ANOVA tables, 227
attribute Y data, 236
center points, 237
curvature, 237
designing, 32, 231, 234-235
Epsilon2, 229-231
noise variables, 236
overview, 224-228, 231
conducting, 208-211
data points, 204-205
designing, 208-211
management, 212
narrow inference, 208
noise variables, 206
optimizing, 203
analyzing, 241, 246-247
central composite designs, 242-244
contour plot of yield, 239
designing, 33, 246-247
optimization algorithm, 244
overview, 237-241
path of steepest ascent, calculating, 238-240, 247-249
overview, 202-208
planning, 206-211
problems, 212
repetition, 208
replication, 208
screening, 203
analyzing, 32, 216, 219-221
attribute Y data, 221
center points, 222

curvature, 222
designing, 32, 216, 219-221
fractional factorials, 214-215
noise variables, 222
overview, 213-216
resolution, 216
sequential experimentation, 206
DOH (Days on Hand), 79
dollar-days, 95
downstream demand problems, 19
downtime for unplanned maintenance, 70, 69
DPMO (Defects per Million Opportunities), 145
DPU (Defects per Unit), 144
DSO (Days Sales Outstanding), 95-98
duration variability, 57

E

effectiveness (KPOVs), 259
effects analyses (control plans), 108
efficiency (KPOVs), 260
entities, 4
entity types
identifying, 198
segregating, 58
too many, 49-51
industry examples, 51
measuring, 52-54
Epsilon2, 229-231
executing processes, 113
external pull system implementation, 351-354

F

failure
attribute MSAs, 282-284
handoffs, 46
intermittent process, 38-41
measuring, 39
modes
control plans,108
FMEA, 326
processes, 19-20
Failure Mode & Effects Analysis. *See* FMEA
final reports, 113
final reviews, 113
finished goods out, 85
first time right, 29
fishbone diagrams, 249-252
creating, 251-252
output, 251
overview, 249-251
unplanned maintenance breakdowns, 70
flag symbol (VSMs), 427
FMEA (Failure Mode & Effects Analysis), 31, 325
creating, 326-329
design, 325
failure modes, 326
market, 325
operations process scheduling, 59
output, 329
overview, 325
primary performance metrics, 31
process, 325
product defects, 74
project, 325
RPNs, 328
standard scoring tables, 327
unplanned maintenance breakdowns, 71
forecasting, 85
defined, 89
deviation problems, 20
eliminating noise from operation process, 87
industry examples, 86

measuring, 86-88
time series analysis, 88-89
fractional factorials, 214
3-factor interaction, 215
resolution, 216
frequency plot check sheets, 264
full factorials (2-factor) characterizing designs
data interaction, 225
design matrices, 223
generic model, 226
graphical representation, 225
runs, 224

G

gage R&R
analytical results, 293
components of variation plot, 293
operator plot, 291
operator-part interaction plot, 291
by part plot, 292
studies, 285-288
Xbar-R charts, 290
gap analysis, 118
global process cycle time
calculating, 394
output, 395
overview, 393
global process problems
accounts payable, 93-95
accounts receivable, 95
accuracy, 96-97
customer non-payments, 97
DSO, 96
industry examples, 95
measuring, 95-98
payments, 97-98
timeliness, 96
accuracy, 29
capacity too low, 27
industry examples, 27
measuring, 27
OEE, 28
pace, 28
rolled throughput yield, 28
throughput, 28
uptime percentage, 28
defects, 29
failure to make product, 72
industry example, 73
measuring defectiveness, 73-75
forecasting, 85
eliminating noise in operations process, 87
industry examples, 86
measuring, 86-88
time series analysis, 88-89
high inventory, 78
industry examples, 79
measuring, 79-82
reducing, 82
rightsizing inventory, 80
setting to demand, 81-82
intermittent failures, 38-41
lead time
industry examples, 41
measuring, 42-44
losses, 82-85
measurement systems, 60
D-Study, 62
destructive testing, 62
in-line testing, 62
industry examples, 61
measuring performance, 61-63
process variation studies, 62
reference materials, 63
meeting Takt, 44-47

on-time deliveries, 23
  capacity too low, 25
  industry examples, 23
  intermittent process failures, 26
  measuring performance, 23-26
  process lead time too long, 26
  variation in demand, 26
pace too slow, 36-38
performance characteristics, 63-64
planned maintenance, 64
quality, 29-33
resource usage, 75-78
rework, 29
sales, 89
  categories for growth, 90
  customer decisions, 91
  industry examples, 89
  measuring, 90-91
sales backlogs, 91-92
scheduling, 54
  cancellations, 58
  designed experiments, 60
  duration variability, 57
  improving schedule process, 58
  industry examples, 54
  measuring performance, 55, 59-60
  operations process, 56-58
  segregating entity types, 58
setup/changeover times, 65-67
too many entity types, 51-54
unplanned maintenance, 68
  causes, 70-72
  downtime, 69-70
  industry examples, 68
  measuring, 68, 72
uptime percentage, 33
  changeover time, 35
  industry examples, 33
    measuring, 33-35
    rework time, 35
    scheduled workforce breaks, 35
    unplanned maintenance, 35
  variable demands, 47
    customer usage, 50-51
    industry examples, 47
    measuring, 47-49
good project examples, 5
guide (customer interviewing), 175-177

H

handoff
  failures, 46
  maps
    as-is construction, 254
    creating, 254-256
    failures, 41
    meeting Takt, 46
    output, 256
    overview, 253-254
    to-be construction, 255-256
headcount reduction, 75
  industry examples, 75
  measuring costs, 76-78
headers of project charters, 335
healthcare examples, 5
  accuracy problems/defects, 29
  broken measurement systems, 61
  delivery problems, 23
  high inventory, 79
  intermittent process failures, 39
  lead time problems, 41
  pace problems, 36
  planned maintenance, 64
  process capacity problems, 27
  scheduling, 54
  setup/changeover times, 65

too many problems, 51
unplanned maintenance, 68
uptime, 33
variable demands, 47
high inventory, 78
industry examples, 79
measuring, 79-82
reducing, 82
rightsizing inventory, 80
setting to demand, 81-82
high-level VSMs, 424
human entities, 4
hybrid VSM-swimlane maps, 387

I

identifying
demand buckets, 198
entities
primary, 4
types, 198
problem categories, 11
imbalanced lines (load charts), 270
implementation plans, 112
implementing kanbans, 110
improving
operations process of scheduling, 56-57
schedule process, 57-58
in-line testing measurement systems, 62
in-process kanbans, 345
inanimate entities, 4
individual process problems
cycle time variations, 104-106
not meeting Takt, 99-102
pace, 102-104
individual step cycle time
calculating, 397
output, 398
overview, 395
individuals charts, 164
industrial examples
broken measurement systems, 61
delivery problems, 23
failure to make product, 73
forecasting, 86
high inventory, 79
intermittent process failures, 39
lead time examples, 41
pace problems, 36
performance characteristics, 63
planned maintenance, 64
processes
capacity problems, 27
losses, 83
rework examples, 29
scheduling, 54
setup/changeover times, 65
too many products, 51
unplanned maintenance, 68
uptime, 33
variable demands, 47
industry examples
accounts payable, 93
accounts receivable, 95
accuracy, 29
broken measurement systems, 61
capacity, 27
deliveries, 23
failure to make product, 73
forecasting, 86
high inventory, 79
intermittent failures, 39
meeting Takt, 44
order backlogs, 92
pace, 36
performance characteristics, 63
planned maintenance, 64

process
lead time, 41
losses, 83
sales, 89
scheduling, 54
setup/changeover times, 65
too many products, 51
unplanned maintenance, 68
uptime, 33
variable demands, 47
intermittent process failures, 38-41
internal pull system implementation, 348-351
interpersonal roles, 14
interview teams (customer interviewing), 179
interviewing customers, 116, 174
conducting interviews, 179-182
output, 183
overview, 174
planning/preparation, 175-179
reviewing interviews, 182
validation, 182
inventory, 78
demand, 80
dollars, 79
industry examples, 79
measuring, 79-82
processes
forecasting, 88
problems, 20
reducing, 82
responsive processes, 80
rightsizing, 80
setting to demand, 81-82
turns, 79

J – K

joins, 430

kanbans, 109
implementing, 110
in-process, 345
material, 346
operation, 347
pull systems, 345-347
supplier, 347
withdrawal, 347
kappa, 281
Kolmogorov-Smirnov tests, 309
KPOVs (Key Process Output Variables), 17, 257
attribute measures, 258
collecting data, 262, 266
forms/tracking systems, 263-264
sampling, 262-263
testing, 266
continuous measures, 258
discovery, 117
effectiveness, 259
efficiency, 260
identifying data, 261-262, 266
operational definitions, 260
overview, 257
predictors, 259
results, 259

L

lead time problems, 19
delivery performance, 25
too long, 26
Lean Sigma
process improvements, 6
Roadmap, 7-9
load charts
area applications, 271
creating, 269

imbalanced lines, 270
line balancing, 270
machines/equipment, 271
meeting Takt, 46
output, 270-271
overview, 268
pace, 38
process lead time, 43
resource usage, 77
long-term continuous capability deviation, 149
loop symbol (VSMs), 430
losses from processes, 82
industry examples, 83
measuring, 83-85
LSL (Lower Specification Limit), 152

M

maintenance, 108
breakdown, 410
corrective, 410
planned, 20, 64
predictive, 410
preventive, 410
scheduled, 410
total productive maintenance
overview, 408-409
timeline steps, 409-410
unplanned, 68
causes, 70-72
downtime, 69-70
industry examples, 68
measuring, 68, 72
problems, 20
management
DOE, 212
projects, 18
market FMEA, 325
material kanbans, 346
Mean Time Between Failures (MTBF), 39
Measurement System Analysis. *See* MSA
measurement systems, 60
D-Study, 62
destructive testing, 62
in-line testing, 62
industry examples, 61
performance, 61-63
process variation studies, 62
reference materials, 63
measuring
accounts payable, 93-95
accounts receivable, 96-98
backlogs, 92
capacity (processes), 27-28
customer usage, 50-51
cycle time, 27, 105-106
entity types, 52-54
failures, 39
first time right, 29
forecasting, 86-88
eliminating noise from operation process, 87
time series analysis, 88-89
headcount costs, 76-78
inventory, 79-82
reducing, 82
rightsizing, 80
setting to demand, 81-82
meeting Takt, 44-47
pace, 36-38
performance
characteristics, 64
deliveries, 23-26
planned maintenance, 64
primary performance metrics, 29-33
processes
cycle time, 99-102
lead time, 42-44

losses, 83-85
pace, 102-104
product defects, 73-75
quality, 29
reliability metrics, 39
RTY, 29
sales, 90-91
scheduling performance, 55, 59-60
cancellations, 58
designed experiments, 60
duration variability, 57
improving schedule process, 58
operations process, 56-58
segregating entity types, 58
setup/changeover times, 65-67
throughput, 27-28
unplanned maintenance performance, 68, 72
causes, 70-72
downtime, 69-70
uptime, 33-35
meeting effectiveness tools, 14
meeting Takt, 44-47
methodologies, 6
mistake proofing. *See* Poka Yoke
moving range charts, 166
MSA (Measurement System Analysis), 24
attribute
between appraiser statistics, 281
conducting, 278-279
failures, 282-284
kappa, 281
output, 282-284
overview, 276-278
within appraiser statistics, 279
continuous
conducting, 289-290
discrimination metric, 289
gage R&R analytical results, 293
gage R&R by part plot, 292
gage R&R components of variation plot, 293
gage R&R operator plot, 291
gage R&R operator-part interaction plot, 291
gage R&R studies, 285-288
gage R&R Xbar-R charts, 290
output, 290, 293-294
overview, 284-289
precision to tolerance ratio, 288
%R&R ratio, 289
repeatability, 287
reproducibility, 288
COV, 48
critical parameters, 108
cycle time, 100, 105
delivery performance, 24
DSO, 96
entity types, 52
failures, 39
forecasts, 86
headcount, 76
inventory, 80
measurement systems, 61
overview, 272-273
pace, 37, 103
payments, 94-96
primary performance metrics, 30
process, 111
cycle time, 45
lead time, 42
processing rates, 45
product defects, 74
sales, 90
sales backlogs, 92
scheduling variance, 55
setup/changeover times, 65

throughput, 28
unplanned maintenance, 69
uptime percentage, 34
validity
confirming, 274-275
output, 275
overview, 272-273
weigh scales, 84
yield, 83
MTBF (Mean Time Between Failures), 39
multi-cycle analysis
conducting, 295, 298
meeting Takt, 46
output, 298-300
overview, 294-295
pace, 37
payments, 95-97
process lead time, 43
resource usage, 77
multi-vari studies
analysis, 302
conducting, 301-305
data collection, 301
operations process of scheduling, 59
overview, 300-301
primary performance metrics, 32
unplanned maintenance breakdowns, 72
Murphy's analysis, 15
creating, 307-308
discovery, 116
output, 308
overview, 306-307

N

narrow inference, 208
noise variables
blocking, 222, 236
managing, 206
normality tests
hypotheses, 309
output, 309-311
overview, 308
NVA analysis, 77

O

objectives of customer surveys, 187
OEE (Overall Equipment Effectiveness), 28, 311
applying to people, 317
capacity loss, 315
output, 316
overview, 311-312
pace, 314-315
process capacity too low, 28
process cycle time, 100-101
quality, 315
uptime, 312-314
on-time delivery problems, 19, 23
capacity too low, 25
industry examples, 23
intermittent process failures, 26
measuring performance, 23-26
process lead time too long, 26
variation in demand, 26
one-phase C&E matrix, 153-155
one-tailed 1-Sample T-tests, 416
one-tailed 2-Sample T-tests, 421
one-way analysis of variance. *See* ANOVA
operations
kanbans, 347
planning, 201
process
eliminating noise, 87
improving, 56-58
opportunities, 145
optimizing designs, 203
analysis results, 241
central composite designs, 242-244

contour plot of yield, 239
designing/analyzing, 246-247
optimization algorithm, 244
overview, 237-241, 245
path of steepest ascent, 238-240, 247-249
order backlogs, 20
organizational priorities, 336
output. *See also* analysis
affinity diagram of customer requirements, 133
ANOVA, 138-140
attributes
capability, 146
MSAs, 279, 282-284
box plots, 143
C&E matrix, 155
Chi-square tests, 161-162
continuous capability, 151
continuous MSAs, 290, 293-294
control charts, 168-169
critical path analysis, 173
customer
interviews, 183
requirements trees, 186
surveys, 192
demand
profiling, 194-196
segmentation, 199-201
fishbone diagrams, 251
FMEA, 329
global process cycle time, 395
handoff maps, 256
individual step cycle time, 398
load charts, 270-271
MSA, 275
multi-cycle analysis, 298-300
Murphy's analysis, 308
normality tests, 309-311
OEE, 316
Pareto charts, 319
process
lead time, 400
variables maps, 333
regression analysis, 370-371
replenishment time, 403-404
SIPOC, 376
spaghetti maps, 379
SPC, 382-384
swimlane maps, 387
T-tests
1-Sample, 414-415
2-Sample, 420-421
Takt time, 407-408
test of equal variance, 391
VSMs, 432-433
Overall Equipment Effectiveness. *See* OEE

P

pace
percentage, 314-315
processes, 21, 102
measuring, 36-38, 102-104
problems, 19, 36
too slow, 28
variants, 314
paired T-tests, 422-423
parallel processing symbol (VSMs), 429
Pareto
charts, 318-319
failures, 40
finished goods out, 85
inventory, 80
process losses, 84
total shift time, 35
unplanned maintenance downtime, 70
Parts per Million (PPM), 151
path of steepest ascent, 238-240, 247-249

payments
  accuracy, 96-97
  suppliers, 20
  terms, 98
  timeliness, 94, 97
percentage contribution, 229-231
performance
  accounts payable, 93-95
  accounts receivable
    accuracy, 96-97
    customer non-payments, 97
    DSO, 96
    measuring, 95-98
    payments, 97-98
    timeliness, 96
  baseline, 124
  characteristics, 63-64
  deliveries, 23-26
  forecasting
    eliminating noise from operations process, 87
    measuring, 86-88
    time series analysis, 88-89
  measurement systems, 61-63
  planned maintenance, 64
  problems, 20
  product defects, 73-75
  setup/changeover times, 65-67
  unplanned maintenance, 68, 72
    causes, 70-72
    downtime, 69-70
physical process maps, 379
physical transfers of materials, 84
planning
  customer interviewing, 175-179
    guide, 175-177
    interview teams, 179
    practicing, 179
    purpose, 175
    sampling plans, 177
    scheduling, 179
  DOE, 206-211
  maintenance, 64
    industry examples, 64
    measuring, 64
    problems, 20
    uptime percentages maintenance, 35
  production/operations, 201
Poka Yoke, 109
  conducting, 323-325
  overview, 321-322
poor project examples, 5
populating VSMs, 431
PPM (Parts per Million), 151
practicing customer interviewing, 179
precision to tolerance ratios, 288
predictive maintenance, 410
preparations for customer interviewing, 175-179
  guide, 175-177
  interview teams, 179
  practicing, 179
  purpose, 175
  sampling plans, 177
  scheduling, 179
  SMED, 356
prerequisites, 2-3
preventive maintenance, 410
primary entities, 4
primary performance metrics, 29-33
probing, 181
problems
  categories
    identifying, 11
    single process step, 21
    whole processes, 19-21
  customer demand, 19
  DOE, 212

global process
  accounts payable, 93-95
  accounts receivable, 95-98
  accuracy, 29
  capacity too low, 27-28
  defects, 29
  failure to make product, 72-75
  forecasting, 85-89
  high inventory, 79-82
  intermittent failures, 38-41
  lead time, 41-44
  losses, 83-85
  measurement systems, 60-63
  meeting Takt, 44-47
  on-time deliveries, 23-26
  pace, 36-38
  performance characteristics, 63-64
  planned maintenance, 64
  quality, 29-33
  resource usage, 75-78
  rework, 29
  sales, 89-92
  scheduling, 54-60
  setup/changeover times, 65-67
  too many entity types, 51-54
  unplanned maintenance, 68-72
  uptime percentage, 33-35
  variable demands, 47-51
individual processes
  cycle time variations, 104-106
  not meeting Takt, 99-102
  pace, 101-104
maintenance, 20
on-time deliveries, 19
order backlogs, 20
performance, 20
  capacity, 19
  cycle time, 21
  defects, 19
  downstream demand, 19
  failure, 20
  forecast deviation, 20
  intermittent failures, 19
  inventory, 20
  lead time, 19
  pace, 19-21
  Takt, 21
  uptime, 19
  waste/loss, 20
resource usage, 20
sales, 20
setup/changeover, 20
supplier payments, 20
too many products, 19
process
  accounts payable, 93-95
  accounts receivable, 95-98
    accuracy, 96-97
    customer non-payments, 97
    DSO, 96
    industry examples, 95
    measuring, 95-98
    payments, 97-98
    timeliness, 96
  accuracy problems, 29
  capacity too low, 25-28
  competences, 112
  customer requirements, 15
  cycle time, 25
  defects, 29
  defined, 3
  defining, 110
  deliverables, 4
  documenting, 111
  entities, 4
  execution, 113

failure to make product, 72
  industry example, 73
  measuring defectiveness, 73-75
final reports, 113
final reviews, 113
FMEA, 325
forecasting, 85
  eliminating noise in operations process, 87
  industry examples, 86
  measuring, 86-88
  time series analysis, 88-89
global problems, 25-26
high inventory, 78
  industry examples, 79
  measuring, 79-82
  reducing, 82
  rightsizing inventory, 80
  setting to demand, 81-82
intermittent failures, 26, 38-41
laying out, 110
lead time
  calculating, 400
  delivery performance, 25
  industry examples, 41
  lost sales, 91
  measuring, 42-44
  output, 400
  overview, 399
  sales backlogs, 92
  too long, 26
Lean Sigma improvements, 6
losses, 82
  industry examples, 83
  measuring, 83-85
maps, 108
measurement systems not working, 60
  D-Study, 62
  destructive testing, 62
  in-line testing, 62
  industry examples, 61
  performance, 61-63
  process variation studies, 62
  reference materials, 63
meeting Takt, 44-47
not meeting Takt, 99-102
on-time deliveries, 23-26
pace, 36-38, 102-104
performance characteristics, 63-64
planned maintenance, 64
problems
  capacity, 19
  categories, 19-21
  cycle time, 21
  defects, 19
  downstream demand, 19
  failure, 20
  forecast deviation, 20
  intermittent failures, 19
  inventory, 20
  lead time, 19
  pace, 19-21
  Takt, 21
  uptime, 19
  waste/loss, 20
quality, 29-33
resource usage, 75
  industry examples, 75
  measuring costs, 76-78
rework problems, 29
roles, 111
sales, 89
  backlogs, 91-92
  categories for growth, 90
  customer decisions, 91
  industry examples, 89
  measuring, 90-91

- scheduling, 54
  - cancellations, 58
  - designed experiments, 60
  - duration variability, 57
  - improving schedule process, 58
  - industry examples, 54
  - measuring performance, 55, 59-60
  - operations process, 56-58
  - segregating entity types, 58
- settings, 111
- setup/changeover times, 65-67
- signing off, 113
- Six Sigma improvements, 6
- too many entity types, 51-54
- training plans, 112
- unplanned maintenance, 68
  - causes, 70-72
  - downtime, 69-70
  - industry examples, 68
  - measuring, 68, 72
- uptime percentage, 33-35
- variable demand, 47
  - customer usage, 50-51
  - industry examples, 47
  - measuring, 47-49
- variable maps
  - creating, 330-332
  - operations process of scheduling, 58
  - output, 333
  - overview, 330
  - primary performance metrics, 31
  - unplanned maintenance breakdowns, 70
- variation
  - control charts, 169
  - demands, 26
  - studies, 62

production planning, 201

products
- backlogs, 91-92
- portfolios, 201
- problems, 19
- processes unable to make, 72-75
- too many, 51-54

profiling demands, 25

project
- charter
  - benefits, 339
  - creating, 335-342
  - departments affected, 340
  - discovery, 115
  - header information, 335
  - metrics, 338-339
  - organizational priorities, 336
  - overview, 333-335
  - process, 335
  - project descriptions, 336
  - schedule, 341
  - scope, 337
  - support, 341
  - team members, 339
- FMEA, 325
- good/poor examples, 5
- management, 18
- practical questions, 11
- sequence of completion
  - communication plans, 112
  - competences, 112
  - controls, building, 109
  - documenting, 111
  - execution, 113
  - final reports, 113
  - final reviews, 113
  - implementation plans, 112
  - laying out processes, 110
  - process triggers, 110

roles, 111
settings, 111
signing off, 113
skills matrix, 112
standard of work, 110
swimlane maps, 109
training plans, 112
validation, 113
pull systems
continuous improvement, 354
graphical representation, 343
implementing
external, 351-354
internal, 348-351
kanbans, 345-347
location processing, 344
overview, 342, 345-347
process requirements, 344
push systems, compared, 344
purpose (customer interviewing), 175
push systems, 344

Q – R

quality
measuring, 29
percentage, 315

%R&R, 289
random components (time series analysis), 89
range charts, 170
rapid changeover. *See* SMED
raw materials in measuring, 84
reaction plans, 108
red tags, 123
reducing
changeover times, 56
headcount, 75
industry examples, 75
measuring costs, 76-78
inventory, 82
reference materials, 63
regression analysis
ANOVA example, 365
conducting, 370
fitted line plot example, 363
full analysis example, 364
linear, 371-372
output, 370-371
overview, 362-368
residual evaluation, 367-368
signal-to-noise ratios, 366
reliability metrics, 39
replenishment, 200-201, 401-404
replenishment time
calculating, 403
output, 403-404
overview, 401-402
rescheduling cancellations, 58
resource usage, 75
industry examples, 75
measuring costs, 76-78
problems, 20
response surface methodology. *See* optimizing designs
results of KPOVs, 259
reviewing customer interviews, 182
rework time, 35
reworking processes, 29
rightsizing inventory, 80
Risk Priority Numbers (RPNs), 328
roadmaps
affinity diagram of customer requirements, 131-132
ANOVA, 135, 138
attribute capability, 145

C&E matrix, 153-155
characterizing designs, 234-235
Chi-square tests, 160
continuous capability, 148
continuous MSAs, 289-290
control charts, 167
critical path analysis, 172
customer interviewing
  conducting interviews, 179-182
  planning/preparations, 176-179
  reviewing interviews, 182
  validation, 182
customer requirements
    trees, 184-185
customer surveys, 186-192
  analysis plans, 192
  conducting, 192
  creating, 186-192
  data collection methods, 189
  designing, 187-189
  following-up, 192
  objectives, 187
  output, 192
  overview, 186
  sampling plans, 190-192
demand
  profiling, 194
  segmentation, 198-199
fishbone diagrams, 251-252
FMEA, 326-329
global process cycle time, 394
handoff maps, 254-256
individual step cycle time, 397
KPOVs
  collecting data, 262, 266
  identifying data, 261
Lean Sigma, 7-9
load charts, 269
MSA
  attribute, 278-279
  validity, 274-275
multi-cycle analysis, 295, 298
multi-vari studies, 301-302, 305
Murphy's analysis, 307-308
normality tests, 309
optimizing designs, 246-247
Pareto charts, 319
Poka Yoke, 323-325
process lead time, 400
process variables maps, 330-332
project charters, 335-342
  benefits, 339
  departments affected, 340
  header information, 335
  metrics, 338-339
  organizational priorities, 336
  process, 335
  project descriptions, 336
  schedule, 341
  scope, 337
  support, 341
  team members, 339
pull systems
  external implementation, 351-354
  internal implementation, 348-351
regression analysis, 370
replenishment time, 403
screening designs, 216-221
SIPOC, 374-376
SMED, 357, 360-362
spaghetti maps, 377-379
SPC, 381-382
swimlane maps, 386-387
T-tests
  1-Sample, 411-415
  2-Sample, 417-420

Takt time, 406-407
test of equal variance, 389-391
total productive maintenance, 409-410
VSMs, 426-432
roles
interpersonal, 14
processes, 111
RPNs (Risk Priority Numbers), 328
RTY (Rolled Throughput Yield), 28-29, 39
runners, 88
Ryan-Joiner tests, 309

S

Safety, Purpose, Agenda, Code of Conduct, Expectations, and Roles (SPACER), 14
sales, 89
backlogs, 91-92
categories for growth, 90
customer decisions, 91
industry examples, 89
measuring, 90-91
problems, 20
sampling
KPOVs, 262-264
plans (customers)
interviews, 177
surveys, 190-192
scheduling
customer interviews, 179
maintenance, 410
processes, 54
cancellations, 58
designed experiments, 60
duration variability, 57
improving schedule process, 58
industry examples, 54
measuring performance, 55, 59-60
operations process, 56-58
segregating entity types, 58
project charters, 341
workforce breaks, 35
scope
5S events, 123
project charters, 337
screening
designs, 203
analyzing, 216, 219-221
attribute Y data, 221
center points, 222
curvature, 222
designing, 216, 219-221
noise variables, 222
overview, 213-216
resolution, 216
DOE, 214-215
seasonal patterns (time series analysis), 89
segregating entity types, 58
selecting control charts, 167
sequence of completion
communication plans, 112
competences, 112
controls, building, 109
documenting, 111
execution, 113
final reports, 113
final reviews, 113
implementation plans, 112
laying out processes, 110
process triggers, 110
roles, 111
settings, 111
signing off, 113
skills matrix, 112
standard work, 110
swimlane maps, 109

training plans, 112
validation, 113
sequential experimentation, 206
service/administrative examples, 5
service/transactional examples
accuracy problems, 29
broken measurement systems, 61
delivery problems, 23
high inventory, 79
intermittent process failures, 39
lead time problems, 41
pace problems, 36
performance characteristics, 63
planned maintenance, 64
process capacity problems, 27
scheduling, 54
setup/changeover times, 65
too many products, 51
unplanned maintenance, 68
uptime, 33
variable demands, 47
setup
problems, 20
reduction, 66-67
times, 65-67
Shine tool (5S), 128
short-term continuous capability deviation, 149
Sigma Breakthrough Technologies Web site, 15
signing off, 113
Single Minute Exchange of Dies. *See* SMED
SIPOC (Suppliers, Inputs, Process, Outputs, Customers), 15
COV, 48
creating, 374-376
discovery, 116
output, 376
overview, 372-373
process cycle time, 45
processing rates, 45
Six Sigma process improvements, 6
skills matrix, 112
SMED (Single Minute Exchange of Dies), 354
implementing, 357, 360-362
overview, 354-355
preparations, 356
time analysis charts, 358
smoothing, 88
SOPs (Standard operating Procedures), 108
Sort (5S), 124, 127
SPACER (Safety, Purpose, Agenda, Code of Conduct, Expectations, and Roles), 14
spaghetti maps
creating, 377-379
meeting Takt, 45
output, 379
overview, 376
setup/changeover times, 67
SPC (Statistical Process Control), 380
output, 382-384
overview, 380-381
setting up, 381-382
splits, 430
stability of control charts, 165, 170
Standard Operating Procedures (SOPs), 108
standard scoring tables, 327
standard work, 110
Standardize tool (5S), 129
star points, 242
start symbol (VSMs), 427
Statistical Process Control. *See* SPC
stop symbol (VSMs), 427
Store tool (5S), 127
store symbol (VSMs), 429
strangers, 88
subgroups (data points), 170
summaries of control plans, 108

suppliers
Suppliers, Inputs, Process, Outputs, Customers. *See* SIPOC
surveys (customer), 117
  analysis plans, 192
  conducting, 192
  creating, 186-192
  data collection methods, 189
  designing, 187-189
  following-up, 192
  objectives, 187
  output, 192
  overview, 186
  sampling plans, 190-192
Sustain tool (5S), 129-130
swimlane maps
  cross-functional, 384
  cross-resource, 384
  failures, 40
  hybrid VSM-swimlane maps, 387
  meeting Takt, 45
  output, 387
  overview, 384
  process lead time, 43
  sequence of completion, 109
  setup/changeover times, 66
  VSM conversions, 386-387
symbols of VSMs, 427-430

T

T-tests
  1-Sample, 411-416
    conducting, 411, 414
    graphical representation, 411
    one-tailed, 416
    output, 414-415
    overview, 411
    two-tailed, 415
  2-Sample, 417-421
    conducting, 417-420
    graphical representation, 416
    one-tailed, 421
    output, 420-421
    overview, 416
    two-tailed, 421
  paired, 422-423
Takt time
  calculating, 406-407
  delivery performance, 25
  graphical representation, 402, 405
  meeting, 44-47
  output, 407-408
  overview, 404-406
  processes not meeting, 99-102
Target Training International Web site, 14
team formation tools, 14
team interaction, 14
test of equal variance, 389-391
testing
  destructive, 62
  equal variance, 389-391
  in-line, 62
  KPOV data collection, 266
throughput, 27-28
time
  global process cycle time, 393-395
  individual step cycle time
    calculating, 397
    output, 398
    overview, 395
  process lead time, 399-400
  replenishment time
    calculating, 403
    output, 403-404
    overview, 401-402
  series analysis, 88-89

Takt
calculating, 406-407
delivery performance, 25
graphical representation, 402, 405
meeting, 44-47
output, 407-408
overview, 404-406
processes not meeting, 99-102
time series analysis, 88-89
timeliness
accounts receivable, 97
deliveries, 24
measuring, 96
Total Productive Maintenance (TPM)
overview, 408-409
timeline steps, 409-410
tracking systems, 263-264
training
materials, 108
plans, 112
transport symbol (VSMs), 428
traveler check sheets, 264
trends (time series analysis), 89
trigger symbol (VSMs), 428
triggers. *See* kanbans
troubleshooting DOE, 212
two-phase C&E matrix, 155-157
two-tailed 1-Sample T-tests, 415
two-tailed 2-Sample T-tests, 421

U

units, 144
unplanned maintenance, 68
causes, 70-72
downtime, 69-70
industry examples, 68
measuring, 68, 72
problems, 20
uptime percentages, 35
uptime, 312-314
problems, 19, 33-35
process capacity too low, 28
variants, 314
USL (Upper Specification Limit), 152

V

VA (Value Added) uptime percentage, 33
validating
control plans, 113
customer interviews, 182
MSA studies
confirming, 274-275
output, 275
overview, 272-273
VSMs, 431
Value Stream Maps. *See* VSMs
variables
demand, 47
customer usage, 50-51
industry examples, 47
measuring, 47-49
noise
blocking, 222, 236
managing, 206
variances
defined, 55
pace, 314
test of equal variance, 389-391
variations in demand, 25
VOC (Voice of the Customer), 130
5 Whys, 17
affinity diagramming, 16
creating, 131-132
output, 133
brainstorming, 15

customer interviewing, 16, 174
conducting interviews, 179-182
output, 183
overview, 174
planning/preparation, 175-179
reviewing interviews, 182
validation, 182
customer requirements trees, 17
creating, 184-185
output, 186
overview, 183-184
customer surveys, 16
analysis plans, 192
conducting, 192
creating, 186-192
data collection methods, 189
designing, 187-189
following-up, 192
objectives, 187
output, 192
overview, 186
sampling plans, 190-192
Murphy's analysis, 15
tools, 15
VSMs (Value Stream maps), 423
converting to swimlane maps, 386-387
creating, 426-432
cycle time, 105
discovery, 117
high-level, 424
hybrid VSM-swimlane maps, 387
meeting Takt, 45
output, 432-433
overview, 423, 426
pace, 37, 103
payments, 94, 97
physical transfers, 84
populating, 431
process lead time, 42
resource usage, 77
symbols, 427-430
validating, 431
variable demands, 49

W

wait/delay symbol (VSMs), 429
Web sites
Belbin, 14
Sigma Breakthrough Technologies Inc. (SBTI), 15
Target Training International, 14
withdrawal kanbans, 347
within appraiser statistics, 279

X – Z

x-bar charts, 170

yield, measuring, 83

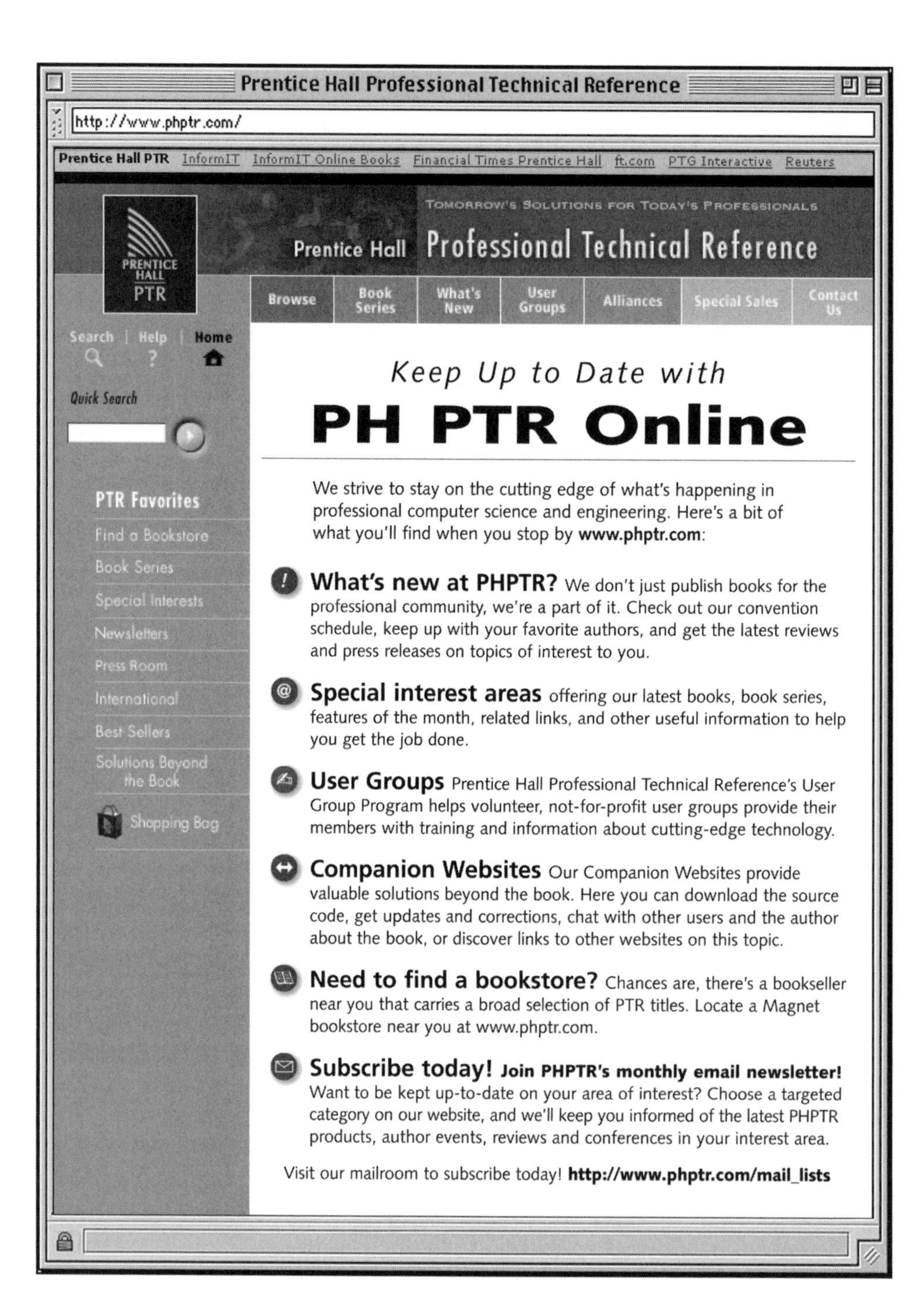

Prentice Hall Professional Technical Reference
http://www.phptr.com/
Prentice Hall PTR
InformIT
InformIT Online Books
Financial Times Prentice Hall
ft.com
PTG Interactive
Reuters
Tomorrow's Solutions for Today's Professionals
Prentice Hall Professional Technical Reference
PRENTICE HALL PTR
Browse
Book Series
What's New
User Groups
Alliances
Special Sales
Contact Us
Search | Help | Home
Quick Search
PTR Favorites
Find a Bookstore
Book Series
Special Interests
Newsletters
Press Room
International
Best Sellers
Solutions Beyond the Book
Shopping Bag
Keep Up to Date with
PH PTR Online
We strive to stay on the cutting edge of what's happening in professional computer science and engineering. Here's a bit of what you'll find when you stop by www.phptr.com:
What's new at PHPTR? We don't just publish books for the professional community, we're a part of it. Check out our convention schedule, keep up with your favorite authors, and get the latest reviews and press releases on topics of interest to you.
Special interest areas offering our latest books, book series, features of the month, related links, and other useful information to help you get the job done.
User Groups Prentice Hall Professional Technical Reference's User Group Program helps volunteer, not-for-profit user groups provide their members with training and information about cutting-edge technology.
Companion Websites Our Companion Websites provide valuable solutions beyond the book. Here you can download the source code, get updates and corrections, chat with other users and the author about the book, or discover links to other websites on this topic.
Need to find a bookstore? Chances are, there's a bookseller near you that carries a broad selection of PTR titles. Locate a Magnet bookstore near you at www.phptr.com.
Subscribe today! Join PHPTR's monthly email newsletter!
Want to be kept up-to-date on your area of interest? Choose a targeted category on our website, and we'll keep you informed of the latest PHPTR products, author events, reviews and conferences in your interest area.
Visit our mailroom to subscribe today! http://www.phptr.com/mail_lists

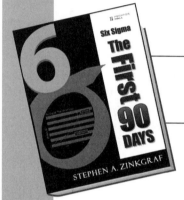